PRAISE FOR
HOW GREAT THINKERS TRANSFORMED OUR IDEAS

C. C. Hagan's *How Great Thinkers transformed our Ideas* is one of the finest syntheses of history, science, philosophy, and theology written in the last thirty years. Page after page is a revelation of the unfolding of early-modern and modern scientific thought and the implications it all has for our deepest questions regarding the universe, meaning, and God. As well as being a first-order work of intellectual history, Hagan makes first-order contribution to these questions himself. A virtuoso contribution.

Dr. Stephen Chavura, Ph.D, Lecturer in History

How Great Thinkers transformed our Ideas deals with some of the most important scientific discoveries in history, the philosophy behind the science and even the struggles, hopes and beliefs in the lives of the great scientists. Questions ... spanning the Enlightenment, the Age of Reason and the Cosmos give vivid impressions of the times in which they lived. .... The author's ingenious approach was to present each scientist's ... achievements across separate chapters to relate their lives, world, science and realities. A big advantage of such approach allows a reader not only to read and learn but also to see and analyze how the scientists' background impacted their personal and scientific growth providing valuable insights into these great thinkers.

Dr Andrew Firek M. Sc., Ph.D., FAusIMM, FAIE

How Great Thinkers transformed our Ideas investigates reality..[and] .. the theories and discoveries of five giants of science, Newton, Leibniz, Curie, Einstein and Hawking, ...[for example, Marie Curie's] work led to further scientific discovery and opened up the new world of nuclear medicine as well as forging new role models for women by winning two Nobel Prizes. …Standing on the shoulders of giants, these five great scientists developed unimagined knowledge of our world. … I was amazed by the insights the author gives us into the interaction between scientists and society and how our world both impedes and promotes their discoveries. An educational adventure and a wonderful combination of social theory, philosophy and science."

Elizabeth Ruth Cohen BA(Hons I), JD (Sydney), GDLP, LLM (Applied Family Law), LLM (Applied Business Law), Solicitor

The volcanic eruption which was the Age of Enlightenment, the advent of reason and experiment, the rejection of religious domination occurred in every field of human endeavour and hence the Modern was born. Mr Hagan masterfully shines a light on one aspect of that eruption -science- and major thinkers who since Sir Isaac Newton have propelled scientific thought and process to where it is today. A marvelously researched, energizing and profound examination of the lives and world changing ideas of scientists from Newton to Hawking!

Michael Ellicott BA LLB – Barrister-at- Law

Chris Hagan has written a very profound book about some of history's most famous scientists namely Newton, Leibniz, Curie, Einstein and as a coda Hawking's space, time and black hole cosmos. ... The book is easy to read for anyone with a smattering of science knowledge as Hagan explains it all in easy to understand language. And for those not so familiar with calculus a chapter explains it all from need to discovery to practical use. *How Great Thinkers transformed our Ideas* is a must read to better understand these titans of scientific discovery.

Wes Harder, Geoscientist & former Corporate CEO

How Great Thinkers Transformed our Ideas by C.C. Hagan transcends the boundaries of a typical popular science book. Apart from historical facts behind scientific breakthroughs and the personal lives of the corresponding scientists, highlighted are their feats of endurance in overcoming, frequent personal tragedies before rising to scientific triumphs and recognition. Also featured are philosophical diversions related to the meaning of physical reality and existence, and at the same time containing a scientific and mathematical anthology (with an impressive number of references and citations - counting over one of their academic breakthroughs and their evolving implications over more than 300 years.

C.C. Hagan not only describes the work, accomplishments, and life of scientific giants such as Newton, Leibniz, Curie, Einstein, and Hawking (perhaps the greatest scientific thinkers and founders of numerous scientific revolutions) but also gives the context and the reason for these breakthroughs which took place by examining the "local reality" of each scientist. I find that quite interesting.

As an educator, I believe that this book has the potential to stimulate interest both in maths and in science not only for the maths/science enthusiast but also for both school-age and university-age students as well as the general public, by taking the reader on a cosmic trip that spans almost three and a half centuries from the foundation of Classical Physics and Calculus to Quantum Physics and String Theory.

Dr Stavros Mouslopoulos , BA, PhD, Dip Ed , Theoretical Particle Physicist

HOW GREAT THINKERS TRANSFORMED OUR IDEAS

SHARE THE INSIGHTS OF NEWTON, HAWKING, CURIE, LEIBNIZ AND EINSTEIN

C.C. HAGAN

Troubador Publishing Ltd
Unit E2 Airfield Business Park
Harrison Road, Market Harborough
Leicestershire LE16 7UL
Tel: 0116 279 2299
Email: books@troubador.co.uk
Web: www.troubador.co.uk

ISBN 978 1 80514 254 6

British Library Cataloguing in Publication Data.
A catalogue record for this book is available from the British Library.

Printed and bound in Great Britain by 4edge Limited
Typeset in 11pt Gill Sans by Troubador Publishing Ltd, Leicester, UK

Cover background image of spacescape by Reimund Bertrams

To my wife, family, friends and all those who gave me valuable input and support.

CONTENTS

PART II: MASTERS OF THE COSMOS

PREFACE

It had been a tumultuous day. BBC News announced the sad death of Professor Stephen Hawking in the early hours of yesterday morning – Wednesday, 14 March 2018. ABC News in Sydney, Australia described him as the most brilliant scientist of our time – famous for his work on gravity, black holes and quantum mechanics. The BBC also interviewed bestselling author and theoretical physicist Michio Kaku about the significance of Stephen Hawking, to which he said: "Einstein and Hawking were seen as 'messengers from the stars' who popularised science." He commented that Stephen Hawking had applied quantum mechanics to black holes which were thought not to emit light. He had found they did emit faint light. Importantly, he saw Hawking as setting the agenda in physics to develop a unified theory between Einstein's relativity and quantum mechanics, giving us a 'huge piece of the gigantic jigsaw puzzle of unification'. Einstein set this quest in motion and Hawking developed it.

Yet it was not just physics that he spoke of – people would often ask him what the meaning of life was. One theme he followed was contained in an interview with veteran talk show host Larry King on CNN in 2010, which is summarised below:

King: You say that science can explain the universe without the need for a creator – what is that explanation… why is there something rather than nothing?

Hawking: Gravity and quantum theory causes universes to be created spontaneously out of nothing.

King: You write that because there is a law such as gravity the universe can create itself from nothing – can you tell me how that law came into existence?

Hawking: Gravity is a consequence of M-theory, which is the only possible unified theory – it is like saying 'why does 2 + 2 = 4?'

King: Simply put, do you believe in God?

Hawking: God may exist, but science can explain the universe without the need for a creator… the scientific account is complete. Theology is unnecessary.

King: One of your colleagues out of Cambridge says that science provides the narrative of how existence happened but theology addresses the meaning of the narrative.

Hawking: The scientific account is complete – theology is unnecessary.

This book collects the symphony of ideas from the study of five of the greatest thinkers who ever lived in the hope that they can be understood by anyone who might not have gone beyond the science they learnt at school. Yet the book has the depth to satisfy the scientific mind. Moreover, it extends to the complexities of M-theory (string theory – see Chapter 26), relativity and quantum mechanics and much more, including the worlds of science, the philosophy behind the ideas and the histories that made possible the unmasking of many of the riddles of the cosmos. Most inspiring are the stories behind the science – Hawking and Curie overcoming their personal tragedies to rise to great heights and the intellectual battles fought by Einstein, Newton and Leibniz.

Newton stands in stark contrast to Hawking on the meaning of life. Not only did Newton dedicate himself to science but also to theology. Newton saw theology as a necessary fabric of reality because he believed those laws were the harmonious work of a creator. Newton's quest was to discover the inner workings of those laws. In fact, Newton believed he was rediscovering laws of nature known by the ancients but lost over time. Newton's quest for knowledge of the laws of the universe succeeded in such a magnificent way that it remains a major foundation of science and mathematics today. Newton worked in the framework of reality known as the Clockwork Universe but, as we shall see, his insights were probably more modern.

Science writer Edward Dolnick[1] explains what the Clockwork Universe was:

> ***...at some point in the 1600s, a new idea came into the world. The notion was that the natural world not only follows rough-and-ready patterns but also exact, formal, mathematical laws. Though it looked haphazard and sometimes chaotic, the universe was in fact an intricate and perfectly regulated clockwork... God was a mathematician, seventeenth-century scientists firmly believed. He had written His laws in a mathematical code. Their task was to find the key.***

Both Isaac Newton and Gottfried Leibniz had displayed their genius in inventing calculus but fought over who had invented it. Dolnick describes the outcome: "Because calculus was the ideal tool to study the natural world, the debate spilled over from mathematics to science and then from science to theology. What was the nature of the universe? What was the nature of God, who had designed the universe?"[2] Today, one of these key questions is all-encompassing – what is the nature of reality?

Einstein's theories of relativity shattered the Clockwork Universe, but despite limiting Newton's laws of motion, they remain valid except at the extremes of reality. Newton's surprise would be that gravity operates in curved space in contrast to his concept of 'absolute space'. More surprising might be the answer to the question 'What is gravity?' after Einstein. Einstein was educated under the German system of *Bildung* (literally 'formation'), where physics and mathematics are taught with philosophy.[3] Clearly, the benefit of 'metaphysics' allowed Einstein's lateral thinking to overview physics and reconstruct it.

This book reveals the astonishing story of five people who exemplify the most extraordinary achievements of the human mind in history. Their stories class them as true founders of scientific revolutions who experienced rags to riches, challenged the bounds of endurance and give courage and inspiration to all of us. The book paints a rich scope across each life in contrast to history, ideas against philosophies and science built with mathematics in a quest to broaden interest for popular science enthusiasts, students and anyone who wishes for a gateway into what the riddles of the cosmos are all about. Science, personal stories, history of ideas and philosophies intertwine in the hope that further insights emerge from these great people. These insights include not only their mindsets but the life philosophies that gave them the perseverance and capacity for work to achieve great things.

The news bulletins also carried the famous episode 'Descent' (*Star Trek*), where Newton, Hawking and Einstein were playing an imaginary card game. Let us hope that they are all in the draw now for their weekly 'celestial' card game. On 10 April 2019, just over a year since Hawking's passing, the first direct visual evidence of a supermassive black hole (dubbed M87) and its 'shadow' was photographed through the Event Horizon Telescope (EHT). The black hole was found at the centre of the Messier 87 galaxy, in the Virgo galaxy cluster, 55 million light years away from Earth. Both Einstein, whose theory of relativity predicted black holes, and Hawking, whose theories explained them, would have been delighted.

Newton's approach to 'natural philosophy' reads like a playbook of the scientific method, discovery and a symphony of mathematical analysis. His work underpinned the Scientific Revolution of the 17th century and led to a vast body of science and mathematics using Newtonian physics and mathematics as its foundation. He had discovered and produced valid theories for gravity, laws of motion, fluid mechanics, optics and new techniques in mathematics such as infinite series and the binomial theorem. He had created the first major building blocks of *reality*. Yet at the time he had conceptualised all this he was too shy to publish and remained unknown many years before he published his great works. He was also an alchemist, historian, theologian and member of parliament, finally becoming Master of the Mint of England – a life journey worth anyone's time in exploring.

So how has Newton's reality changed after the sad passing of Hawking?

FOREWORD

Richard Feynman used to say that physics is one scientific field that has not admitted any evolutionary question: "Here are the laws! we say. We don't even think about how they got that way..." Luckily, we are in a position today to view such a statement as an exaggeration (although it still holds true for the way of thinking of the majority of theoreticians). This book spans the history of physics from the classical (Newton) to the modern (Hawking), providing a glimpse into the evolution of ideas over the centuries since the beginning of modern science. However, here let me take the top-down perspective.

Amidst the search for the nature of reality, the quest for the Theory of Quantum Gravity has led us to question even the most deeply rooted assumptions of the physicist's dream – the ultimate unified theory, namely, the expectation that the theory itself will have the form of an immutable set of equations governing all the physical laws and thus determining and describing the world around us.

The need for such a shift in the scientific paradigm did not only come from theories such as Loop Quantum Gravity[4] with its emergent laws of nature and the discrete nature of space-time, itself at the fundamental level. Perhaps, unexpectedly, this shift was also shared by its seeming competitor, String

Theory, over the last two decades or so. Given that String/M theory, which is formulated in ten and eleven space-time dimensions, has to eventually reproduce the 4D phenomenology as we know it, the compactification of the additional dimensions is inevitable. It is exactly this process of 'fixing of the moduli' of the Calabi-Yau compactifications that brings us to the string theory 'landscape'. The 'landscape' is a direct consequence of the large (and perhaps infinite) number of Calabi-Yau manifolds, and a huge number of possibilities of stabilising them, that has brought forward a vast number of solutions (vacua) for the theory:[5] "The resulting four-dimensional vacua have all possible physical laws with all possible constants, and this has led to a radically new view of physics in which one argues that the constants in the physical laws that we measure in our universe do not come from an underlying unified theory, but are environmental (anthropic) variables that are determined by where we are in this multiverse."[6]

The previous comment is not mentioned as necessarily a flaw of String Theory but certainly as one of the most obvious ways that the Theory has fallen short of its promises. At the same time, it may be interpreted as an indication that we should be looking towards fundamentally probabilistic and emergent laws springing from the 'continuum' of possible vacua. One should have in mind that previous questions may not be possible to completely address without a satisfactory resolution of the problem of measurement in quantum mechanics. The previous comments are naturally related to Black Hole physics, which is another arena where the clash between classical and quantum physics appears and where both aforementioned theories have not only made progress on their own but also seem to converge surprisingly.[7]

I find this ongoing shift in scientific thought fascinating, and *How Great Thinkers Transformed Our Ideas,* by C.C. Hagan, encaptures not only the scientific elements of it but also the corresponding philosophical and even perhaps social contexts. The study of the history of science should provide a profound understanding of the way that scientific thought evolved over time, either gradually or in the form of a major scientific revolution. One hopes that this understanding may help bring home the next leap in scientific thought in each given discipline. Perhaps one of the most effective ways that this study can be done is by looking into both the details of the actual scientific breakthroughs and the local reality of the scientists that delivered them.

How Great Thinkers Transformed Our Ideas by C.C. Hagan on the one hand describes this evolution of scientific thought from the clockwork deterministic universe and cassical physics (Newton) and the associated maths and philosophy (Newton, Leibniz) to the realm of quantum theory (Curie and Einstein) and gravity (Einstein) to the modern era developments of string theory and loop quantum gravity related to black hole physics and cosmology (Hawking). On the other hand, this book transcends the boundaries of a typical popular science book as it involves not just the historical facts in regard to scientific breakthroughs but also the personal lives of the corresponding scientists and philosophical influences related to the meaning of physical reality and existence, and at the same time consists of a scientific and mathematical anthology of the related scientific breakthroughs.

How Great Thinkers Transformed Our Ideas is certainly an exciting reality to experience and I am confident it will encourage readers to seek deeper knowledge.

Dr Stavros Mouslopoulos, BA, PhD, Dip Ed, Theoretical Particle Physicist

NOTES TO PREFACE AND FOREWORD

1. Edward Dolnick – *The Clockwork Universe* – Harper Perennial, 2011, p xvii to xviii.
2. Ibid
3. *Oxford Illustrated History of Science* (edited by I.R. Morus), Oxford University Press, 2017, p357
4. In loop quantum gravity, the smooth background of Einstein's theory of gravity (general relativity) is replaced by nodes and links to which quantum properties are assigned and space is built up of discrete entities – the loops.
5. Here, we refer to the possible 'flux compactifications' of string theory to four-dimensions. For further reading, please refer to the review article by Iosif & Bena Mariana Graña, *Comptes Rendus Physique*, Volume 18, Issues 3–4, March–April 2017, pp200–206
6. See Note 5.
7. In the context of loop quantum gravity, the black hole entropy can be done approximately using spin networks and for large enough cases does give the area law for the entropy. In the context of string theory, in specific cases, the area law can also be derived by counting the string microstates. Furthermore, in string theory, the black hole formation can be seen as similar to a phase transition when the Calabi-Yau shape goes through a space-tearing conifold transition.

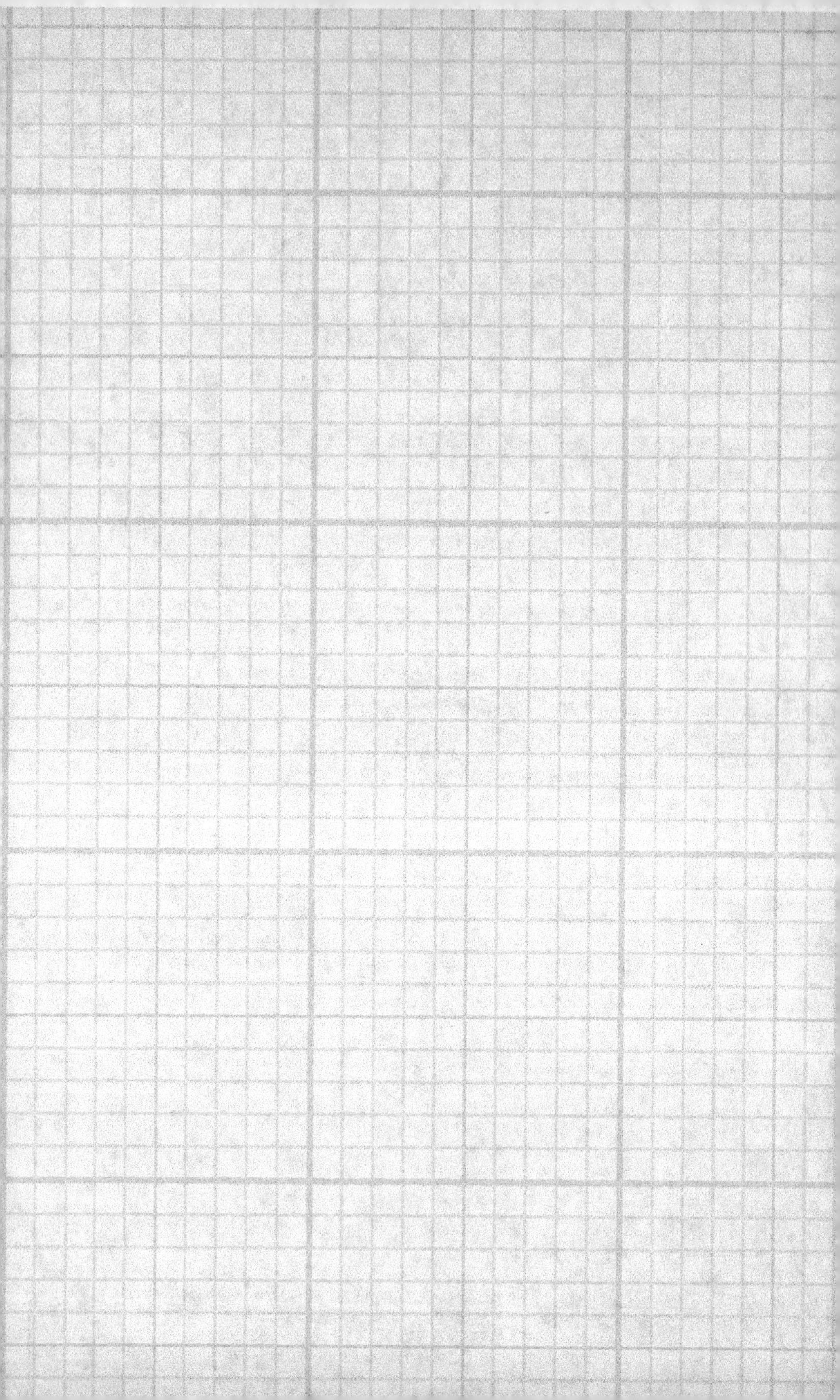

PART ONE

MASTERS OF THE ENLIGHTENMENT

A genius emerges only once in every thousand years
Voltaire

ISAAC NEWTON

NEWTON

Chapter One Plate

Isaac Newton performing his crucial prism experiment – the 'experimentum crucis' – in his Woolsthorpe Manor bedroom. "I placed my prism at its entrance, that it might be refracted to the opposite wall. It was at first a very pleasing divertisement, to view the vivid and intense colours produced thereby; but after a while applying myself to consider them more circumspectly, I became very surprised."

ONE

ISAAC NEWTON

Isaac Newton lived at a time when England was full of political and religious upheaval. Newton was said to be the greatest thinker of his time – the leading scientist and mathematician in the 17th century, regarded in awe since. Newton is famous for having seen an apple fall from a tree before having a new idea – that an invisible force guided everything from apples to planets. The theory of gravity was born. In 1642, Newton was born into a relatively prosperous family. His father was a farmer but died three months before Newton was born. His uncle and stepfather were rectors of local parishes. Newton's mother had wanted him to be a farmer but this hope faded when one day the sheep wandered away whilst Newton was reading under a hedge.[1] Fortunately, his prowess for maths and science was recognised by his schoolmaster or uncle persuading his mother to drop her vision of Isaac as a farmer from her wishes: "What a loss it was to the world, as well as a vain attempt, to bury so extraordinary a talent in rustic business."[2] And so he was sent to Cambridge. Newton was admitted to Trinity College, Cambridge in 1661 as a sub-sizar.[3] (A sizar was an undergraduate who received assistance with meals, lodging, etc. A sub-sizar was required to earn his living by acting as a servant for wealthier students.) Newton was also expected to become a priest after his Cambridge studies but was exempted. In fact, Newton has been called a 'Priest of Nature' by Oxford Professor Rob Iliffe, who wrote a book bearing that description.[4]

Newton's first years at Cambridge were interrupted when the great plague hit, and so in 1665 he was forced to return to the family home at Woolsthorpe in the country and didn't return to Cambridge until 1667. But this interlude of two years turned out to be his making, when his major discoveries occurred – the apple falling from the tree and his theory of universal gravitation set out some years later in his masterpiece known as *The Principia,* published in 1686–87. In his own words, when reminiscing on these two years:

> ***In the beginning of the year 1665 I found the method of approximating series and the rule for reducing any dignity of any binomial into such a series. The same year… I had the method of fluxions & the next year in January had the theory of colours… And the same year I began to think of gravity extending to the orb of the moon… All this was in the two plague years of 1665 and 1666. For in those days I was in the prime of my age for invention & minded mathematicks & philosophy more than at any time since.***[5]

Newton had discovered and invented much more than gravity – such as the laws of motion and the invention of calculus, the core of mathematics. Later, he made discoveries in optics and even built a new telescope. After his return to Cambridge, his genius at mathematics was recognised by the then (first) Lucasian Professor of Mathematics, Isaac Barrow. Newton had written *De Analysi per aequationes numero terminorem infinitas (On analysis by Means of Equations with an Infinite Number of Terms)*[6], which contained the binomial theorem he had discovered in 1665.[7] Barrow had sent this paper in July 1669 to an amateur mathematician of the day, John Collins, who published mathematical books relating to algebra and navigation[8]. Newton had a habit of sitting on things without revealing them and had a peculiar shyness for his work. To take one example, he invented the binomial theorem, which was extremely useful in expanding equations into infinite series such as in trigonometry, where he could expand sine and cosine functions beautifully, and of immense importance in science. He went down to London to meet Collins in November 1669 and discussed his telescope, harmonics and many mathematical things. Two years later, when Collins was to publish work relying on some insights Newton had given him, Newton wanted his contribution to remain anonymous, telling

Collins he had no desire "to gain the esteeme of one ambitious among the croud to have my scribbles printed."[9] This bizarre shyness prevented Newton becoming known as the most brilliant mathematician of his age.[10] By this time, Barrow had become aware of Newton's incredible ability and so in September 1669 he was appointed Lucasian Professor of Mathematics – a position held more recently by Stephen Hawking.[11]

Newton's shyness to publish was galvanised a year later when he presented his optics paper. Before he became a Fellow of the Royal Society in 1672,[12] he had sent the Royal Society (only founded twelve years earlier in 1660) a telescope which he had made (and which King Charles II is said to have looked through) and said he had a far richer gift to give to the Royal Society, which he termed "the oddest, if not the most considerable detection, which hath hitherto been made in the operations of nature."[13] Some weeks later, he sent his optics paper to the Royal Society and in the opening passage said:

> *To perform my late promise to you, I shall without further ceremony acquaint you, that in the beginning of the year 1666 (at which time I applyed myself to the grinding of the optick glasses of other figures than* **Spherical***), I procured me a triangular glass-prism, to try therewith the celebrated Phaenomena of colours. And in order thereto having darkened my chamber, and made a small hole in my window-shuts, to let in a convenient quantity of the sun's light, I placed my prism at its entrance, that it might be refracted to the opposite wall. It was at first a very pleasing divertisement, to view the vivid and intense colours produced thereby; but after a while applying myself to consider them more circumspectly, I became very surprised…*[14]

It seems Newton was inspired to experiment on light after reading Descartes's book of colours – *Dioptrique* (1637) (Descartes was another leading scientific figure of the age), which he did after buying two prisms at a market (although Newton had assisted Barrow with lectures on optics and mathematics).[15] The key moment described in the above quote from Newton was when he had drilled a hole in his 'window-shuts' and noticed not a circular (refracted) white spot on his wall but an oblong spectrum of colours.[16] This refuted the existing theories of Descartes and Hooke that the colours were merely transformations of the white light. Rather, the white light was *composed* of the colours.[17] Moreover, in a decisive step, Newton refocused the spectrum of coloured rays emanating from a prism back into a spot of white light[18], presumably through the second prism he had purchased. His work also led to Newton's formula for a lens.[19] Despite both his papers on light and the telescope making him famous, within a week of the Royal Society having received the paper, Newton's soon-to-be nemesis objected to Newton's theory of light – Robert Hooke. Hooke had been elected as assistant to the Earl of Cork, Robert Boyle (a scientist famous for his air pump and other works and a founder of the Royal Society[20]), and in 1662 Hooke was appointed Curator of Experiments, a salaried post carrying a handsome £80 per year plus £50 per year for lectures.[21]

So, in 1672, after the Royal Society's delight in receipt of Newton's paper, Robert Hooke, by then a key figure in the Royal Society, criticised the theory of light, saying there was no proof without a comprehensive hypothesis of why white light split into these colours – indeed, this could be just an anomaly causing light to appear multicoloured – an alteration of light caused by refraction.[22] Newton was furious at this point, demonstrating his distaste for a hypothesis in favour of a theory backed by rigorous experiment. In doing so, Newton showed himself to be a truly modern scientist utilising the scientific method. This dispute galvanised Newton to be meticulous with what he showed the public and also to keep all his documentation, which would later prove decisive in a dispute over his invention of calculus in mathematics. However, this may have been the beginning of Newton becoming overly suspicious of others competing with his work. For example, in 1677, a year after Newton formally sent his binomial theorem to the Royal Society,[23] he decided to publish his optical work and during preparations a letter was passed on to him by Hooke from a Jesuit, Francis Linus, who had criticised Newton's work but who had died in 1675. The letter had been passed on to Hooke by a colleague of the Jesuit, Anthony Lucas. When Newton saw the letter given to him by Hooke was slightly different from the original, Newton became furious and described Lucas and his friends as mounting a Jesuit conspiracy against him.[24] Psycho historians believed that these character traits emerged from the damage he suffered during the years of separation from his mother after she remarried to move into her new spouse's home without young Isaac – reminiscent of a situation from a Dickens novel.[25]

Newton had achieved considerable status from his work on optics and mathematics and in a sense was 'resting on his laurels' when in 1679

Robert Hooke sought to 'wake' Newton from his 'slumber', urging him to think about planetary motions and to impart anything philosophical to Hooke. In Chapter 3, it will be shown in detail how this critical exchange of ideas between Newton and Hooke and others led to the publishing of *The Principia* in 1687, which contains his groundbreaking theory of gravity, the laws of motion, his mathematics and incredibly detailed worked examples of a great range of planetary and other problems which he had solved using his new physics and mathematics.

Newton's fame grew in the years after the publishing of *The Principia* and indeed was championed in the next century by Voltaire in Europe. Voltaire saw Newton as 'the greatest man that ever lived' and a genius the likes of which emerges only once in every thousand years.[26]

In the next chapter, we will see the incredible transformation of Newton from academic to member of parliament after the invasion which saw King James II flee and William and Mary of Orange take power. Remarkably, he also became the 'streetwise' Warden of the Mint to tackle coin counterfeiters – a long way from the equilibrium of Cambridge. In fact, when the Master of the Mint, Thomas Neale, died in 1699, Newton succeeded him as the new Master of the Mint himself. Then followed a series of further successes – the election again as a member of parliament in 1701 and upon Hooke's death, his appointment as President of the Royal Society in 1703, which must have been a sweet success indeed, given his turbulent dealings with it in previous years. He was also knighted by Queen Anne in 1705.[27]

With his appointment as President of the Royal Society and his meticulous record keeping, he was able to win a Royal Society dispute in 1713 fought against German mathematician Gottfried Leibniz as to who was to take credit for the invention of calculus.[28] (However, today's historians give joint credit to both men for the invention of calculus. Professor of Mathematics William Dunham, after summarising the events of the dispute, concludes: "The manuscript trail establishes that Leibniz, his contacts with Newton notwithstanding, had discovered the principles of the calculus independently and rightfully shares the glory of discovery."[29]) We will revisit the unfolding of the Newton-Leibniz dispute in the later chapters on Leibniz.

By the end of the two miracle years of 1665 and 1666, Newton had conceptualised his great theories which grounded his great works in later life, and thus the daunting realisation must have occurred to him that he was leading the known world in science and mathematics. Yet the 'academic school of hard knocks', which he endured over many years, saw him reluctant to publish his great concepts and without the urging of others, his works may have remained hidden. As historian Richard Westfall observed: "In 1665, as he realised the full extent of his achievement in mathematics, Newton must have felt the burden of genius settle upon him, the terrible burden which he would have to carry in the isolation it imposed for more than sixty years."[30]

Chaper Two Plate

Newton's work at Cambridge was carried out in his rooms located to the right of the gate house on the first floor.

TWO

NEWTON'S WORLD

As Newton was born in 1642, he found himself in a century of religious and political turmoil, but it is the 200 years that led up to this period that properly explains the complex world views, political thinking and religious change. In 1417, the humanist movement of the Renaissance began with Italian scholar Poggio Bracciolini's 'rediscovery' in a German monastery of certain writings from ancient Rome. These manuscripts included Vitruvius's *De Architectura*, Lucretius's epic poem *De Rerum Natura (On the Nature of Things)* – important for the coming Scientific Revolution[31] – and some of Cicero's famous orations.[32,33] Vitruvius was a Roman author, architect and engineer. He studied perfect proportions in architecture and the human body. After Bracciolini sent these manuscripts to Florence, Da Vinci's drawing, *Vitruvian Man* began to exemplify the new 'humanism' movement of the Renaissance at that time. It was essentially a shift in focus from the world to focus upon the human being. The humanist movement symbolised by the famous *Vitruvian Man* drawing demonstrated the "relationship between the microcosm of man and the macrocosm of the earth".[34] Lucretius was a pupil of Epicurus, a Greek philosopher, who, like Democritus, adopted a purely materialist viewpoint and saw humans as composed of atoms, and so he surmised that upon death we dissipate and thus there is no afterlife, regardless of any gods.[35] Thus, in Lucretius's epic poem *On the Nature of Things,* Epicurus's natural philosophy was the theme of his verse "denying divine design and depicting the universe as a whirl of atoms, in a perpetual state of flux".[36] Epicurus had said, "The atoms move continuously forever, some… standing at a long distance from one another, others in turn maintaining their rapid vibration whenever they happen to be checked by their interlacing with others or covered by the interlaced atoms".[37] We see here one of the great conflicts between different versions of the nature of reality. Newton had helped pioneer one version – the Clockwork Universe (mentioned in the Preface) whilst another version – the more modern concept of the chaotic universe was yet to come. As this book unfolds, these competing versions will be seen as fundamental to understanding the nature of reality.

The re-emergence of Greek thinking had been suppressed by religious thinking, and it was only the work of Aquinas in making the Greek natural philosophy of Aristotle acceptable to Christian doctrine[38] that allowed Greek natural philosophy to be taught at university in Newton's day. The critical change with Renaissance humanism was the concept of Lucretius of the universe existing in a natural state and not governed by 'nature Gods' or by a divine driving force.[39] Epicurus, Lucretius and others represented materialism, which in that age led to atheism. However, unlike the 20th-century secular humanism of Sartre and others, these Renaissance humanists did not reject Christianity or religion. In fact, the Catholic hierarchy supported humanism.[40]

In England, apart from the influence of the Renaissance, the 16th century saw Henry VIII, famous for his many wives, take England away from the Catholic Church, proclaiming himself head of a new church – the Church of England, which occurred during a larger movement in Europe – the Protestant Reformation. Famously, Martin Luther had nailed ninety-five theses to a Wittenburg church door in 1517 objecting to the offering of indulgences by the Catholic Church for sins. These 'after-life reductions of punishment' for sins were 'sold' in return for church donations which had contributed to the building of St Peter's Basilica in Rome.[41] This protest saw the founding of Lutheran and other Protestant religions.[42] This move by Henry VIII was exemplified in the movie *A Man for All Seasons*, which saw Sir Thomas More (1478–1535), advisor to King Henry VIII and Lord High Chancellor of England, executed on a charge of high treason. He had refused to support the King's legal steps to divorce his wife, which could only be done if England broke away from the Catholic Church. He was regarded as a humanist and Catholic martyr who was canonised as St Thomas More in 1886.[43]

The Catholic Church reacted to this great period of change by organising the Council of Trent (1545–1563) on how to respond to the Protestant Movement. (This council proved to be a landmark, with the next most important council – Vatican II – held in 1962–65.)[44] Of important consequence for the Scientific Revolution, of which Newton was to be a part, was a reform to improve the education of priests with the formation of an intellectual group within the Catholic Church. This group, known as the Jesuits, was founded by a former soldier, St Ignatius of Loyola, who had been hit by a cannonball and determined his life on a (perhaps) more challenging course. The Jesuits founded hundreds of schools and not only taught church doctrine but mathematics and sciences based on Aristotle. As Professor of Humanities, Lawrence M. Principe explains: "Jesuit schools were often some of the first to teach some of the new scientific ideas of the Scientific Revolution, and educated many of the thinkers responsible for them… The Jesuit attitudes in studies of science and mathematics expresses their motto 'to find God in all things'. While Jesuits emphasised this incentive, it was not unique to them – it undergirded virtually the entire Scientific Revolution."[45] By the 1700s, more than half of mathematics professorships in Europe were held by Jesuits.[46]

Newton himself had disputes with Jesuits with respect to his theory of light[47] but more importantly he too saw God in science and mathematics and saw the universe as fine tuned and designed by God. These concepts remain active viable theories in modern ideas in blending science and religion. The trend that occurred in medieval times has been a shrinkage in the scope of God's intervention in the universe from an active creator to a passive creator who merely created and 'wound up' the Clockwork Universe to run autonomously. Indeed, although the Clockwork Universe allegory has subsided with the Heisenberg Uncertainty Principle and the discovery of chaos – both discussed later in this book – this same trend continues today.

In Newton's time, it became evident that "The fashion of the age is to call everything into question."[48] One such challenger of the age was Thomas Hobbes, who was an influential philosopher and writer who mixed with Europe's leading intellectuals of the day, including Descartes, Galileo and Gassendi.[49] He had travelled to Venice and was well acquainted with this Renaissance humanism and its revolutionary thought. He accepted the Greek atomism in the form of material atoms of motion and therefore argued that the immaterial substances, such as the soul, were illogical.[50] Material processes governed everything, including the mind, which governed (selfish) human nature. His materialist philosophy led to Hobbes publishing his landmark work, *Leviathan*, in 1651, which proposed that a social contract is necessary due to the selfishness of human beings.[51] At one stage in his life, Newton is thought to have suffered a mental episode of some sort and insulted his friend, philosopher John Locke, by calling him a 'hobbist', which meant a "materialist" or in those days could even mean an atheist. Newton had also accused Locke of endeavouring "to embroil [him with women.".Newton later apologised to Locke when he recovered from his condition.[52]

Moving into the 1600s, after the reign of Elizabeth I (daughter of Henry VIII), saw the threat of the return of Catholicism – the threat of 'Restoration' – which resulted in the execution of Mary Queen of Scots.[53] King Charles I was lenient towards Catholics and, being politically naïve, made several blunders leading to civil war and his ultimate execution in 1648.[54] Oliver Cromwell, a general and statesman, then succeeded him and in 1653 became Lord Protector of the Commonwealth (a newly created title for him to govern instead of a monarch in conjunction with Parliament). He did govern and introduced puritan reforms to the Church of England.[55] The Restoration did occur when Cromwell died in 1658 and the new Catholic king, Charles II, was installed as monarch. He reigned from 1660 to 1670 during the time Newton was at Cambridge and throughout his return to Woolsthorpe for the two 'marvellous' years in which he conceived of his landmark theories. Writer Jenny Uglow wrote about these ten years in her book *Charles II, A Gambling Man.* Charles II lived his life 'poker-faced' in the sense that during these turbulent times no one was able to gauge his attitude to the numerous issues of state that he presided over. She wrote of this decade: "It was an extraordinary decade, marked by struggles for power in state and church and by blows like the plague, the Great Fire and the Dutch war… It saw the founding of the Royal Society, the return of the theatre, the glamour, fashion, gossip and scandal of the court, and the resurrection of London rising like a phoenix from the ashes."[56]

Heresy – challenging church doctrine in some way – was a real threat to anyone who published ideas that could conflict with the church and indeed could be used as a political weapon against perceived enemies. Sir Thomas More himself had convicted heretics in England, who were burned at the stake for heresy. Two main sources of heresy which endangered Newton were (1) whether Christ was God and man or just divine, and (2) the apparent conflict between God's purpose and human free will.[57] This may have partly explained Newton's shyness to publish. In fact, Newton was a heretic because he secretly wrote extensively on the view that whilst Christ was divine, He was not God. Newton's view was that the politics of the early church had undergone a corruption of church doctrine, primarily by a Bishop Athanasius[58] for his own purposes, to make Christ both divine and God.[59] He spent years of intense scholarship studying ancient texts, seeing it as his mission to restore a proper theology to the church. As Professor Iliffe explains: "Not only could one discern important truths about scripture and nature, but by using the intellect properly one could see exactly how attributes of a divine being were adumbrated in mere mortals… The claims that *The Principia* had revealed the language and blueprint of the divine creation, and that (in Halley's famous words) no mortal could approach the gods more closely, should not be read as mere hyperbole. If Newton's achievements said a great deal about the human intellect, so they said much about the mind of God."[60]

Newton was forty-four years old when *The Principia* was published, and in the same year the then King James II, who had converted to being a Catholic, was keen to permit Catholics to be elected to office or to attend university. Newton became embroiled with the religious turmoil of the times when a particular controversy brought him and others before the notorious Judge Jeffries. The Duke of Monmouth, a Protestant, had led a rebellion against his uncle James II (against whom he held a grudge), and Judge Jeffries had sentenced to death hundreds of the Duke's supporters (the Duke was executed as well). Indeed, some years before the rebellion, a plot to kill the previous King Charles II saw the Duke of Monmouth (his own son) accused of being part of the plot, and Judge Jeffries had sentenced to death various plotters.[61] When a proposal to admit Father Alban Francis to an MA degree was refused, to maintain the Protestant integrity of Cambridge, Newton and seven others had been called before Judge Jeffries sitting on the Ecclesiastical Commission and eventually made orders that they refrain from refusing admission.[62] Newton had written essays on the importance of safeguarding the Protestant religion and the conflict of interest of the king in supporting the Anglican Church in his official role as defender of the faith – ie. the Anglican Church – and his own Catholicism.[63] In November 1688, William of Orange landed in England with a force of 500 ships and 21,000 men who had sailed from the Dutch republic. The Duke of Monmouth's invasion, a few years earlier, consisted of three ships and eighty-three men, who had also sailed from the Dutch Republic.[64] King James had fled in the face of this overwhelming force (called the Glorious Revolution) and indeed Judge Jeffries was caught attempting to flee, disguised as a sailor at Wapping.[65]

During the period when the Duke of Monmouth had been accused of plotting to assassinate his own father, Charles Montagu, the Earl of Halifax, had acted as a liaison between the duke and the kng to avoid execution.[66] Newton thrived under the new regime of William and Mary and was elected one of two members of parliament representing Cambridge University and in allegiance to the new regime voted in favour of an important resolution – the acknowledgement that James had abdicated his throne – presumably fleeing has sufficient legal force.[67] A few years later, the Earl of Halifax, then Chancellor of the Exchequer (the Royal Treasury) and President of the Royal Society, had acted as liaison in a quite different role – to procure the appointment of Newton as Warden of the Royal Mint – quite a transformation for a new scientist![68,69] Newton's half niece Catherine Barton was later to enter into a romantic liaison with the earl, cementing Newton's association with him.[70] At that time, Newton estimated that approximately 20% of coinage in England was either counterfeit or clipped. Illegal clipping of coins involved the counterfeiter taking enough silver or gold to then melt down and make up counterfeit coins from the clippings.[71] Newton applied his scholarly skills to investigating how to stop the practice and this revealed another side to Newton – a man of the world. Newton investigated illegal counterfeiting with vigour and launched prosecutions after paying informers for leads and witnesses to dress well when they appeared for him in court. Newton had himself declared a justice of the peace and secured more than twenty-eight convictions after cross-examining more than one hundred witnesses[72,73]. It seems that Newton, the academic-cum-parliamentarian, had become the inveterate prosecutor with zeal. His litigious posture gained after the scientific challenges against him had left their mark.

In an article in *National Geographic History* (May/June 2018 issue), 'The Alchemy Paradox', the author described a further insight into Newton: "One 17th century manuscript handwritten by Isaac Newton came up for auction in 2016. It contained cryptic instructions for preparing 'sophick mercury', an alchemist's potion… It turns out, this scientist was also an alchemist… Today the word 'alchemy' evokes magical images of wizened men surrounded by dusty books and bubbling potions, but in Newton's time, it fell in the realm of legitimate scientific enquiry."[74] King Charles II was very interested in alchemy and possessed alchemical works including translations of John Dee's alchemical works.[75] John Dee was an advisor to Elizabeth I and was also an alchemist to the Holy Roman Emperor, Rudolf II in Prague, who was also very keen on alchemy.[76] Alchemy was in its heyday during the Scientific Revolution. It is popularly known for the pursuit of the Philosopher's Stone' which could transmute base metals like lead into gold, but in Newton's age, it was also known by some as '*chymistry*', to include not only the 'magical' but chemistry, the study of matter, production of pharmaceuticals, dyes, salts, perfumes, oils, etc.[77] Robert Boyle, a contemporary of Newton, founder of the Royal Society, famous for his air pump and responsible for the scientific equation enshrined in Boyle's Law, was also active in alchemy or, as he called it, *chymistry*. Boyle was a mechanical philosopher who saw alchemy as a way of studying the mechanisms of the universe.[78]

Given this backdrop, we are in a better position to answer a challenging question – who was Newton? Author Niccoló Guicciardini describes a division between two schools of thought by historians of science – between "the scientist and the alchemist, the experimental philosopher and the heretic", which arose in connection with the release of Newton's private papers in 1936.[79]

When the 5th Earl of Portsmouth, a descendant of Newton's half niece Catherine Barton, donated Newton's private papers which related to science'to the University Library of Cambridge in 1888, the Earl requested that any papers concerning theology, chronology, history and alchemy be returned to him. Some years later, in 1936, his family must have put these 'non-scientific' papers up for auction and they were subsequently purchased by a semitic scholar, Abraham Yahuda, and the economist John Maynard Keynes. Keynes later wrote a bombshell paper *Newton the Man* in 1945 asserting that Newton was not the enlightenment scientist but the 'last of the magicians' and follower of ancient scholars. As Niccoló Guicciardini asks, "Who then, was Newton? The cold mathematician who calculated the orbits of the planets, espousing a deterministic view of nature… Or rather a heretic who, denying the tenets of orthodox Christianity, conceived of natural philosophy as a search for the providential action of a single God *pantocrátor*?"[80] Since the Keynes essay of 1945, historians had proposed that Newton was still of a 'single mind', but Professor Iliffe's view is that "the recent publication of his religious, historical, and chronological papers has provided no support" for such notion.[81] Indeed, he sees his work in religion emerging in "a parallel intellectual universe" to that of mathematics and science, yet he used his 'formidable reasoning powers' on all of these to seek truth.[82]

Yet Newton saw his work on scripture as connected to his scientific work because he subscribed to the *Prisca Sapentia* – the idea popular in the humanism of the Renaissance that original wisdom was known to the ancients but corrupted over time through war, chaos, negligence or mere human manipulation.[83] Niccoló Guicciardini's view seems to be that despite these apparent contradictions, "we should not think that the 'traditional' Newton, the mathematician and physicist, is less remote to us than Newton the alchemist and theologian. In Newton's England, astronomy, optics and even mathematics were disciplines whose aims and methods were defined in ways considerably different from the ones accepted nowadays. What matters [in terms of Newton and his work] is how these scientific disciplines were conceived by Newton and his contemporaries as deeply connected to issues of great religious and philosophical reach, contrary to what happens today."[84] So the combining of activities such as science and alchemy (with its 'magic') that Newton, Boyle and others pursued and even historical and religious research should not diminish these pioneers as the first of a generation of modern scientists. Professor Lawrence M. Principe warns us not to misjudge any of the great figures of this period when he says: "…applications of *magia naturalis* and the whole idea of an interconnected world of sympathies and analogies are sometimes dismissed as irrational or superstitious. But this harsh judgement is faulty. It results from a certain smug arrogance and failure to exercise historical understanding. What our predecessors did was to observe various mysterious and apparently similar phenomena in nature and extrapolate thence into a more universal statement – a law of nature – about connections and the transmission of influences in the world… They were trying to understand the world… and to make use of the powers of nature."[85]

Chapter Three Plate

A replica of Isaac Newton's new design of a second reflecting telescope that he made in 1672; from the Whipple Museum of the history of science, Cambridge

THREE

NEWTON'S SCIENCE

Newton's discoveries occurred during a turning point in history which has been called the Enlightenment. In essence, a new approach of science and reason started to prevail over the previous religious-dominated mindset of the age, resulting in the Scientific Revolution. In Newton's early years at Cambridge, this was the overarching backdrop to medieval thinking but something was changing. Aristotle's science was the standard 'natural philosophy' taught at Cambridge. In those days, natural philosophy *was* science, being a blend of religion, philosophy and naturalistic studies. Aristotle had promoted a way of thinking that predated the dominant Christian era, and this was the method of dialogue – almost a court process to establish a correct principle. This use of reason to find truth was contrary to the method of revelation to discover knowledge revealed through analysis of scripture or theory. This new way of thinking led to a new science where nature had to be studied differently. Pioneers of science such as Francis Bacon suggested that science be investigated directly through observation and experiment. This is now called the scientific method rather than Aristotle's philosophical analysis. This new thinking and approach interested Newton.

Rather than focusing on Aristotle, Newton was more interested in this new way of thinking and, amongst others, read works by Galileo, Kepler and Descartes, who were contributing to what was really a 'new science'. One of the most famous examples of this new science was Galileo's fight with the Catholic Church. The church saw the earth as the centre of the universe until in 1514 Copernicus proposed that the sun was the centre and the seven known planets revolved around the sun – a heliocentric system. In the next century, Galileo wrote a treatise to demonstrate that Copernicus was right, but the church disagreed and he was convicted for heresy by the Roman Inquisition. He was imprisoned under house arrest until his death in 1642. Galileo had transformed astronomy with the telescope and Kepler had discovered some stunning mathematics of astronomy, including intricate equations and a very significant discovery – that the planets did not revolve in perfect circles as was expected by religious thinking but revolved in ellipses.

But it was Descartes, a pioneer of the Scientific Revolution and a founder of the Enlightenment, who proposed a mechanical theory of planetary motion and gravity. Newton first accepted but by the early 1680s had rejected this theory by the publishing of his masterpiece on the theory of 'invisible' gravity[86] set out in *The Principia*.[87,88] *The Principia* has been described as 'the most admired scientific treatise of all times'.[89] Descartes's major influence in this new science had changed the mindset of the age with the publishing of his *Discourse on the Method* (1637), which proposed that reason rather than revelation be used to gain knowledge of the world through both analysis and experiment. This eventually led to the scientific method along the lines that Francis Bacon had also proposed. Descartes's work was dangerous for the times (as seen by the plight of Galileo) and thus it was safer for him to work in the Dutch Republic (now the Netherlands) than his native France. In his *Principia Philosophae (Principles of Philosophy*, 1644) and *Le Monde (The World* – suppressed but published after his death in 1650), Descartes had proposed that the world was created ready-made, like a clock wound up to exist and operate with ready-made laws of the universe – known as the Clockwork Universe.

Descartes, unnerved by the harsh treatment of Galileo Galilei by the church, feigned to deny the Clockwork Universe. In the opinion of two historians, "This thinly veiled attempt to eliminate God as a necessary element in the creation of the world brought Descartes considerable notoriety. In relegating God to the remote and seemingly minor task of establishing the laws of nature he had overstepped the bounds of 17th-century tolerance. His fellow countryman Blaise Pascal (1623–1662) could never forgive such blatant impiety. 'In all his philosophy', wrote Pascal, Descartes 'would have been quite willing to dispense with God. But he had to make Him give a fillip to set the world in motion; beyond this he has no further need for God'..."[90]

Although this thinking is more than 300 years old, it echoes comments by the late Stephen Hawking, who has referred to the laws of nature being sufficient to account for the origin of the universe – as seen in the Preface. Hawking saw no need for a creator in the scientific theory of the universe. Yet, over 300 years before that, Isaac Newton provided an opposing view in *The Principia* when he said:

> ***"It is not to be conceived that mere mechanical causes could give birth to so many regular motions, since the comets range over all parts of the heavens in very eccentric orbits... The most beautiful system of the sun, planets, and comets could only proceed from the counsel and dominion of an intelligent and powerful Being..."[91] and as to the capability of laws of nature to create worlds, Newton said in a letter to Thomas Burnet (c. 1635–1715): "Where natural causes are at hand God uses them as instruments in his works, but I do not think them alone sufficient for ye creation."[92]***

In fact, Newton's thinking departed from the literal Genesis story in the Bible in that he envisaged our solar system emerging from 'common chaos' with subsequent condensation into solid globes by gravitational attraction. However, in 'compliance' with Genesis, Newton declared that the duration of creation was 'as long as you please'. This was because time was not yet running during the first two days of the Genesis creation week. Why not? The earth was yet not created for time to exist.[93] It would seem that Newton would be quite comfortable with today's Big Bang theory, with no time existing before the Big Bang. In fact, Newton had remarked that he saw the Bible as an instrument to explain creation to those not versed in the skilled art of natural philosophy and not as a scientific record of events.[94]

Science historian Lawrence M. Principe summarised how Descartes's mechanical theory of the universe declined after Newton:

> ***The mechanical philosophy waned by the end of the 17th century… If God is a clockmaker, did He start the world running and then abandon it, or must He regularly readjust it as if He were less than a master mechanic? … Newton's forces of attraction, a kind of action-at-a-distance, were not reducible to mechanical explanation. The triumph of Newtonianism in fact meant the defeat of strict mechanism.***[95]

In Chapter 1, we saw the dramatic presentation of Newton's *Opticks* paper in which he revealed his 'Theory of Colours', showing white light to be composed of the 'celebrated Phaenomena of Colours'. The significance to the science of the day was that Newton established a theory of the corpuscular theory of light consisting of a stream of particles disturbing the 'aether'.[96] Later, the wave theory of light would result in viewing light as wave-like or particle-like depending on the context. In 1905, Einstein proposed light as a stream of photons in his work on the photoelectric effect which resolves itself with quantum theory.[97] Newton had encountered chaos in astronomical calculations and presumably just saw it as anomalous. However, we will see later that chaos is a real mathematical feature of reality.[98] Thus, it is interesting to note also here that the 'quantum chaos' that can arise in the classical treatment of light might be viewed as a type of 'bridge reality' between the chaotic anomalies or uncertainties encountered in nature with the system of quantum theory, which can avoid these anomalies.[99]

In 1679, Robert Hooke had written to Newton urging him to impart anything 'philosophical' (meaning the scientific natural philosophy) and asked Newton about how to analyse planetary motions by means of an inertial path coupled with a force directing one body to the centre of an attracting body. Professor of Intellectual History and History of Science Rob Iliffe describes his response to Hooke as follows: Newton "offered a small 'fansy' concerning the earth's daily motion. If an object fell to earth… it would fall *in front of* its original position… If an object were dropped from a tall tower [daily], rotation might thereby be proved and on the assumption that the earth offered no resistance he drew a diagram of the spiral path of the object towards its centre."[100] Professor Iliffe comments that in view of Hooke's reply to Newton (that instead of a spiral path an elliptical path would result – which Newton rejected). Hooke has justifiably been given more credit by historians and indeed Newton's analysis here falls short of his brilliant analysis in *The Principia* which was about to take shape.[101] Hooke wrote again to Newton that his understanding had been that "gravity was always inversely proportional to the square of the distance from the centre of attraction."[102] To explain proportionality, let us say that x is directly proportional to y. If so, as an example, 'if x doubles then y doubles', which means the relation is x = y as distance varies. Note it could also be more complex, eg. x = ay where a is any constant value which in our first example means a = 1. Now inverse/indirect proportionality would mean, eg. if x doubles then y halves or x = $1/y$ and again it could be more complex, such as x = a/y which in our example means a =1 again. However, there are many more formulae that can fit this inverse equation and in the case of gravity the inverse formula was suspected by Hooke as applicable but the exact formula or equation remained elusive. Hooke posed what could have been a key catalyst for Newton's *Principia* – a crucial question – in effect, what was the mathematical inverse relation defining a planetary orbit (known to be an ellipse under Kepler's first law) – Hooke explained to Newton he had no doubt that "you will easily find out what that curve must be and its proprietys [sic] and suggest a physical reason of this proportion."[103]

This episode seemed to rekindle Newton's interest in 'matters philosophical' because Newton later confessed to Edmond Halley (of Halley's Comet fame) that this correspondence with Hooke had prompted him to think again about planetary motions. Newton later used Kepler's second law to show that with an elliptical orbit a body not only undergoes an inverse-square law of attraction but that the mathematical relation is enshrined in an equation. This shows distance between the two bodies does vary in inverse proportion to each other, as Hooke had supposed without knowing the exact equation. This was in fact an inverse-square law relating to the reciprocal of the distance between the two bodies squared (a then unknown equation involving 1/ distance x distance or force of attraction between the two bodies F = $1/r^2$ where r is distance so that if r doubles then the force F become four times weaker. Now, we mentioned inverse proportionality can be more complex and, in fact, Newton's law of gravitation is more complex and is not simply F = $1/r^2$ or F = a/r^2 which we will see shortly).

Newton was also influenced by a separate train of correspondence with John Flamsteed, the first Astronomer Royal, who was observing a comet in about November 1680 known as the Great Comet. Flamsteed had told Edmond Halley that the sun had attracted the comet, which he thought was a dead planet within the sun's vortex. When the sun drew the comet into its anti-clockwise vortex, it would get closer but then be twisted by the vortex so much that it showed its opposite side upon reaching a tipping point where the attractive force of the sun became a repulsive force sending the comet back again.[104]

Newton disagreed that the comet that went was the same as the comet that came back, stating it would not follow the same path back, but if it were the same comet, he surmised that the path of the comet had been attracted into an orbit by the sun and that it might have travelled around the opposite side of the sun. Newton saw that Flamsteed had not explained why the sun's vortex would change from attraction to repulsion. It was generally thought that the sun's attraction was like a magnet attracting iron but that by changing the pole of a magnet it would repel, and so there was some theoretical basis for this thinking. However, Newton rejected the magnetic model, stating that once the comet was within the influence of the sun, it would continue to be attracted and this would explain its orbit around the sun and back again. He added in a letter to Flamsteed that the competing forces of the sun's attraction and the comet's centrifugal force resulted in the comet decelerating but overpowering and breaking free of the sun's attraction. Eventually, continuous attraction rather than centrifugal force was his final analysis in *The Principia.*[105] Newton's brilliance was beginning to shine through in the years before *The Principia* was published.

Further correspondence in the period 1684–1685 with Flamsteed shows the evolution of Newton's thinking. One fascinating turning point in his thought was the insight that since the tail of comets did not dissipate – as a fiery ball in flight on earth might do (as the air tended to put it out) – this suggested there was no 'aether' in space offering resistance analogous to air that would do likewise.[106] This was a major plank in Newton's coming theory of space that formed the basis of his rejection of the reigning theory at the time, namely, Descartes's theory of the 'aether' composed of vortices providing a mechanical 'crowding' of vortices that objects like comets would have to 'push' through to move. Indeed, this reigning theory relied on this pivotal mechanism to explain gravity as that 'crowding' of vortices pushing objects like planets or comets.

Newton continued to formulate his laws of motion and crucially came to the conclusion that the bulk of matter was 'usually identical' to its gravity and redefined his definitions in the coming *Principia* by defining the 'quantity of matter' (or 'mass') as basic and undifferentiated matter. In fact, he extended his insight to stating that where there was matter there was mass, and where there was not matter there was nothing at all ('space'). Newton also pondered the relationship between *weight* and *mass*. The weight measured the gravitational pull of a body whilst the mass measured the amount of matter. Mass and weight must therefore be precisely proportional to each other.

Recall that Galileo discovered that all bodies fall to earth with the same speed. Newton observed that a body that is twice the weight or mass of a smaller body would be pulled with twice the force; otherwise, the smaller body would fall faster than the larger body (assuming the same gravitational force applies to both bodies).[107] Like Galileo, Newton saw pendulums as excellent tools for understanding both gravity and motion. Galileo had shown that a falling body increases speed with evenly uniform acceleration. Newton's invention of calculus (which he called the method of fluxions) was instrumental in his analysis of gravity for he suddenly realised that gravity was not a centrifugal force as he had surmised during the exchange of discussion on the comet but a 'centripetal' force (a term Newton himself had coined).

This new way of looking at force or gravity prompted Newton to think deeply about a body's power at resisting motion at rest or changes in direction when in motion (recall the comet's resistance against the sun's attraction that Newton had explained). Newton called this new concept of resistance 'inertia'. This led to Newton realising that forces must be proportional to the changes they cause to inertia of a body.[108] As we saw above, if something is proportional to another thing, a formula can be framed and experiments such as with pendulums or dropping weights from heights can be used to arrive at a workable equation. In fact, his analysis led to Newton's famous three laws of motion, where inertia is a central principle to those laws. Newton would then ponder – does such a formula apply to everything? It started to grow on Newton that such new laws with formulas and equations not only applied to the earth and the moon or to comets and the sun but to every body in the universe. Thus, since his renewal of interest in 'matters philosophical' prompted by the letter in 1679 from Hooke and later correspondence with Flamsteed through 1685 regarding comets and planetary motions, Newton discovered his Law of Gravitation and the elusive equation that Hooke had pressed him for. This became Newton's famous Law of Gravitation between any two massive bodies in the universe as enshrined in the equation $F = M_1 x M_2 x G/d^2$ (or using shorthand better known as $F = M_1 M_2 G/d^2$ – so, in fact, using our previous examples, it is not simply $F = a/r^2$ or, if you like, it is if $a = M_1 M_2 G$) where F is force, M_1 and M_2 are the respective masses of two bodies such as the earth and the moon, G is called the gravitational constant and d is the distance between the two bodies (ie. distance from the earth to the moon using their centres).[109] His great analysis had resulted in Newton's Laws of Motion, which are summarised in the *Oxford Dictionary of Physics* as follows:

1. A body continues in a state of rest or uniform motion in a straight line unless it is acted upon by external forces.
2. The rate of change of momentum of a moving body is proportional to and in the same direction as the force acting on it, ie. $F = d(mv)/dt$ where F is the applied force, v is the velocity of the body, and m its mass. If the mass remains constant $F = mdv/dt$ or $F = ma$, where a is the acceleration. $F = d(mv)/dt$.
3. If one body exerts a force on another, there is an equal and opposite force, called a reaction, exerted on the body by the second.[110]

Thus, armed with his Law of Gravitation and three Laws of Motion, Newton was able to explain falling bodies, motion of fluids, the tides and the shape of the earth and soundly refute Descartes's reigning gravitational theory of vortices mechanically pushing everything through the 'crouding of the aether'. Thus, in April 1686 he presented *The Principia* to the Royal Society. *The Principia* is a vast work, but one striking stroke of genius is his analysis of the shape of the earth. In *The Principia*, Newton had described a discovery by explorer Jean Richer, who had discovered that the net downward force of gravitation varies with latitude. This was because a pendulum clock ran slower in Cayenne in French Guiana than in Paris. Newton ruled out changes in temperature as the cause of this variance and deemed that it must be the net force of gravity. The reference to 'net force' is the addition of the gravitational pull to earth and the centrifugal 'lift-off' that any object on earth 'feels' because of the earth's rotation. Clearly, the gravitation overpowers centrifugal force; otherwise, every object would fly off into space and thus the net force of gravity is positive.

Newton observed that the earth is mostly water so he imagined a tunnel built from the North Pole to the centre of the earth flooded with water – call it the Arctic water column. He then imagined a second tunnel from the equator, also to the centre of the earth – call it the African water column. Theoretically, *if* the earth were a perfect sphere then both columns *would be* equal radii and thus *would be* of the same length. For the earth to rotate in equilibrium, both these columns of water would need to weigh the same; otherwise, there would be an imbalance, prompting a wobble in rotation. However, gravity was known to be less in Africa than in Paris or presumably the Arctic, and thus the two columns could not weigh the same if the earth was a perfect sphere. Therefore, the earth is not a perfect sphere. It followed that the African column must be longer than the Arctic column so that both columns weighed the same to maintain earth's equilibrium in rotation. This was because the African column needed more water to compensate for the weaker gravity in Africa than the Arctic. If so, the earth bulged at the equator and flattened at the Arctic (and thus the Antarctic), giving it an oblate shape. Newton even calculated the extra bulge at the equator that was needed to compensate for the weaker gravity there.[111]

However, controversy followed Newton again because Hooke accused Newton of stealing the inverse-square law of gravity from him (despite Hooke not having an equation – Hooke seemed ignorant of the difference between an idea and a proper theory, yet he must be recognised positively for his influence on Newton). Newton was perhaps reluctant to publish the crowning achievement, the 'System of the World' section – including his Law of Gravitation – probably due to fear of more controversy, yet Halley persuaded Newton to publish it in *The Principia*. Newton commented: "Philosophy is such an impertinently litigious Lady, that a man had as good be engaged in lawsuits, as to have to do with her".[112] Newton eventually presented his manuscript for *The Principia* to the Royal Society in April 1686, whose full name was *Philosophae Naturalis Principia Mathematica* (it was written in Latin in which Newton was fluent), which means *Mathematical Principles of Natural Philosophy*. (Newton's friend Edmond Halley paid for it to be published as the Royal Society could not afford to print it.) Former Professor of History and Philosophy of Science, P. M. Rattansi ,described this momentous event as follows: "The great work was published in May 1687 and created an immediate sensation. Even the followers of Descartes, who rejected its central idea of an attractive gravitational force, came in time to recognise its achievement. The dream of a mathematical science of nature was at last set on firm foundations..."[113]

Newton had not only explained Galileo's experiments on free-falling objects in mathematical terms but of equal importance he had 'clothed' Kepler's laws of planetary motion with his new mathematics and physics in one giant scientific coup, demonstrating the universality of his new laws of gravity and motion.

Chaper Four Plate

Newton as divine geometer by William Blake

FOUR

NEWTON'S REALITY

Newton's discovery of gravity was a revolutionary idea, but it contained risks to religion due to the mindset of the age. People then believed that the planets moved in perfect circles according to the harmonics from the music of the spheres – those spheres being the planets. All this was consistent with religious thinking. Newton was able to draw on Galileo's demonstration of Copernicus with the sun at the centre of the solar system and use Kepler's insight that the planets moved in elliptical not circular orbits, which had corrected the mathematics which predicted the positions of the planets. However, there was a lot left to be explained and mathematical equations to be invented.[114] With Hooke's clue that somehow there was an inverse-square law to be found governing all this, the stage was set for Newton's major discovery – his equation governing the mysterious invisible force – gravity, as we saw in Chapter 3. Astronomers could make predictions but they started to look for the causes for these anomalies not with God but with mathematics. But this required a quantum leap – that God used mathematics. Why would God need mathematics? Today, we can quote Nobel Prize winner Paul Dirac, who said that "God is a mathematician of the highest order" but in that day this type of thinking was not accepted. In medieval times, everything had a purpose. If something moved, there was a reason for this movement. Everything had a reason for its existence – birds were designed to fly and therefore had wings given to them by God.[115] Darwin's theory of evolution was yet to come, another 200 years later. So Newton had to walk this delicate line between religion and science.

The Principia contained laws and principles of motion based on Newton's theory of universal gravitation in extraordinary detail. So whereas previously scientists made predictions based on observations, Newton was able to give causes to these predictions and allow people to calculate and make predictions. This was a major turning point in the Enlightenment because previously the mindset had been that knowledge was there to be revealed through scripture but scripture was not accessible to reason. However, this had started to change. The realisation that man is not missing a rib as in the story of Adam and Eve produced the insight that the Bible should not be taken literally but rather as a book of principles. Thus, science and reason (using God's mathematical designs of creation) became tools where the principles of the universe could be discovered.

In Chapter 2, mention of Newton calling Locke a Hobbist was made in relation to philosopher Thomas Hobbes, who had embraced Greek atomism and Renaissance humanism. Hobbes was a materialist who shunned religion. In that age, a materialist was almost an atheist and was often called a Hobbist. In actual fact, Hobbes today would be called a deist because he believed in a first cause (a passive creator God) but, like Aristotle, believed the universe just works on its own steam despite being created by a first cause. So, the changing reality of 'natural philosophy' with its fusion of science and religion saw Newton and other scientists of his day view the world changing from the religious 'everything has a purpose' universe to the naturalistic 'Clockwork Universe', where the universe runs on scientific and mathematically based laws. Religion had left the laboratory. Historian of Science, Richard Westfall refers to tension between Christianity and science being resolved by Newton, as author Peter Watson relates: "[Sixty-five years after a Cardinal Bellarmino of the Catholic Church condemned Copernicus astronomy] Newton engaged in a correspondence with a certain Thomas Burnet, who claimed that the scriptural account of the Creation was a fiction, composed by Moses for political purposes. Newton defended Genesis, arguing that it stated what science-chemistry would lead us to expect. 'Where Bellarmino had employed scripture to judge a scientific opinion, both Burnet and Newton used science to judge the validity of scripture'. This was a huge transformation. Theology had become subordinate to science, the very opposite of the earlier position..."[116]

As we saw in Chapter 3, Newton's study of alchemy (the forerunner of chemistry) caused historical questions to be raised as to whether Newton was a scientist or alchemist or even a theologian. Yet, Newton saw what may be a perfectly valid connection between science and alchemy in some instances. For example, he linked his mathematical concept of mass in *The Principia* with alchemy when he introduced a further theory 'Hypothesis 3'. In Book 3, Proposition 6 of the final version of *The Principia*, Hypothesis 3 stated that the fundamental building blocks of matter were the same. Therefore, all forms of matter could be transformed into each other. This was a prophetic scientific insight, albeit based on early 'chymical' theory.[117]

Indeed, Newton had seen past the Greek concept of *atomism* referred to in Chapter 2. This concept formed part of a crucial principle that helped shape modern science. The earliest Greek thought recorded on this topic came from Thales (585 BCE), of the Greek colony Miletus. Thales advised Pythagoras (circa 570–495 BCE) to travel to Egypt to learn mathematics. Why? The Pythagorean school of thought was that the nature of reality was governed by mathematics, and indeed Newton embraced this concept in investigating reality. The atomism of the Greeks was part of this thinking and a number of Greek philosophers contributed:

(a) Before the atomism of Epicurus and Democritus, Leucippus was probably the first known proponent of atomism who proposed that the universe only consists of atoms or vacuum (although Aristotle argued against it)[118] and in rare cases 'atoms' could 'swerve' unpredictably.[119] This was probably the earliest origin of the modern term *indeterminism*, and even Aristotle had acknowledged the possibility of accidents in nature.[120]

(b) Parmenides and Zeno proposed that matter was immutable in that nothing can be created or destroyed. There is one unchanging reality – our perception of change is an illusion.[121]

However, whereas the Greeks saw atoms interacting or fitting together via their shapes, Newton postulated that the atoms connected through a network of strong forces acting at tiny scales.[122] Newton saw matter as made up of:

(a) mathematical points; or
(b) mathematical points and 'parts'; or
(c) a simple indivisible entity; or
(d) 'individualls' (real indivisible parts – this is like the Greek concept of 'atom').[123]

He saw these mathematical points as imaginary and lacking real magnitude – even an infinite number would sink to a single point. This evokes the Zeno Paradox, which will be discussed in Chapter 10 – Leibniz's Reality. Newton then described his concept of space as the separation between these basic parts (atoms) as 'vacuities interspersed' and that the cosmos was full of these empty spaces.[124] Matter would populate these spaces in the form of that space, eg. a ball would fit into a spherical space. Newton held that space extended infinitely in all directions but with an inherent mathematical potential or structure, such that it was composed of motionless parts or containers being 'eternal in duration and immutable in nature.'[125] This was in contrast to Descartes's theory that extension (the filling of space by matter) had an indefiniteness or undetermined nature.[126] Newton argued against Descartes on the basis that if Descartes's reality were true then this indefiniteness would imply God was only indefinitely perfect.[127] It turns out that this indeterminacy of infinity expressed by Newton may support a more modern version of Leibniz's answer to the famous problem of evil which I explain in Chapter 11, 'Leibniz's 'Best of All Worlds' Theodicy'.

It is important to recognise that Newton's concept of space involved three separate dimensions: up, down and across (or x, y and z for mathematicians) in which events occur to observers at the same time regardless of position (Newton assumed absolute time, space and velocity and an absolute resting frame, which he saw as basic scientific assumptions).[128] In Einstein's later 'tweaking' of Newtonian physics, a great shift occurs, namely that events do not happen at the same time due to differing frames of reference, light being a maximum speed and space having the ability to curve with the effect of gravity (a concept that would have surprised Newton). In fact, in this four- dimensional space (x, y, z and t for time as the 4th dimension), space and time are not separate but linked so that an object moving is given its coordinates x, y, z and t in a system used by Einstein called Minkowski space time, where a point is called a world point and follows a path called a world line so you can pinpoint its essential statistics (being its x, y, z and t coordinates).[129]

So time is a constant quantity flowing at a constant rate in Newtonian physics – what we call part of classical mechanics today. Gravity is quantified and given an equation but the cause is not explained, nor is the essence of gravity as a mysterious force explained. As Newton conceded: "I have not been able to discover the cause of those properties of gravity from phenomena, and I frame no hypotheses."[130] As observed earlier, Newton disliked hypotheses in favour of just presenting a theory based on the experimental results.

What else is unexplained in Newtonian physics? It seems mass is also a mystery. Recall Newton's second law set out in the last chapter:

2. The rate of change of momentum of a moving body is proportional to and in the same direction as the force acting on it, ie. F = d(mv)/dt where F is the applied force, v is the velocity of the body, and m its mass. If the mass remains constant, F = mdv/dt or F = ma, where a is the acceleration.[131]

Physicist Ernest Mach saw a shortcoming in the Newtonian concept of mass which he explained as follows: "With regard to the concept of 'mass', it is to be observed that the formulation of Newton, which defines mass to be the quantity of matter of a body as measured by the product of its volume and density, is unfortunate. As we can only define density as the mass of a unit of volume, the circle is manifest."[132] In other words, references to 'mass' in Newton's equations are circuitous and do not tell us what mass is. Mach's principle that the inertia of an object derives from its interaction with the rest of the universe was made use of by Einstein in his theory of relativity.[133] Indeed, scientists are still struggling with their understanding enhanced by the 'mass-giving' Higg's Boson, whose existence was confirmed by experiment many years after the theory of the Higg's Boson was postulated.[134]

Newton had expressed the view that his work rediscovered what the ancients had known. Thus, on space the ancients had taught that "a certain infinite spirit pervades all space into infinity, and contains and vivifies the entire world" and that "matter was moved within that infinite spirit" in accordance with the principles of physics within his *Principia*.[135] Newton and Leibniz clashed on this conception of space. Leibniz charged that Newton made space into God's spirit or body and needed God's interventions when imperfections occurred. Newton counterclaimed that Leibniz saw the universe as created by an 'absentee landlord'.[136] We will compare Newton's physics to Leibniz's physics in the next chapters on Leibniz.

In fact, in the final part of his then yet-to-be-published *Principia*, he conjectured that as comets approached the sun, their fiery tails were replenished by solar material which in turn rejuvenated the same fluids that provide sustenance to living things, whenever planets passed through their tails.[137] Which indeed has happened since and has caused alarm to some people due to the unfounded fear that the tail could contaminate the earth. This stunning conjecture shows again the prophetic reach of Newton's intellect because scientists today have indeed found evidence that the fluids of life – amino acids – have been found on comets or can form from comet ice. During the Stardust space mission, glycine, an amino acid, was found for the first time on a comet – Comet Wild 2. Origin-of-life theories now include conjectures that such building blocks of life have found their way to earth to contribute to Darwin's postulated pre-biotic soup that spawned life.[138]

As author Peter Watson observes: "The beauty of Newton's solution to the problem of gravity is astounding to modern mathematicians, but we should not overlook the fact that the theory was itself part of the changing attitudes of wider society."[139] Watson saw that more practical problems of formulating a practical astronomy for navigation rather than the workings of the divine mind dominated this drive to solve the mystery of gravity.[140] Today, aspects of gravity remain a mystery to physicists and, as quoted earlier, Newton himself conceded he was not describing the cause or nature of gravity. He was stating mathematical principles derived from empirically based data. Indeed, Kepler's data was based on the meticulous data built up by his former principal Tyco Brahe.[141]

Thus, it is important to look at the main rival gravity-like theory proposed at the time by Frenchman René Descartes using 'mechanical philosophy'. In contrast to Newton's universal gravitation, Descartes had proposed that nature had to be studied like a machine – not as living things full of purpose as the prevailing medieval mindset saw them. Thus, the Clockwork Universe became a major paradigm shift in thinking. Machines only have the purpose instilled by their makers, so it made no sense to look at purposes. Yet, according to Descartes, God still figured in this scheme because he saw God creating the world based on principles of matter and motion. Matter would rub against other matter and the quantity of motion remained in equilibrium so the Clockwork Universe would never run down. He then imagined the universe full of 'subtle matter' of whirling vortices with the slower moving subtle matter congregating to make up the earth, planets, etc. A solar vortex of this matter then formed the atmosphere and the pressing down of this matter formed the earth. Thus, gravity was a *mechanical* process where the motion of matter pushed against a chain of crowding matter.[142] This mechanical philosophy was consistent with the Greek atomism but was contrary to Aristotle's theory of forms and 'sensible qualities' of objects. A mechanical philosopher would hold that an object does not have a particular smell but only that particles of particular size, shape and mobility strike our sensory organ – our nose – producing that smell, whereas Aristotle would have held the object has an inherent smell due to its unique nature.[143] Professor Principe explains the mechanical philosophy as follows: "First, effects were caused only by mechanical contact – like a hammer on stone... There is no room for action-at-a-distance. Second, the world and objects in it... were conceptualized as machines. Mechanical philosophers compared the world to a complex clockwork."[144] (Recall the Clockwork Universe discussed earlier.)

Newton's thinking broke through all this in that his theory of gravitation involved no such mechanical forces of crowding subtle matter – it was an invisible force. This was a controversial theory, contrary to Descartes's more popular theory, so much so that one of the great scientists of the day, Christian Huygens – who had clashed with Newton – called it absurd. Although Newton's invisible force of gravitation was contrary to Descartes's mechanical theory, Newton did ponder whether attractions or repulsions between particles could explain gravity. It could have been the attractive forces he had witnessed in his alchemy that may have inspired this line of thought. In effect, it showed the great span of Newton's intellect because today the explanation of gravity involves the greatest unanswered question in science – the unification of the various forces of nature – gravity, electromagnetic forces and the quantum forces. Newton saw that matter and motion were not enough to explain nature. Newton had done what scientists see as paramount: produce universal laws governed by mathematical equations. Anything more is metaphysical – the realm of hypothesis was the realm that Newton eschewed, unlike Hooke, who first criticised Newton's theory of light for not having a hypothesis for what light was.

Newton's theory of light had been partly inspired by French scientist René Descartes, who saw light as a type of pressure through the ether. He thought space was made of 'luminous matter' which made up the sun and the stars. Descartes saw matter like 'ethereal globules' spinning like tops that obliquely strike an object – different speeds gave different colours.[145] Newton believed in the ether but not an ether composed of mechanically grating microparticles. Newton described his concept of ether in relation to light, gravity and even the life forces or animal spirits or even the soul. This was no criticism because the theory of the ether was still a valid theory until 1887 when Albert Michelson and Edward Morley conducted an experiment with an instrument (an interferometer) supposed to measure changes in the speed of light when it was rotated, which it would only do if an ether existed. When no change in the speed of light was observed, the theory of the ether was abandoned.[146]

Today, we consider space is empty and not filled with an ether, but it is hard not to admire Newton's intuition. He was fascinated with the possible connection between musical harmony and colours.[147] He also saw light as composed of luminous corpuscles travelling across the medium of the ether (which was in contrast to Descartes's luminous matter crowding across the ether). Newton explained refraction by suggesting these light corpuscles accelerate in the less dense ether and slow down through glass, deflecting the pathway of these light particles. Newton measured the 'bigness of the vibrations' of the ether associated with various colours. Elementary lengths between lens and plane coincided relationally with the length of the 'pulse' of the vibrating ethereal medium (corresponding to today's correct scientific classification of colours in light to the respective constituent wavelengths).[148] However, Newton did not claim a complete theory of light including wavelengths as Hooke said he should have had, but his intuition towards the complete theory of light was impressive. It is interesting that Newton did not postulate 'refraction of gravity' due to the ether even though his theory involved gravity travelling through both the ether/space and matter, eg. through to the centre of the earth – the idea that light could bend only occurring to Einstein over two centuries later. This may have been because it was not supported by experimental evidence or perhaps that gravity was not composed of any 'particles', like the vortices suggested by Descartes.

Historians have often questioned whether Newton's conception of ether was of a material substance or of a spiritual substance, but author Niccoló Gucciardini believes it is an open question.[149] It would seem that if Newton believed light's interaction with the ether obeyed mathematical laws and gravity obeyed mathematical laws then he could not have seen the ether as spiritual, for that would have placed it in a 'magical' non-scientific realm. Newton's discovery of the universal laws of gravity and motion rolled back the 'magical fabric' of the universe, exposing it as a mathematical framework of matter and space. It is hard to see him making the ether an exception to this framework. Like time, gravity and mass, Newton went no further to dissect what the ether might be, and for his purposes it was the fabric of space (yet with an underlying structure or potential to contain all the shapes of objects in the universe).

To Newton, God had created a Clockwork Universe but not quite. In Chapter 3, we saw how Newton envisaged our solar system emerging from 'common chaos' with subsequent condensation into solid globes by gravitational attraction. Chaos theory, involving the principle that bodies governed by a deterministic law can behave indeterminately (ie. erratically), was not yet invented, but again, Newton's intuition gave him clues to think in terms of the uncertainty of physical systems. Indeed, he had to cope with the now well-known chaos inherent in astronomical calculations and has been wrongly criticised in the 'curve fitting' he did to make some of his equations 'very nearly' equate. Scientists still have to cope with curve fitting and formulation of constants to make equations work – 'margin of error' is a key working concept to any scientist.[150] Newton had remarked that the human intellect is incapable of dealing with certain planetary motions.[151]

The famous 'three-body problem', which puzzled mathematicians in the 18th and 19th centuries, involves what is now known as a classic example of chaos – that is, chaotic 'unknowable' behaviour arising from a deterministic system of three planetary bodies interacting.[152] In summary, Newton may have conceived of a hybrid chaotic and Clockwork Universe governed by both deterministic and non-deterministic laws.

Laplace, himself one of the great scientists, said that Newton was the luckiest scientist who ever lived because he had discovered the major physical laws which could only be done once. Yet, unlike today's scientists, he wrote *The Principia* to promote the belief in God by demonstrating the harmonious and beautiful way in which the laws of the universe worked. This was an argument for the existence of God and was really an early form of the present-day argument that the universe seems to display fine tuning making the universe bio-friendly enough to originate life itself. One version of this argument has become a full hypothesis known as the *anthropic principle*. This principle postulates that the universe appears to be designed with tailored universal laws whereby tiny changes in those laws would mean that humans could not exist. Put another way, an accident of the forces of nature which varied those laws in some minute way would mean that humans would not exist. Newton was also concerned that the concept of the Clockwork Universe could turn people away from believing in God. Rather, Newton strongly believed that God had planted active principles in nature.

Much of Newton's maths and science remains valid today. Einstein's later theories showed Newton's laws broke down at the extremes of reality, such as in a black hole, but Newton's laws are preserved in most everyday situations. In one of his famous quotes, Newton paid homage to the founders of the new science when he said that he had been able to go further as he had stood on the shoulders of giants or in his old age said, "I don't know what I may seem to the world but as to myself, I seem to have been only like a boy playing on the seashore, and diverting myself in now and then finding a pebble smoother or prettier shell than ordinary, whilst the great ocean lay all undiscovered before me."[153]

Chapter Five Plate

"The artist's concept depicts NASA's Kepler mission's smallest habitable zone planet. Seen in the foreground is Kepler-62f, a super-Earth-size planet in the habitable zone of a star smaller and cooler than the sun, located about 1,200 light-years from Earth in the constellation Lyra. Much like our solar system, Kepler-62 is home to two habitable zone worlds."

FIVE

NEWTON'S FINELY-TUNED COSMOS

In Chapter 2, we saw the origin of Renaissance humanism with Lucretius's epic poem *On the Nature of Things*, including themes of Epicurus's natural philosophy "denying divine design and depicting the universe as a whirl of atoms, in a perpetual state of flux."[154] Newton referred to Lucretius as leading the "crowd of philosophers that preceded Aristotle." Despite Newton's own view of the cosmos, he approved of Lucretius's account of infinite space because, in the absence of space, gravity would cause all matter to clump into one giant mass.[155] Newton's *Principia* was an in-depth analysis of what gravity did to matter. For instance, no celestial body could move in a straight line or in a proper conic section because gravity dragged it one way or another. As Professor Iliffe explained: "Newton laid down a novel and highly original way of mutually fine tuning theory and data. Increasingly sophisticated empirical evidence from a host of independent sources could be fed back into the mathematical theory and then tested against both well-known and unexpected phenomena – thereby corroborating the existence of the general force in question.[156]

Despite Lucretius's and Epicurus's 'materialism' and Newton's fear that *The Principia* might promote atheism, Professor Iliffe observes that Newton and his followers made "entirely novel physical and astronomical contributions to the field of natural theology… [and] paid lip service to …the so-called Argument from Design."[157] Further, in correspondence with Richard Bentley, Newton stated: "…God had ensured that planets travelled at *just* the correct velocities, were at *just* the right distance from the sun (and travelled in just the right orbits), had *just* the right mass (and hence the right amount of gravitational power) and revolved on their axes at *just* the correct rates." Also referring to: "… a perfect (ie. divine) mechanic in the 'Preface to the Reader' of *The Principia*, he insisted that only a being exquisitely skilled in mechanics and geometry could have compared and adjusted all these features; he might have added that only a great philosopher could have discovered the principles underlying their motions."[158]

In further correspondence with Richard Bentley, Newton thought aspects of *The Principia* could shed light on God's actions at the time of creation: "[and]… the long-term effects of universal gravitation on all matter in the universe… [were that] provided space were really infinite, matter initially dispersed evenly across the universe… would never come together into one mass in its centre."[159] As we will see later with the latest cosmological theories, this is remarkably similar to at least one modern account of God creating the universe via a Big Bang being consistent with the present 'spread-out' (inflated) state of the universe.

In fact, scholars have placed Newton's appeal to divine intelligence as the ultimate or 'final cause' as one of the pillars of Newtonianism in contrast to Descartes and his followers' strict support for mechanical causes (the other pillars including both Newton's mathematical and synthetic approach to science and his inductive, experimental approach to establishing the science).[160]

This is Newton's designed and finely-tuned universe (as opposed to the fine-tuning of his theories of physics), but he also referred to biological design and it is remarkable that today's scientists have indeed discovered that the universe not only has the appearance of Newton's fine-tuning but has the scientific laws and constants that permit life. Scientists, by analysing these laws and constants, found that had they been outside certain narrow ranges, life may not have emerged. One recent publication by two scientists entitled *A Fortunate Universe – Life in a Finely Tuned Cosmos,* by G.F. Lewis and L.A. Barnes, explains:

> *Peering more deeply into [the great scientific advances] ...examining the basic make-up of the Universe, reveals that we are not as mediocre as it seems. The fundamental particles from which everything is constructed, and the fundamental forces that dictate interactions, appear to be fine-tuned for life. Minor tinkering with either would leave the Universe dead and sterile... We find ourselves questioning the nature of many of the things we take for granted, from the fabric of space and time, to the mathematical underpinnings of the universe. At every level, we find that our universe's ability to create and sustain life forms is rare and remarkable.*[161]

How finely tuned is the universe? Professor Paul Davies, in his book *The Mind of God*, refers to the laws of nature and why the universe obeys scientific laws and mathematical rules so well and is not more chaotic than it is.[162] The issue of whether the universe is also bio-friendly involves the question of whether there was a supernatural cause. The major question which has puzzled scientists is 'Why does the universe appear fine tuned to support life?' Professor Paul Davies has a good analogy to explain 'fine-tuning' via a thought model:

> *A good way to think about this is to imagine playing God and setting out to design a universe. Suppose you had already settled on the basic laws of physics, but you still had some free parameters at your disposal. The values of these parameters could be set by twiddling the knobs of a designer machine... Turn one knob and the electron gets a bit heavier, turn another and the strong force becomes a bit weaker, and so on... How many knobs are there? The standard model of particle physics has about twenty undetermined parameters, while cosmology has about ten. All told, there are over thirty 'knobs'."*[163]

In another book by Professor Paul Davies, *The Goldilocks Enigma*, the author says:

> *If almost any of the basic features of the universe, from the properties of atoms to the distribution of the galaxies, were different, life would*

very probably be impossible. Now, it happens that to meet these various requirements, certain stringent conditions must be satisfied in the underlying laws of physics that regulate the universe, so stringent in fact that a bio-friendly universe looks like a fix or 'a put-up job', to use the pithy description of the late British cosmologist Fred Hoyle. It appeared to Hoyle as if a super-intellect had been 'monkeying' with the laws of physics. He was right in his impression. On the face of it, the universe does look as if it has been designed by an intelligent creator expressly for the purpose of spawning sentient beings. Like the porridge in the tale of 'Goldilocks and the Three Bears', the universe seems to be 'just right' for life, in many intriguing ways. No scientific explanation for the universe can be deemed complete unless it accounts for this appearance of judicious design. Until recently, 'the Goldilocks factor' was almost completely ignored by scientists.[164]

Probably the most striking example of this fine-tuning is the 'setting' of the cosmological constant – or more correctly, we should ask: 'Was it set or did it just occur by a lucky fluke on formation of the universe at the Big Bang?' The Wikipedia entry for 'Cosmological Constant' provides a good explanation for what the cosmological constant is:

It was originally introduced by Albert Einstein in 1917 as an addition to his theory of general relativity to 'hold back gravity' and achieve a static universe, which was the accepted view at the time. Einstein abandoned the concept after Hubble's 1929 discovery that all galaxies outside the Local Group (the group that contains the Milky Way Galaxy) are moving away from each other, implying an overall expanding universe. From 1929 until the early 1990s, most cosmology researchers assumed the cosmological constant to be zero.[165]

In fact, scientists were baffled when the cosmological constant was discovered not to be zero but one part in $10,^{120}$ which is a number so small that it is hard to describe. Author and former journalist Lee Strobel has given his opinion of how small it is: imagine a ruler stretching from one side of the universe to the other – if the settings of this constant were moved by one inch, the effect on life would be catastrophic and anything larger than a pea would be crushed. Paul Davies described it this way:

A factor of ten would suffice to preclude life... A factor of ten may seem like a wide margin, but one power of ten on a scale of 120 is a pretty close call. The cliché that 'life is balanced on a knife edge' is a staggering

understatement in this case: no knife in the universe could have an edge that fine.[166]

What is being postulated? Clearly, Newton's explanation of a divine intelligence aligns with a number of 'fine-tuning' examples like this. However, many more examples than were available to Newton have emerged. Take for example the conditions required for the building blocks of life, namely, the carbon atom. Using carbon, a model theory may work along these lines:

- Life requires carbon-12
- Carbon-12 is formed in the heart of large stars (star life of 5 to 10 billion years)
- The universe is over 13 billion years old
- It took life 3.8 billion years to emerge
- Thus, to form carbon *and* human life, it took up to 13.8 billion years to see life emerge as we know it today (up to 10 billion star life years *plus* 3.8 billion years to emerge)
- Thus, the universe was created to incubate the building blocks for life *and* life itself.

NASA states we now know of over 3,200 habitable planets after analysis of Kepler space telescope data. There may be millions more as there are billions of stars with planets as yet unknown. With our present level of knowledge, we cannot know the alchemy behind universe and life building or the precise time scales required and why, but there are clues. One of the best clues is the nucleosynthesis of carbon. In an article headed 'Nucleosynthesis. Life, Bent Chains and the Anthropic Principle'[167] by nuclear physicist Dr N.M. Clarke of the University of Birmingham (who worked for many years on the Nuffield cyclotron), a number of interesting points emerge. Looking back at the history of one branch of nuclear physics, the improbability of carbon-12 seemed apparent. Nuclear physics had difficulty accounting for it as the combination of particles to make carbon-12 did not make sense. Although beryllium and helium nuclei have more energy than a carbon nucleus, beryllium breaks into two helium nuclei in a fraction of a second unless a third helium atom intervenes (resulting in the three helium nuclei forming carbon). In fact, the insight for the solution came only because astronomer Fred Hoyle had postulated the solution, namely that a resonance must exist (7.6 MeV) which would explain the unusual nucleosynthesis that must take place to form carbon-12; otherwise, we would not be here (because carbon-12 is necessary for life). His prediction proved correct and the resonance was found. Dr Clarke commented: "…Hoyle's prediction of its existence is possibly the only proven example of a scientific prediction using the 'Anthropic Principle'."[168]

Astronomer Fred Hoyle had made an audacious prediction. He had said that "there must be an excited state of the carbon nucleus… with precisely 7.65 MeV of energy…"[169] Hoyle's prediction turned out to be true for 3 helium-4 nuclei fusing at 100 million degrees inside a red giant because it creates a more stable resonant reaction before the excited nucleus decays into carbon-12. "It was an astonishing triumph, and remains the only successful prediction based on the anthropic argument ahead of experiment – that is, 'I exist, therefore it must exist'".[170] Hoyle concluded from this that the universe was fine-tuned for life because without it there would be no (carbon-based) life.

But Dr Clarke goes further:

> ***If you think that the two-step process that forms the excited state at 7.6 MeV is an amazing coincidence, then the decay of this state is even more remarkable! For every ten thousand decays, four result in the emission of two gamma rays which takes the excited carbon-12 to its stable ground state! So it is that rare two-step process and those 4 in 10,000 decays which enable me to write this piece and for you to read it!"***[171]

This would make Fred Hoyle's comment that the existence of carbon-12 is a lucky fluke an understatement. The fine-tuning theory is also known as the strong anthropic principle, proposed by physicists John D. Barrow and Frank Tipler – that the fundamental constants of the universe are 'just right' and fine-tuned for life to exist. More precisely, they said: "The universe must have those properties which allow for life to develop within it at some stage of its history."[172]

In more general terms, the authors of *A Fortunate Universe – Life in a Finely Tuned Cosmos* have produced an interesting table regarding carbon and its 'neighbours' on the periodic table as to the number of chemical compounds that can be made with each element and hydrogen, which is:

Li	Be	B	C	N	O	F	Ne
4	6	38	29,019	65	21	6	0

From *A Fortunate Universe – Life in a Finely Tuned Cosmos*, namely, G.F. Lewis and L.A. Barnes, p268[173]

On other constants, Sir Roger Penrose, after commenting on confirmation of the Hoyle resonances, said: "It is remarkable that the constants of Nature are so adjusted that such energy level should be in just the right place, so life, as we know it, could come about. Another example of cosmic good fortune is the fact that the neutron's mass is just slightly greater than that of the proton (1,838 and 1,836 electron masses, respectively). The existence of stable nuclei, on which almost the whole of chemistry depends, rests upon this seemingly fortuitous fact."[174]

Australian physicist Brandon Carter had introduced both the weak and strong anthropic principles in 1973 and had defined the weak anthropic principle as follows: "We must be prepared to take account of the fact that our location in the universe is necessarily privileged to the extent of being compatible with our existence as observers."[175]

This meant that the fine-tuning theory suffers from 'selection bias' in that we humans just happen to think the universe is special because it hosts us when, in fact, the whole of our existence is a series of lucky flukes. An objection to the strong anthropic principle is that the universe appears 'unintelligent' or badly designed in that it is really a dangerous, inhospitable place. This issue is fully addressed in Chapter 11, 'Leibniz's Best-of-All-Worlds Theodicy' and Chapter 13, 'Is a Perfect World Possible?'

What has emerged today that might challenge Newton's explanation of a divine intelligence for fine-tuning? Everett's Many Worlds hypothesis[176], the Megaverse and the Multiverse hypothesis – all these different parallel universe hypotheses (discussed in Chapter 25) are now being used by some 'anti fine-tuning' opponents (such as atheists or agnostics) to explain away the astonishing fine-tuning of the universe and the Goldilocks enigma highlighted by Professor Paul Davies. In fact, Barrow and Tipler added a further gloss to Carter's anthropic principle which was:

1. There exists one possible universe 'designed' with the goal of generating and sustaining 'observers'.

2. Observers are necessary to bring the universe into being.
3. An ensemble of other different universes is necessary for the existence of our universe.[177]

Thus, the anti fine-tuning opponents plus Barrow and Tipler are effectively stating that to explain the anthropic fine-tuning, there must be multiple universes with multiple laws and we just happen to live in the universe with bio-friendly laws. This seems to be missing Newton's alternative concept, namely, that a divine designer created the universe with such laws and constants to permit life to evolve and live.

The overarching problem with Barrow and Tipler's interpretation is that the real scientifically based multiple universe theories existing today, such as Everett's Many Worlds theory, the Megaverse or Multiverse, were all proposed for completely different reasons than to counter the anthropic principle. Crucially, Everett's theory involved many universes with *identical* universal laws to ours, which would make its use inconsistent with 'Multiple Universal Laws' relied on to oppose the fine-tuning theory.

What about the Megaverse or Multiverse hypotheses? Megaverses, which we will explain in Chapter 26, arose from Andrei Linde's theory of chaotic inflation producing island or pocket universes branching off from the main expanding universe in a fractal fashion. We will see that the Multiverse hypothesis has arisen from purely mathematical implications based on concepts such as string theory. However, whilst the Megaverse results from one of the leading theories for inflation of the universe, it does not imply different laws – Linde himself comments that it implies only varying application of those laws.[178] It may, however, allow varying constants within those laws, but we just do not know. The Multiverse does possibly imply different laws and constants but at present it has the status of a purely mathematical hypothesis according to its leading proponent, Professor Leonard Susskind (see Chapter 26). On the other hand, Linde's theory of chaotic inflation producing island universes (the Megaverse) does seem to have evidence – but not proof – in its correlation with the fractal patterns of background radiation in the universe (as seen in the COBE and WMAP satellite maps showing imprints of quantum fluctuations in the early universe).[179]

The authors of *A Fortunate Universe – Life in a Finely Tuned Cosmos*, namely, G.F. Lewis and L.A. Barnes, seem to be divided as to the implications of fine tuning, as shown in this extract of their 'Socratic dialogue' towards the end of their book:

> ***Geraint:*** *Well, I'm sure you'll agree that we won't settle these debates today. But I think I can summarise your thoughts on this. To you, the apparent fine tuning of the properties of the universe, the properties that allow you to exist as a living, thinking, active creature, is not an accident, not some random role of the dice in an inflating cosmology. To you, the conditions were chosen, the dials were explicitly set to allow for your existence. The universe contains good things, like free moral agents and all that they can do and learn and appreciate. The presence of these qualities is not accidental, but reflects the intent of the creator, the person that set the dials. How's that?*
>
> ***Luke:*** *That's about right. But I feel that you don't buy the argument.*
>
> ***Geraint:*** *Alas, I don't. I think that moral beliefs have arisen during our evolution and allowed us to survive and thrive in communities and clans. Anyway, all these people appear to be a little amoral sometimes!*[180]

This reflects two scientists' points of view, but what of the philosopher's modern viewpoint in view of the lack of a verifiable 'multilaw' Multiverse theory? Richard Swinburne, Oxford philosopher, has said: "To postulate a trillion trillion other universes, rather than one God in order to explain the orderliness of the universe seems the height of irrationality."[181]

Richard Swinburne has clashed with another leading philosopher, Herman Philipse, the author of *God in the Age of Science?: A Critique of Religious Reason.*[182] In this book, Herman Philipse tackles many arguments for and against the existence of God, such as the cosmological argument, which is the ancient argument that all natural things depend for their existence on something else, ie. are caused – thus, the universe exists implying a first cause which is God. However, the topic relevant here that he tackles is 'The fine-tuning argument from logically possible universes', which he addresses with reference to Richard Swinburne as follows: "… if the constants in the laws of nature holding for our universe fall within these narrow intervals, the universe is said to be 'fine-tuned' for (human) life… Then it is argued that *a priori* these facts of fine-tuning are very improbable if God does not exist, whereas they are much more probable if God exists. Consequently… [w]e have here a powerful C-inductive argument for the existence of God as Swinburne concludes."[183] Philipse then challenges the argument:

> *"But why are the facts of fine-tuning thought to be a priori very improbable if there is no God… Since there are many of these (alleged) facts, which can be understood properly only by those who possess the relevant scientific knowledge, I shall mention (only the cosmological constant)".*

Recall that this is one of the thirty or so constants referred to by Paul Davies above. He then presents the argument as simply the calculation of a probability as the usual proportion of 1., eg. a one chance in eight probability equals 0.125. Thus, the proportion of fine-tuned universes compared to all logically possible universes is very small and therefore,

> *"Because this proportion is very small, it is concluded that the prior probability of a life-permitting universe is very low or the prior probability (in a Godless universe) that a universe will be tuned is a function […] of all possible theories and boundary conditions… for any universe at all'. Swinburne assumes that this probability is small indeed, nor – I suspect – within the ability of any present-day mathematician."*

Philipse then argues that the problem with the argument is that each constant could take on any real number in the interval $[0,\infty]$ or perhaps be more limited to some finite interval for each constant, eg. $[a_0, a_n]$ but a_0 and a_n are unknown. Furthermore, fine-tuning proponents will support a narrow range of constants and anti fine-tuning opponents will support a large range of constants. He also raises the technical issue of countable additivity, namely, that if you are calculating a probability distribution, it must add up to 1 and the continuum of infinities between 0 and 1 causes calculation problems. This does not seem a serious problem for physicists who are expert at renormalisation of infinities which they meet often and in fact Philipse candidly concedes:

> *"Readers may not be impressed by these technical objections to Swinburne's first fine-tuning argument [because mathematical techniques may exist to overcome the difficulty]. However, the burden of proof rests on the proponents of the fine-tuning argument."*[184]

Thus, Philipse has effectively invited solutions to this 'technical issue' and so to accommodate him a qualitative rather than a quantitative approach is suggested:

1. To illustrate the issue, assume that a member of the public asked NASA for the probability of finding a habitable planet in the universe. NASA then replied that they can only compare their estimate of 3,200 habitable planets against their 'guesstimate' of trillions of possible planets and concluded the probability was extremely low. Is that qualitative conclusion reasonable? If NASA had put a quantitative estimate on the trillions of planet and gave a figure of 0.00001 probability, would that be a more reasonable conclusion?
2. The technical problem of countable additivity only arises if one attempts to calculate the probability for each universe with the myriad permutations and combinations of ranges for the thirty or so constants that are 'fine-tuned' to make a universe fit or unfit for life. Stephen Hawking described applying the anthropic principle as a 'counsel of despair', and it would seem that this quantitative attempt to calculate the anthropic 'fine-tuning probability' is worse. Is there is a qualitative pathway to scientific or philosophical insight?
3. It is suggested that a qualitative result is sufficient and reliable. Unlike the NASA 'guesstimate' example above, the two scientists referred to in this chapter, namely, G.F. Lewis and L.A. Barnes, have done an extensive analysis of the narrow ranges of the constants that permit life in their book *A Fortunate Universe – Life in a Finely Tuned Cosmos* and late in the book discuss assigning probabilities to possible universes and without assigning a figure state: "If all we knew was (a) that a certain universe obeyed the laws of nature as we know them, without specifying the values of the constants of nature and initial conditions, and (b) mathematical knowledge, what is the probability that the universe would contain life forms? We have seen in previous chapters that this probability seems to be extremely small."[185]

The reason why we normally feel more confident about a calculation is that the assumptions must be subjected to more rigorous hypothesis and calculation. Thus, Philipse has suggested that the probabilities should be calculated. However, in this situation, we would only use mathematics where we would anticipate surprising or strange results occur which are contrary to our intuition. The calculations of narrow ranges over all possible ranges are linear and simple and instead of precise calculations, 'order of magnitude' calculations – common in science – provide a reliable alternative to reach qualitative conclusions.

Thus, the question becomes whether G.F. Lewis and L.A. Barnes's conclusion in point 3, that their unquantified 'fine-tuning' probability is extremely low is reasonable . It appears the result is reliable because:

(i) We have no reason to doubt their assumptions drawn from standard scientific sources.
(ii) Nor is there anything that indicates a possible surprising or strange result that might spring from a calculation because of the nature of the calculations – moreover, due to the problems that Philipse raises with the contentious choice of ranges of constants, a calculation may not be as reliable.
(iii) It would seem that there is a compelling case for anthropic fine-tuning which can be appreciated by viewing not only G.F. Lewis and L.A. Barnes's book but also such books as *The Goldilocks Enigma* by Paul Davies and *Just Six Numbers* by Martin Rees.

The harder question is whether a parallel universe explanation is a reasonable explanation for the fine tuning.[186]

(iv) Two opponents of fine-tuning who have challenged some of the assumptions of fine-tuning (before Lewis and Barnes' book was published) are Nobel Prize winner Stephen Weinberg and Physicist Victor Stenger. The particular 'for and against' arguments are better left for books just listed but in Stenger's article 'The Fallacy of Fine-Tuning', he states in the abstract: "…I will focus first on the five parameters that have the most significance because, if interpreted correctly, they seem to pretty much rule out almost every conceivable kind of life without fine tuning: ratio of electrons to protons; ratio of electromagnetic force to gravity; expansion rate of the universe; mass density of the universe; cosmological constant. I will admit that the features a universe would have, for slightly different values of these parameters, all other parameters remaining the same, would render unlikely any form of life even remotely like ours, that is, one that is based on a lengthy process, chemical or otherwise, by which complex matter evolved from simpler matter."[187] Stenger has various philosophical objections, including that other forms of life may exist, meaning we just do not know whether any chaotic universe would permit life. Furthermore, in 2011, he referred to an opinion of cosmologist Don Page, an evangelical Christian, namely, "…that the apparent positive value of the cosmological constant is somewhat inimical to life, because its propulsion acts against gravitational attraction needed to form galaxies. If God fine-tuned the universe for life he would have made the cosmological constant slightly negative". And generally in response to fine-tuning, Victor Stenger said: "The currently favoured solution to the problem among physicists is called the 'multiverse' in which our universe is just one of a great many others having a wide variation of values for the cosmological constant as well as other physics parameters. We happen to live in a universe suitable for us."[188]

In fact, a demonstration of how close we have come to not existing is with the settings for the cosmological constant. Physicist Max Tegmark describes its fine tuning (to support the multiverse):

> ***"…the actual knob setting for our universe is about 10^{-123} of a turn away from the halfway point. That means if you want to tune the knob to allow galaxies to form, you have to get the angle by which you rotate it right to over 120 decimal places! Although this sounds like an impossible fine-tuning task, some mechanism appears to have done precisely this for our universe."***[189]

In Chapter 26, we summarise Max Tegmark's four categories of multiverse and their experimental basis and meaning.

Newton's fine-tuned cosmos also included biological design. In this regard, there seems to be an 'elephant in our universe' not contemplated by the anthropic principle, and this is represented by a black hole – not in space but at the cornerstone of biology – origin of life. All discussions on physics just refer to life evolving in our 'fine-tuned' universe as a given, but it is far from that. In fact, leading biologist Professor Nick Lane refers to a second 'black hole at the heart of biology'[190] with respect to a paradox – the phylogenetic void that generated complex life from simple life such as a microbe.[191]

The leading scientific theory for origin of life is the RNA world theory published in 1968 that life began in a primordial mud pool with a type of self-replicating primitive RNA molecule or something like that self assembling (ribozymes have been suggested). However, the 'father' of RNA world, Professor Leslie Orgel, who with Nobel Laureate Francis Crick proposed the theory, pronounced in 1993 that the very existence of this ideal mud pool required such a peculiar set of substances, catalysts and self assemblies that he called it 'The Molecular Biologist's Dream'. Professor Orgel posthumously went further and a year after his death he published a paper concluding:

> ***"It is likely that such catalysts could be constructed by a skilled synthetic chemist, but questionable that they could be found among naturally occurring minerals or pre-biotic organic molecules… it is not completely impossible… but the chance of a full set of catalysts occurring at a single locality on the primitive earth… seems remote in the extreme."***[193]

The state of the art in origin-of-life science is also reflected in this comment by leading biologist Eugene Koonin of the National Center for Biotechnology Information in Maryland, USA, who said in 2007 that the origin of the first cell carrying genetic instructions for making proteins is "a puzzle that defeats conventional evolutionary thinking". He has now appealed to cosmology and the Multiverse to explain life itself and dispense with the 'problem of fantastic probability'.[194]

What is this fantastic probability? No one really knows but we can do some 'order of magnitude' calculations on the way to complex life. There are probably seven great hurdles necessary for human life – (1) a life molecule (2) the code in the life molecule (3) energy for the life molecule (4) the first complex cell, 15,000 times larger than a microbe, eg. a bacteria (5) photosynthesis – life as we know it could not exist without it, eg. oxygen, food chain, etc. (6) sexual reproduction (a mystery as to why life graduated away from 'least costly' asexual reproduction) and (7) consciousness (the evolution of 'gleams of consciousness' or higher consciousness is another mystery). Taking the 'Molecular Biologist's Dream' referred to by Orgel and Crick, with some biologists' estimates for populations of self-replicating molecules, we can get an 'order of magnitude' probability. The assumption here is a transition from a replicating chemical molecule to a candidate life molecule (whatever that may be) in a process known as 'chemical evolution', where molecules are 'selected' by replications. Thus, let us choose a model 29 chain polymer for our 'order of magnitude' calculation for a 'life molecule' to emerge from a primordial pond or underwater volcanic vent on the primitive earth 4 billion years ago.

The probability that one '29-mer RNA life molecule' with twenty-nine nucleotides accidentally self assembled from a population of 10^{14} – 10^{16} RNA molecules and vast other molecules within the soup or vents can be calculated on an assumption of 4.19 x 10^{43} molecules in the lakes, swamps and vents on earth. Leaving out much of the tedious calculation beyond these numbers gives a fraction of 4.4 /10^{53} molecules – ie. insufficient to produce one life molecule in, say, one hour. Then, assuming iterations of life molecule replications amongst trillions or more of these self replicators for 500 million years reduces the probability of getting a candidate for (1) – the life molecule (not yet life) to 1.54 / 10^{24} which is another very tiny fraction (ie. only pre-life molecules would be produced during 500 million years). Thus, you would need ~ 2.35 x 10^{22} universe lifetimes (say, ~13.8 billion years each) for one pre-life molecule to emerge. We do not know what the minimum chain length is for the first candidate life molecule, but noting that codons are three nucleotide units, a minimum of twenty-nine nucleotides seems reasonable (and twenty-nine would grow immensely as evolution progressed and the life molecules 'morphed' to a DNA molecule).

At stage (2), for a code to emerge in our 29-link life molecule, another calculation gives a bigger fraction, namely, 3.96 / 10^{3} 'life molecules' emerging (still only one pre-life molecule during one hour), having a pre-DNA-type code during a second 500 million years. This means you need a relatively slight increase of over ~ 2.35 x 10^{22} + 9 universe lifetimes for one life molecule to emerge with the code in it. Of course, if we are missing some fine-tuning factor in nature that we are unaware of, this period may shorten. If one objects that the transition from chemical-replicating molecule to life-replicating molecule includes gaining a code then this removes the nine universe lifetimes. This leaves ~ 2.35 x 10^{22} universe lifetimes which is still more than trillions of universe lifetimes, indicating we are missing the clever mechanisms in knowing how life began. It must be stressed that anyone doing the above calculation can arrive at different results because a biologist may wish to dispense with one assumption or add another or calculate with different probabilities, but the aim of the exercise was to demonstrate the type of fantastic probabilities to which Eugene Koonin was referring.

In reflecting on the seven stages for complex life above, consider stage (4) where somehow microbes increased in size by some 15,000 times to 'morph' into the complex cell – a process that perhaps resulted from microbes replicating (iterating) over 2 billion years.[195] Biologist Nick Lane describes the event:

> ***"After simple cells first appeared, there was an extraordinary long delay – nearly half the lifetime of the planet – before complex ones evolved. What's more, simple cells gave rise to complex ones just once in four billion years of evolution: a shockingly rare anomaly, suggestive of a freak accident. If simple cells had evolved into more complex ones over billions of years, all kinds of intermediate cells would have existed and some still should. But there are none. Instead there is a great gulf."***[196]

So continuing and again relieving the reader of many tedious 'order of magnitude' calculations and consideration of replication times, assembly, reaction and mutation events and taking all seven threshold events for human life, we reach 5.26×10^{202} universe lifetimes being needed for real 'life' to emerge. The reason that such an astronomical number is reached is the great increase from a 29-link life molecule chain to much higher (DNA) molecule chain links. For example, at threshold event (3), for a microbe to have an energy mechanism requires the cytochrome c protein which has 300 'chain links' (bases) and thus implies one hundred codons. Thus, in probability, the permutations and combinations rise from 29 up to possibly 300, which produces extremely low probabilities just at stage (3). As the genome molecules grow in 'chain length' so do the probabilities shrink infinitesimally. This implies that some clever mechanism is hidden within evolution that can beat the fantastic odds.

There is a law used in probability known as Borel's Law which, if applied here, would state that such a fantastic improbability for the existence of life would be deemed impossible in human terms. This is the 'order of magnitude' for the type of fantastic probability that Eugene Koonin was referring to which explains his need to turn to the Multiverse hypothesis to explain the existence of life as we know it. The Multiverse hypothesis itself will be dealt with in more detail in the Hawking chapters towards the end of this book. However, there is a fatal flaw with the Multiverse explanation.

Harvard professor Leslie Valiant won the Turing Award (the 'Nobel Prize of Computer Science') for inventing a form of analysis for testing rules with algorithms which is set out in his classic book *Probably Approximately Correct*

(2013).[197] In his book, Professor Valiant analyses evolutionary algorithms and shows that some of the typical evolutionary algorithms that nature uses cannot solve biological tasks even by using brute force/numerical computing to overcome the fantastic odds that Eugene Koonin refers to. He has demonstrated that a general evolutionary algorithm for a range of functions required for evolution of typical life features using non-uniform distributions of input data will be unsolvable.[198] Professor Valiant comments on the results: "A weakness of the current conventional wisdom in evolutionary theory is that it gives centrality to competition per se without proving that that mechanism is sufficient… We need to explain how evolution is possible at all, how we got from no life, or from very simple life, to life as complex as we find it on earth today. This is the BIG question."[199]

But even our most basic assumption above – self-replicating 'life molecules' in primordial ponds or vents – may be ambitious. Philosopher of science Samir Okasha has said, "We know that about 4 billion years ago, molecules with the ability to make copies of themselves appeared in the primeval soup, and life evolved from there. But we do not understand how these self-replicating molecules got there in the first place."[200] In view of Eugene Koonin's views, perhaps even the existence of such molecules is not certain. The present state of origin-of-life science suggests a new principle of cosmology may emerge – call it the *anthrobiopic principle* – defined as any principle explaining what laws of the universe, constants or even processes/mechanisms could generate not only self-replicating molecules but all the multifaceted features of reality that need to be present for evolution. This is a particularly fascinating problem which is challenging origin-of-life scientists. (See 'The Origin of Chirality in the Molecules of Life – a revision of awareness to the current theories and perspectives of this unsolved problem', Guijarro and Yus –2009.[201]) This last reference refers to an unsolved problem of symmetry relating to one striking phenomenon in life – inexplicable to chemistry – which is the chirality of both DNA and amino acids. DNA's bases – A, G, T, C molecules on the DNA chain are right-handed only, yet in nature they are both right-handed and left-handed. The same problem occurs with amino acids in proteins which generally have left-handed chirality. There is no obvious scientific process to solve the chirality problem.

A surprising insight also comes from former atheist, Oxford don and philosopher Antony Flew. His paper supporting atheism, 'Theology and Falsification' (1950), refuting the existence of God, was 'the most widely reprinted philosophical work of the last half century'. Yet in an intellectual upheaval which involved him gaining new scientific insights, he had a change of heart and wrote a book *There is a God* (2007) and explained why he changed his mind. One insight he had was: "In 2004 I said the origin of life cannot be explained if you start with matter alone. My critics responded triumphantly announcing I had not read a particular paper in a scientific journal… In doing so, they missed the whole point. My concern was not with this or that fact of chemistry or genetics but with the fundamental question of what it means for something to be alive… To think at this level is to think as a philosopher. And at the risk of sounding immodest, I must say this is properly the job of philosophers, not of scientists as scientists; the competence specific to scientists gives no advantage when it comes to considering this question…"[202]

Flew also quotes the reaction of one scientist who seemed to share Flew's change of heart. Nobel Prize winner and physiologist George Wald said: "We choose to believe the impossible: that life arose spontaneously by chance." But in later years he seemed to embrace a strange new physical reality when he said:

> ***"How is it that, with so many other apparent options, we are in a universe that possesses just that peculiar nexus of properties that breeds life? It has occurred to me lately – I must confess with some shock at first to my scientific sensibilities – that both questions might be brought into some degree of congruence. This is the assumption that mind, rather than emerging as a late outgrowth in the evolution of life, has existed always as the matrix, the source and condition of physical reality – that the stuff of which physical reality is constructed is mind-stuff. It is mind that has composed a physical universe that breeds life, and so eventually evolves creatures that know and create: science-, art- and technology-making creatures."***[203]

Another Nobel Prize winner, physicist Freeman Dyson, saw quantum particles such as electrons as 'choice-making agents' and that experiments force choices such as seen in the quantum wave collapse, which we will consider in Chapters 12 and 21.[204] Freeman Dyson then considers quantum processes in brains:

"[Brains] ...*appear to be devices for the amplification of*...*quantum choices made by the molecules inside our heads... Now comes the argument from design. There is evidence from particular features of the laws of nature that the universe as a whole is hospitable to the growth of mind* [defined as the capacity for choice]. *Therefore it is reasonable to believe in the existence of a third level of mind, a mental component of the universe. If we believe in the mental component and call it God, then we can say we are a small piece of God's mental apparatus*" (Worthing 1996, 40).[205]

Newton's reality reflected in the age of Hawking.

GOTTFRIED LIEBNIZ
LEIBNITZ

LEIBNIZ

Chapter Six Plate

Statues of Isaac Newton and Gottfried Leibniz at the Oxford University Museum of Natural History Photo by Andrew Gray

SIX

GOTTFRIED LEIBNIZ

His big achievement was the invention of calculus independently from Newton – both are credited with the invention of calculus despite the intense dispute between them fought out in the corridors of the Royal Society. Newton himself was then President of the Royal Society – today's modern legal concept of 'conflict of interest' did not exist. However, if you pick up any mathematics book today, you are more likely to see Leibniz's symbols because they were widely adopted in Europe together with some key methods, notations and principles. Thus, both Newton and Leibniz were true giants in mathematics in establishing calculus as the core of modern mathematics. His scientific discoveries and contributions included contributions to physics, geology and mechanics. His philosophy concerned metaphysics which, at least, established principles that science benefitted from – although some discount metaphysics as beyond science (in fact, I will argue that it is of more importance because we will see that it is difficult to determine where physics ends and metaphysics begins, raising the issue of what is known as Hume's Fork, to be discussed in Chapter 10). For example, his metaphysics laid foundations for the laws of conservation of energy, and even his treatise Monadology (akin to atoms discussed in Chapter 10) may assist quantum modelling.[206] He liked being around people and was a great conversationalist. He served various dukes of the House of Hanover, which provided him a secure position within the Court of Hanover not only as a paid advisor but a position where he could pursue his noble aims of the improvement of the human condition, the advancement of science and to fulfil these noble goals by contributing to the public good and to the glory of God 'by means of useful works and beautiful discoveries'.[207]

Leibniz began his studies at university in 1661 and became interested in the conflict between modern philosophy and the ancient Greek philosophy of Aristotle and Plato under the tutelage of his instructor and mentor Professor Jakob Thomasius. This also led to interest in Renaissance humanism (recall the poem of Lucretius and *Vitruvian Man* referred to in Chapter 2). Jakob Thomasius supervised his first philosophical treatise.[208] In 1666, Leibniz completed another paper – 'Dissertation on the Art of Combinations' ('Dissertatio de arte combinatoria'), where he proposed the 'universal characteristic' and the logical calculus, both of which were to transform into his new system of mathematics which we still use today.[209] His brilliance was reflected in the offer of a professorship at the age of twenty-one which he rejected because he wanted to be a 'man of the world'[210] or because he had 'very different things in view'.[211] He completed his doctorate in law in 1667.[212]

After his rejection of a professorship in favour of being a 'man of the world', he met a baron who was able to secure a job for him in the court of the Elector of Mainz. There he produced a series of works in theology before turning to natural philosophy – ie. science – which included his first attempts at scientific theory, though not impressive. He dedicated his new work on motion to the *Académie des Sciences de Paris* and the Royal Society in London, but he had not yet made a name for himself. However, his big opportunity emerged in 1672 when the Elector of Mainz sent him to Paris – the centre of the Scientific Revolution underway – on a diplomatic mission. He must have overstayed longer than his employer expected because he stayed for four years (and indeed he would overstay long trips for subsequent employers also). In Paris, he met Antoine Arnauld and Nicholas Malebranche, but the giant in his field was the Dutch mathematician and physicist Christiaan Huygens, who had corresponded with Isaac Newton.[213]

The *Stanford Encyclopaedia* describes the experience vividly:

> ***"It was he, 'the great Huygenius' (as John Locke would call him in the Dedicatory Epistle to his* Essay Concerning Human Understanding*), who took Leibniz under his wing and tutored him in the developments in philosophy, physics, and mathematics. Not only was Leibniz able to converse with some of the greatest minds of the 17th century while in Paris, he was also given access to the unpublished manuscripts of Descartes and Pascal. And, according to Leibniz, it was while reading the mathematical manuscripts of Pascal that he began to conceive what would eventually become his differential calculus and his work on infinite series."***[214]

One example of Huygens's guidance is explained by Leibniz himself: "When I arrived in Paris in the year 1672, I was self-taught as regards geometry, and indeed had little knowledge of the subject, for I had not the patience to read through the long series of proofs." Apparently, Descartes's *Géométrie* was even too difficult for him.[215] Huygens, acting as his mentor, gave him a problem. Professor of mathematics William Dunham summarises the problem that Huygens gave him: "[Huygens] …thus challenged Leibniz to determine the sum of the infinite series:

$$1+\frac{1}{3}+\frac{1}{6}+\frac{1}{10}+\frac{1}{15}+\frac{1}{21}+\frac{1}{28}+\frac{1}{36}+\ldots$$

(Here the denominator of the nth fraction is the sum of the first n whole numbers)

Leibniz, having to rely on raw intelligence rather than past training, experimented a bit before rewriting the series as:

$$1+\frac{1}{3}+\frac{1}{6}+\frac{1}{10}+\frac{1}{15}+\frac{1}{21}+\frac{1}{28}+\frac{1}{36}+\ldots=$$

$$2[\frac{1}{2}+\frac{1}{6}+\frac{1}{12}+\frac{1}{20}+\frac{1}{30}+\frac{1}{42}+\frac{1}{56}\ldots]$$

Then, expressing each fraction within the square brackets as a difference of two others, he transformed the right-hand side into:

$$2[\left(1-\frac{1}{2}\right)+(\frac{1}{2}-\frac{1}{3})+(\frac{1}{3}-\frac{1}{4})+(\frac{1}{4}-\frac{1}{5})+\left(\frac{1}{5}-\frac{1}{6}\right)+(\frac{1}{6}-\frac{1}{7})\ldots]$$

$$=2\,[1]=2$$

because within the brackets all terms after the initial '1' cancel. In this fashion, he remarkably concluded that:

$$1 + \frac{1}{3} + \frac{1}{6} + \frac{1}{10} + \frac{1}{15} + \frac{1}{21} + \frac{1}{28} + \frac{1}{36} + \ldots = 2$$

The mathematical novice had passed Huygens's test. Historian Joseph Hoffman, commenting on the critical role played by this problem in Leibniz's career, observed that "another example only slightly more difficult (and hence for Leibniz insoluble) would no doubt have quenched his enthusiasm for… mathematics.[216] Instead, success ignited him."[217] He would later reminisce on this period: "…it seemed to me, I do not know by what rash confidence in my own ability, that I might become the equal of [contemporary maths geniuses] if I so desired" and also after this period he was "ready to get along without help, for I read [mathematics] almost as one reads tales of romance."[218]

By the time of his return to Germany in 1676, he had become a Fellow of the Royal Society[219] and during his visit to London in 1673 had presented his designs for a novel calculating machine to the Royal Society.[220] (See picture at the beginning of Chapter 8.) He had also stopped off to meet the philosopher Spinoza in November 1676, three months before Spinoza's death.[221] It is one of the rare occasions that two great philosophers have actually met.[222] By that time, he had developed his famous calculus in which he would share fame with Newton despite their later quarrel. He was later employed by a series of dukes in the Court of Hanover (Hanover is the capital of Lower Saxony in present-day Germany) and was favoured by some powerful women (see next chapter), one of whom urged him to write his famous *Theodicy* in later life (see Chapter 11).

The dispute with Newton arose many years after Leibniz published his calculus first in 1684 entitled *Nova methodus pro maximus et minimis,* which appeared in a prominent journal of the time.[223] The first date of Newton's calculus being published was 1687, some three years after Leibniz published his calculus.[224] (I majored in mathematics at university, coached students in maths and pursued a career in law where everything must be simplified for court evidence. Thus, I hope to present a simplified version of calculus in Chapter 9 to those readers who do not know what calculus is. I am further encouraged by the fact that some great mathematicians were lawyers!) Newton's calculus was not the topic of *The Principia* but was applied in it. For example, in Lemma 2, Book 2, Section 2, Newton talks about the variation of quantities as "indeterminate and variable, and increasing and decreasing as if by a continual motion of flux" and mentions "their instantaneous increments or decrements". He then refers to "velocities of increments and decrements (which it is also possible to call motions, mutations, and fluxions of quantities)."[225] Historian Niccolò Guicciardini states: "It should be stated clearly that the study of manuscripts that Leibniz penned during his stay in Paris (1672–6) has proven that he formulated his calculus independently from Newton. Further, Leibniz's calculus has foundational as well as algorithmic features that set it apart from Newton's method of series and fluxions."[226] In 1699, Fatio de Duillier, Newton's friend, published a paper accusing Leibniz of 'borrowing' Newton's calculus.[227] Leibniz refrained from retaliation when he was told that Fatio had acted out of subterfuge in connection with the Royal Society.[228] However, in 1708, Scottish mathematician John Keill made an accusation[229] in the Philosophical Transactions of the Royal Society of plagiarism of Newton's calculus – a serious accusation. Leibniz called for a retraction but none was given. Two publications in 1711 and 1713 of Newton's calculus heightened the controversy. The 1711 publication consisted of a little book by a maths teacher (William Jones) collecting Newton's early mathematical writings. For example, as mentioned in Chapter 1, Newton had given a copy of his 1669 paper 'De Analysi'[230] (introducing his method of series), which contained his famous binomial equation, to the amateur mathematician John Collins. In fact, Collins had shown a copy of 'De Analysi' to Leibniz on a second visit to London in 1676, but this would have been just stimulation of Leibniz's interest in developing something more than the binomial theorem and methods of using series.[231] We will see in Chapter 9 that Leibniz himself produced some brilliant mathematical discoveries that caused Newton to praise Leibniz as a genius.[232] For those mathematically minded readers, it should also be mentioned that Newton's brilliance in that paper also produced the famous 'Newton's method'.[233]

The second publication, in 1713, was the 'Commercium Epistolicum' – a publication edited by a Royal Society committee to respond to Leibniz's earlier request of 1708 for Keill to withdraw his accusations – bureaucracy moved slowly in those days also! Newton himself was said to have studied this publication in detail and clearly his influence was obviously in view as it sided with Keill against Leibniz.[234] This heralded the beginning of an academic war between Newton and Leibniz, with claims and counterclaims – ridicules and counterridicules. Correspondence between the Newtonians such as Samuel Clarke and Leibniz occurred by way of a 'go-between' – Princess Caroline, wife to the Prince of Wales. The dispute dragged on right up to Leibniz's death in 1716 when the dispute terminated.[235] However, Newton waged a war of attrition, relying on his great propensity to keep everything he had ever written, whereas Leibniz had better things to do than spend months wading through his huge archive of unpublished papers – he wrote 522 papers or works totalling over 3,000 pages between 1677 and 1690.[236] We will return to some of these exchanges in Chapter 10 as they provide some great insights into the differing 'realities' of both Newton and Leibniz despite their common ground – to reveal the cosmological works of God via mathematics and science.

Since 1673, Leibniz had served the House of Hanover, serving as librarian, historian, lawyer and advisor on many things, including the many projects he championed during his lifetime.[237] Thus, even during the dispute over calculus, he had kept himself busy on official business in Vienna (seeking patronage from the Habsburg Emperor)[238] despite the displeasure of his employer, the Hanoverian monarch George I, who left Germany for England to become king.

George I had referred to Leibniz's "insatiable curiosity", his "delight in endless correspondence" and his inability to bring his many projects to fruition (although this may have been due to the lack of funding that he so eagerly sought for his projects).

In fact, after the death of Queen Anne and the arrival of George I in England, Leibniz hurried back from Vienna in the hope of following his employer to continue as his advisor in London, but King George ordered that he not cross the Channel until he had completed one of his large projects – the genealogical lineage of the Guelf and Este families to give proper foundations to their status as dukes. Sadly, with the large project unfinished, Leibniz never made it to London. He died on 14 November 1716, with only his secretary and coachman by his bedside.[239]

SEVEN

LEIBNIZ' S WORLD

Historians Will and Ariel Durant put Leibniz in context: "If the achievements of science in that revolutionary era were not as startling as those in the preceding century from Galileo and Descartes to Newton and Leibniz, they entered more powerfully into almost every phase of European History".[240]

Leibniz's Germany was not as it is today but was made up of hundreds of independent states. *Germania* was the name of a book written by Roman historian Tacitus in the first century AD[241], who spoke of the 'tall blond and virile' Germans as against the 'debauched and decadent' Romans. After the fall of the Roman Empire and rise and fall of other empires, the Emperor Charlemagne [768 to 814], who ruled the Frankish Empire from France to Austria, adopted the Catholic faith and in 800 AD the then pope Leo III

Chapter Seven Plate

Leibniz's residence in Schiemedestrasse, Hanover (he lived in an apartment in the tallest building as shown from 1698 to his death on 14 November 1716) [Ref. M.R. Antognazza, Leibniz, Oxford University Press, 2016, at p116]

*Axel Hindemith – source Foto aufgenommen von Benutzer Benutzer:AxelHH, February 2008 https://creativecommons.org/licenses/by-sa/4.0/deed.en**

**or such deed or licence instrument as may be applicable to the licence*

crowned Charlemagne head of the Holy Roman Empire. Historians view this curiously as this is not the 'Roman Empire' that had fallen centuries beforehand but a new empire, part of which broke off into hundreds of these small independent states that would later become Germany. A German emperor would then appoint heads of these states called 'electoral princes'. By the 1500s, the Habsburg family, ruling from Vienna, had supplied a lineage of emperors controlling these states comprising the Holy Roman Empire.[242]

In 1517, Martin Luther – a friar, priest and university lecturer who was the son of a Saxon miner – nailed ninety-three theological theses challenging the practices of the Catholic Church to the main church in Wittenberg. After being summoned by the Vatican in Rome and refusing to recant, Luther renounced his religious vows and married a nun before establishing the Lutheran religion in these formative German states. Various rulers of German states with Denmark and Sweden used Luther's principles to reform their churches (the Reformation) freeing the Holy Roman Empire from the control of Rome. This caused various factions or offshoots of religion called 'confessions' to emerge, such as the Calvinists, who followed the strict religious doctrine of Calvinism founded by John Calvin. Calvinism exercised wide influence despite less humane and more severe practices than Lutheranism and even caused religious wars.

Thus, Leibniz lived in one of the 300 or so independent states ruled by a sovereign electoral prince presiding over a court, army and kingdom.[243] Religious wars amongst Christian religions and confessions had followed the Reformation. Leibniz had been born towards the end of the Thirty Years' War (1618–48) and so grew up in the shadow of the stark reality of war and burning religious issues. This series of wars was caused by the Calvinist movement, which unsettled the balance of power between the electoral princes. Catholics fought Protestants, and fights over 'secularised' lands prompted further hostilities. Despite intervening peace, outbreaks of war continued until the Westphalia Treaty saw the then emperor make peace, recognising Calvinism and granting territorial gains to Sweden, France and Brandenburg with independence to Holland and Switzerland.[244] The Peace of Westphalia Treaty allowed rules for the peaceful co-existence of the opposing religious factions and confessions – namely, Catholics and Reformers (such as Calvinists and Lutherans). Leibniz's city of Leipzig in

Saxony was a stronghold of progressive Lutheranism whose inhabitants despised Calvinism.[245]

Leibniz was a key advisor in the courts of these fledgling German states serving dukes. These courts were the centres of cultural and political life in the Holy Roman Empire – the future Germany. It was emerging as a land of great palaces, great homes and great castles. Germany's multitude of states lacked national pride, yet its impressive development saw Hanover with a magnificent opera house and Leipzig as a 'little Paris'.[246] So it was that Leibniz wrote his famous *Theodicy* (see Chapter 11), explaining how God had created the best of all worlds when in fact Leibniz, knowing what he knew, might have thought that his was the worst of all worlds, with wars, religious tortures and executions, plundering, political turmoil and all the upheaval of the age.

Leibniz's presence at the court of the House of Hanover was bolstered by the patronage of Sophie von der Pfalz, whose husband was Ernst August. They were the parents to George I, mentioned in the last chapter, who became King of England and who left Leibniz behind due to his dispute with Newton. The other renowned woman was the daughter of George I – Caroline von Ansbach, who became the Princess of Wales and who acted as go-between for Leibniz in the Newton-Leibniz dispute.[247] His quest for knowledge involved a vast network of contacts in Europe numbering 1,300 correspondents, including such people as Bernoulli, De Volder, Des Bosses, Huygens, Spinoza, Newton, Clarke and his protégé Christian Wolff on a wide range of topics – even the nature of space and time.[248] Leibniz's letters to them showed an astonishing depth and coherence. Professor Antognazza defends Leibniz from a charge by Bertrand Russell that Leibniz disclosed the mythical double-headed 'Janus face' of the appeaser of dukes in public and a different philosophy in private (perhaps Lord Russell, being an earl, was unfamiliar with the day-to-day struggle of earning a living, which normally includes impressing one's employer as Leibniz had to do). In summary, she says that although metaphysical publications were not within the preferred aims of his employers, "the value of each publication was measured with reference to his main aim and objectives. Leibniz was a man of synthesis and reconciliation. His overarching goal was the improvement of the human condition. To his mind theoretical reflections on logic, mathematics, metaphysics, physics, ethics and theology were ultimately in the service of life and aimed at the happiness of mankind."[249] The culmination of all his scientific, mathematical and philosophical work saw him complete his *Theodicy* in 1710 towards the end of his life – a major work exceeding 250 pages.[250]

These objectives can be seen in his letter to the Russian Tsar, Peter the Great:

> *"Although I have very frequently been employed in public affairs and also in the judiciary system and am consulted on such matters by great princes on an ongoing basis, I nevertheless regard the arts and sciences as a higher calling, since through them the glory of God and the best interests of the whole of the human race are continuously promoted. For in the sciences and the knowledge of nature and art, the wonders of God, his power, wisdom and goodness are especially manifest; and the arts and sciences are also the true treasury of the human race, through which art masters nature and civilised people are distinguished from barbarian ones. For these reasons I have loved and pursued science since my youth... The one thing I have been lacking is a leading prince who adequately embraced this cause... I am not a man devoted solely to his native country or one particular nation: on the contrary, I pursue the interests of the whole human race because I regard heaven as my fatherland and all well-meaning people as its fellow citizens... To this aim, for a long time I have been conducting a voluminous correspondence in Europe and even as far as China, and for many years I have not only been a fellow of the French and English Royal Societies but also direct as president of the Royal Prussian Society of Sciences."*[251]

I think this letter sets Leibniz apart from other great scientists, mathematicians and philosophers in that he had a true quest for improvement of the human condition and the promotion of human happiness. He not only wrote over 3,000 works but strived in the hope of support for his many human improvement projects.

Leibniz's quest for the betterment of the human condition and greater glory of God would sit comfortably with the universal goals of today's United Nations, whose noble goals include human rights, religious freedom, education and at times promotion of the arts, sciences and other 'wonders of God'. It is not surprising that some leading philosophers see Leibniz as a truly modern mind, not only inventing new science, mathematics and logic but inventing a new synthesis of his new ideas with religion.

Chapter Eight Plate

Prototype mechanical calculator invented in 1674 by Gottfried Leibniz and built in 1694 and believed to be the first calculator capable of addition, subtraction, division and multiplication – now held in the National Library of Lower Saxony

Source: J.A.V/ Turck, The Origin of Modern Calculators (1921) p133

EIGHT

LEIBNIZ'S SCIENCE

"Leibniz stood out as one of the most important figures in the development of the Scientific Revolution" (*Stanford Encyclopedia of Philosophy*[252]). Leibniz was 'arguably Newton's only intellectual equal in the period' according to Newton historian Professor Rob Iliffe.[253] Philosopher Stephen Law, in Chapter 2, 'History of Philosophy', of his book *Philosophy,* referred to Leibniz and Spinoza as "the great 18th century system builders".[254] In fact, the leaders of the great romantic movement in history which defied the Scientific Revolution built on Leibniz. One of these leaders of this romantic ideological revolt, Herder, admired Leibniz,[255] whose science blended with an ideal romantic view of man as opposed to the cold materialists like Hobbes, Spinoza and Locke. Newton's friend Locke was very influential in introducing empiricism to England, which even extended to underpinning his political philosophy. The empirical observations in nature saw mankind needing to defend itself, and this resulted in an implied contract between the majority to appoint a supreme authority to impose the rule of law to protect mankind.[256] This became the philosophy of our present-day democracy.

Leibniz was the opposite of the empiricists, whose reality was the material world. As one author put it, "Leibniz is a hundred leagues removed from any type of empiricism…"[257] Leibniz's reality was the underlying structure that displayed what we see as phenomena detected by our senses. Descartes had really led the way to the revolutionary thought of using reason to analyse the world (thus, Descartes, Leibniz, Spinoza and others were classified as 'The Rationalists'.[258]) Descartes had become the ultimate sceptic – raising questions like "how do I know that the world exists at all? My senses could be deceiving me. There could be a malevolent force deceiving me." Finally, his breakthrough was something like, "I think therefore I am," or I at least know that I am an entity. A great principle developed. There were two different types of truth. Analytic truths – or Leibniz would call them necessary truths (although 'what is a necessary truth?' will become a complex issue as we shall see in the next chapter) – and empirical truths. Descartes said, "Whether I am awake or asleep two and three added together are five…"[259] This was an analytic or necessary truth.[260] Empirical truths are what we observe, and that becomes a problem because observations can be deceiving. It has become a greater problem in science today, known as the observer problem, associated with the Copenhagen Interpretation in quantum mechanics, namely, that some scientific facts can only exist, eg. a particle has no position or spin until an observation is made when it undergoes 'wave collapse'.[261] This leads to a probabilistic reality which can cause paradoxes.

In fact, in the movie *The Matrix,*[262] this precise situation formed the factual basis of the movie, ie. everyone in our world was being deceived by a malevolent force of robots who had created a world of phenomena – our visible world – which was fake. The underlying structure of the world – what Leibniz would call the underlying reality causing the phenomena – was actually a vast matrix of humans plugged into a network of life support systems with neurological apparatus creating illusions for each human, who was placed into a deep mental state to be deceived on a daily basis into believing they were living in a world that looked precisely like our world. The purpose of this grand deception was merely to provide energy from each plugged-in human metabolism to power the robots – a bleak meaning of life for humans. In fact, Leibniz saw our world as something akin to the matrix (but with a nobler meaning of life!).

Voltaire, whose contributions to literature, philosophy and history brought him great fame, was much younger than both Newton and Leibniz but his travels to England convinced him of Newton's genius. Thus, he and his mistress, Émilie du Châtelet, championed Newton's work in Europe. Yet Voltaire was opposed to Leibniz, unlike his mistress, Émilie du Châtelet, who used Leibniz in her own work. Émilie du Châtelet was not only very capable in mathematics but became the translator of Newton's *Principia,* which popularised it in Europe. During the years of her work, she sheltered Voltaire at the du Châtelet château at Cirey in the province of Champagne in the north east of France because he was exiled from Paris.[263] Émilie's husband, the marquis du Châtelet, was away for king and country on military duties, allowing this sublime, enlightened freedom of thought (and action) to take place at the château. An idyllic location to foster new ideas.

Voltaire's exile followed his publishing *Letters Concerning the English Nation* – also called *[Philosophical] Letters* – described by one author as "one of the most influential and quintessential Enlightenment texts of the eighteenth century, with its emphasis on religious and political tolerance, and its support for philosophical and scientific empiricism as the most objective means of finding truth."[264] Unfortunately, unauthorised French versions fell into the wrong hands. Upon his exile, the French authorities deemed the *[Philosophical] Letters* "scandalous, contrary to religion, morality and the respect due to the authorities." Voltaire replied, "Truly, they cry so much about those wretched Letters that I wish I had said more."[265]

At the du Châtelet château, marvellous scientific and philosophical discussions were held during the days and nights as they both worked on their various projects. Both became familiar with Newton's and Leibniz's work but with mixed views. Voltaire had taken Newton's side in ideological terms, which meant he was against Leibniz. Science should not be about 'them and us' but as philosopher Thomas Kuhn has seen, science is, contrary to appearances, governed by sentiments. Thomas Kuhn is responsible for part of our vocabulary – the term 'paradigm shift' resulted from his landmark book *The Structure of Scientific Revolutions*. His insight was that each successive scientific paradigm is supported by the sentiment of scientists that it is right. They work to support the paradigm until a revolution in thought occurs where the paradigm is broken and replaced with a new one. Newton's paradigm of absolute space and time was broken by Einstein's theory of relativity.

So when Émilie du Châtelet decided to write a new book called *Foundations of Physics*, she not only planned to present Newton's physics but also both Leibniz's science and philosophy. Like Newton, Leibniz's theories encompassed space, time, motion, mass, force, energy and so on. However, unlike Newton, Leibniz sought to 'drill down' to the nature of reality. To empiricists like Locke, Newton had been able to provide laws governing our reality, and that was enough to spark a revolution in thought despite not knowing the underlying realities such as the nature of gravity, mass or energy. Leibniz sought a unifying theory and in fact in Chapter 10 I explain how modern science may benefit from Leibniz's work.

So what was Leibniz's synthesis of science and philosophy? Philosopher Roger Scruton describes Leibniz as producing "a philosophical system of astonishing power and originality" and at the centre of it is what Scruton terms "his philosophical masterpiece – the Monadology."[266] It is good to start with Voltaire's reaction to it in a letter to another philosopher, Claude-Adrien Helvétius. Isaiah Berlin ranked Helvétius alongside Voltaire and other philosophers as those who "believed that reality was ordered in terms of universal, timeless, objective, unalterable laws which rational investigation would discover..."[267] In contrast, philosophers of the Romantic movement, such as Herder, said that attempts to reduce worldly phenomena to uniform elements "tended to obliterate crucial differences which constituted the specific object of the object under study."[268] Indeed, in our present age of chaos theory and quantum mechanics, not even a dialectic approach produces easy answers. In any event, in October 1739, Voltaire wrote to Helvétius stating: "Émilie has brought with her to Paris her König, who has no imagination or sense, but who, as you know, is what they call a great metaphysician... He swears, following Leibniz, that extension is made up of non-extended monads, and that impenetrable matter is composed of tiny penetrable monads. He believes that each monad is a mirror of his universe. When one believes all that, one believes in... miracles." In May 1739, Émilie and a Swiss mathematician Samuel König had moved with Voltaire and her two children to Brussels, but presumably they spent time in Paris.[269]

Ever since Leibniz introduced the concept of 'monads' to describe his reality, they have been the subject of intense analysis and criticism yet continue to be analysed and critiqued amongst today's philosophers and scientists.

In Chapter 10, we will see how monads could be a conceptual model used to create the grand unified theory – the holy grail of physics. The short description is that they come from the Greek *monas*, meaning 'one' or signifying unity.[270] The Greeks had used the concept of 'atoms' as the building blocks of the universe but these had 'extension', ie, length, breadth, height – metrics. The Pythagorean monad had long been used to signify the first existing being – the 'Indivisible One', which generated 'numbers, points, lines, shapes…' etc., which was later adapted to include monads as having a 'vital active force'.[271] Monads were not atoms but were incorporeal.

How were they to be described then? The Greeks also used the word 'soul' for incorporeal substances, which took on a wider meaning than it does today, and Leibniz, not having sufficient scientific jargon in his day, also adopted 'soul' to refer to incorporeal monads, as Professor Copleston explains: "The ultimate constituent of things must, therefore, be 'points' though not mathematical points. They must be, then, metaphysical points, distinct from physical points, which do not exist and cannot together form bodies. Further these metaphysical points, which are logically prior to body, must be conceived after the analogy of souls. There must be some internal principle of differentiation, and Leibniz decided that these substantial units are distinguished from one another by the degree of 'perception' and 'appetite' which each possesses."[272]

The key phrase here is 'substantial units'. 'Substance' was the underlying essence of reality. If you see an object then this is a phenomenon, and underlying it in reality is a substance that could be an aggregate of monads. This led to Leibniz's now accepted scientific principle of dynamics. Professor Copleston continues: "Each substance or monad is the principle and source of its activities: it is not inert but has an inner tendency to activity and self-development. Force, energy, activity are of the essence of substance,"[273] and as Leibniz explains: 'The idea of energy or virtue, called by the Germans *Kraft* and by the French *la force*, and for the explanation of which I have designed a special science of dynamics, adds much to the notion of substance".[274]

This concept then involved active or passive energy in the monad which eventually became known as kinetic energy. It would not imply any concept of Einsteinian energy within nuclear particles because to Leibniz the monads were indestructible and indivisible, unlike the particles arising from splitting the atom resulting in $E=MC^2$ representing the energy released from fission of a nucleus – destruction of matter. Yet this does not mean that monads do not possibly allow conceptualisation of particles because Leibniz would not see the nucleus as a monad nor protons and neutrons. You would have to keep drilling beyond quarks to get to monads and then they have no extension; they are in another dimension. The views of some modern scientists on this 'monadology' will be explored further in the next chapter.

Recall the concept of monads having a 'vital active force' developed by philosophers from the Pythagorean concept of the monad. Leibniz adopted this when he made his landmark extension to physics, which went beyond Newton. Descartes had talked of a quantity of motion and Newton had talked about a 'quantity of matter' (mass) and so Leibniz conceptualised the conatus, impetus or force of a moving object to be a type of energy not contained in a stationary object which was inert. Yet stationary objects had a passive force as they were inert – they had inertia. But the force that was needed to overcome the inertia and move the object could be measured and quantified and this force became kinetic energy. Leibniz suggested it could be measured by multiplying mass by velocity squared, ie. mv^2, which today we know as Kinetic Energy = $\frac{1}{2} mv^2$. Note, however, that modern science considers energy and force as distinct physical quantities but connected by the concept of work. Work equals displacement times force, which in turn requires energy.[275] Unlike Newton, Leibniz was not an extensive experimentalist but he did draw on Galileo's experimental results in deducing that the energy of a falling object was proportional to velocity squared. His concept of force and dynamics was presented in a number of papers, including 'Dymanica de Potentia et Legibus Naturae corporea' (1689) or 'Essay of Dynamics on the Laws of Motion'[276] – Leibniz invented the term 'Dynamics'. So both Leibniz's and Newton's legacies include the great scientific breakthrough of codifying equations of physics to reflect the various relations governing gravity and motion as they relate to other quantities such as mass, force, energy, velocity, momentum or distance.[277] In proposing his theory of dynamics, he had dismantled Descartes's 'quantity of motion' because it violated the conservation of energy, ie. force equalled the quantity of vis viva (living force) – Descartes had not seen the ½ mv^2 equation. Leibniz had also superseded his earlier abstract theory of motion.[278]

Although Newton took the lion's share of fame with his exposition of the physics of motion, gravity, etc. in *The Principia,* with Leibniz's dynamics filling another cornerstone of physics, Leibniz's span of work was wider, including the fields of geology, engineering, history, philosophy, logic and mathematics. Newton too was both a mathematician and historian and had great engineering prowess both early in his life and with his telescope. Leibniz was given more major works such as the modernisation of silver mines – the Harz mines – by Duke Ernst August. Leibniz spent months of solitude in the Harz Mountains whilst continuing to write physics, philosophy and logic.[279] He constructed a calculating machine – a forerunner of the calculator – and presented it to the Royal Society in London during his first trip to London in 1673, following which he was elected a Fellow of the Royal Society.[280] He developed other mechanical devices such as carriage mechanisms and a water pump run by a windmill.[281] In geology, he was one of the first to postulate that the world had a molten core.[282]

A concept that proved very fruitful was his aim to produce an 'algebra of thought' through a logical system. In his article 'De Arte Combinatoria' ('On the Art of Combination'), he had proposed a device which could use such an algebra of thought using numbers, words, colours, etc. that is believed to be embryonic for the modern computer.[283] This 'algebra of thought' led Leibniz to a type of logical calculus – the mathematisation of logic.[284] All this led to his great breakthrough and milestone in the development of modern mathematics with the presentation in October 1684 of his infinitesimal calculus with the publication of *Acta Eruditorium* of the *Nova Methodus pro Maximus et Minimis* and in 1686 his *De Geometria Recondita.*[285]

His conceptualisation of mathematical calculus was deeply blended with his science as seen in his objections to Descartes's theory of gravity and 'aether' where he said: "What Descartes says here is most beautiful and worthy of his genius, namely, that every motion in filled space involves circulation and matter must somewhere be actually divided into parts smaller than any given quantity."[286]

As the online *Stanford Encyclopaedia of Philosophy* describes Leibniz's conceptualisation of matter: "Dismissing Descartes' cautious 'indefinite division' as 'being not in the thing, but in the thinker', he takes the argument to show that every part of matter is actually infinitely divided (A VI.ii.264). Similarly part of matter is everywhere moving... Thus, for Leibniz not only are some parts of matter infinitely divided, but every part of matter is divided to infinity (Levey 1998)!"[287]

Thus, even though it is now acceptable in mathematical calculus to make calculations that go to infinity – ie. to infinitesimal levels – Leibniz saw matter also as capable of such a feat. I will return to this concept later when dealing with Leibniz's landmark *Theodicy,* for it may not be as puzzling as it appears. Another interesting aspect of this quote appears to be that he anticipated the movement of atoms or quantum particles without having any experimental evidence (other than looking at microbes under the microscopes of the day), so again his foresight is surprisingly modern. Or as Magee comments: "...one of the fundamental doctrines of 20th century physics is that all matter is reducible to energy... Now it seems to me that Leibniz was trying to express something astonishingly close to this idea. He was saying that all matter was made up of propensities for activity which are not themselves material – and indeed this is something we now know to be true."[288] Magee's interviewee response to this included: "The predominating view of many people in his time, and indeed since, while admitting that nature consisted of matter in motion, was that motion was not intrinsic to matter itself but has to be imparted to the material world from an external source. Leibniz did not make that assumption".[289] It is no surprise therefore that Magee himself responded: "I think that Leibniz is in many striking ways a startlingly modern thinker..."[290]

Chapter Nine Plate

Isaac Newton [left] and Gottfried Leibniz [right]

NINE

LEIBNIZ AND NEWTON'S MATHEMATICS

Readers can avoid the 'delights of mathematics' by skipping this chapter, although it does seek to explain the key discovery that Leibniz and Newton made – the calculus (and drawing on the author's past maths coaching experience, it has been kept as simple as possible so that a reader with no mathematical knowledge might here have the opportunity to see something wonderful).

Some readers may wish a greater understanding of the mathematics invented by Newton and Leibniz because it forms the cornerstone of their ideas. Many books on mathematics show mathematicians have a blind spot – the inability to perceive what the novice knows and does not know. So to overcome this challenge it is intended to present for the reader an understandable 'crash course' of calculus by explaining it in simple steps. It is then hoped the reader can marvel at understanding Leibniz's fascinating method for transforming π into a series of numbers later in the chapter.

Newton and Leibniz worked with speed, velocity and acceleration, but we will use a running tap filling a bucket of water to analyse the 'speed' of the water or the 'speeding up' of the flow of water ('acceleration' of water). Even velocity can be thought of in terms of water as the water being pumped in (positive velocity) or water being pumped out (negative velocity). In this way, we can understand what calculus does in a way that is more understandable.

To take the most simple example, if water flows into a bucket via a hose at a rate of 1 litre per minute we could graph the behaviour of the water thus, as Figure 9a.

Thus, note that the hose is flowing at a constant rate as shown by the gradient of the bold black line. This bold black line represents what is known as the 'curve' of a function (because it is not always straight). This means that each litre poured into the bucket is a function of time, which simply means that for each minute another litre is poured into the bucket. Let us call it our water function. In maths, they have a notation for this which here

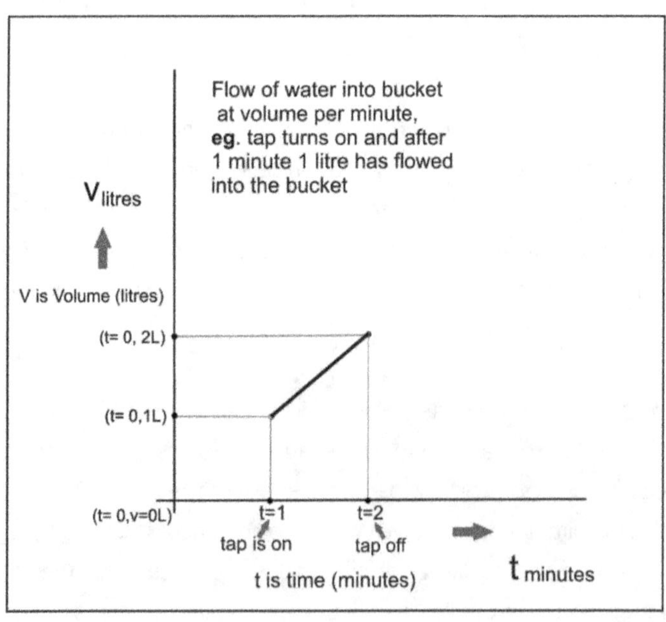

Figure 9a

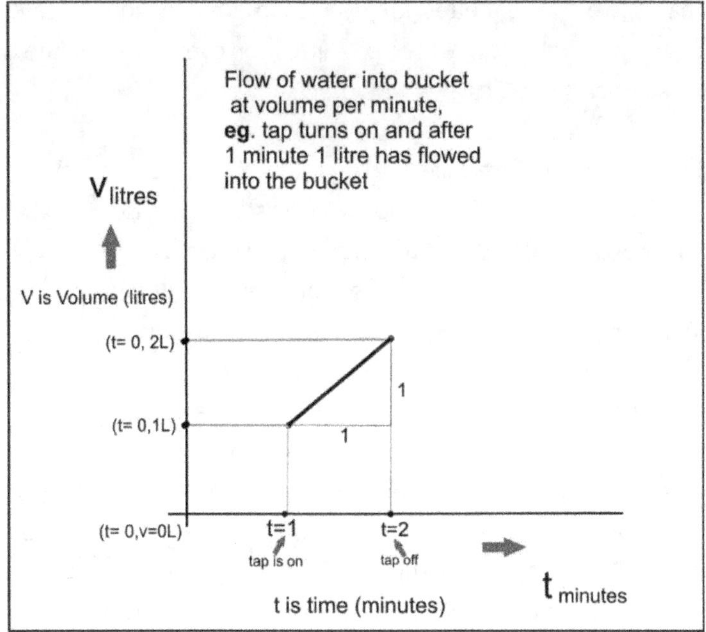

Figure 9b

would mean V= f(t) – the more usual notation is y = f(x). Some functions are much more complicated than this. Now this gradient becomes our focus because what calculus does to it is very clever. So consider a triangle that we construct superimposed below the bold black line representing our water function. This is represented in Figure 9b.

You will see here that our black 'curve' starts at t = 1 and finishes at t = 2, and we have analysed its behaviour using the triangle which shows us that it rose 1 unit and ran 1 unit. Compare this to climbing a ladder of a house of 10 metres height and the ladder being 10 metres from the house you would, in effect, 'run' at the house by 10 metres and then rise by 10 metres (even though your actual climb – the bold line on the graph – was less than 10 metres run followed by 10 metres climb). The gradient of your ladder is calculated by your rise divided by your run – that is 10 divided by 10 which is 1. A gradient of 1 is that for every 1 metre you travelled, you rose 1 metre. On our graph you can see the gradient is 1 divided by 1, ie. the gradient is 1 which means the rate of water into the bucket is 1 unit (litre per minute). If our ladder were 5 metres from the house then the gradient would be a rise of 10 metres divided by a run of 5 metres, giving a gradient of 2. That is, we rise 2 metres for every 1 metre horizontal movement. With water, we might flow water at 2 litres for every one minute, which means a steeper curve (ie. line) would be shown.

Now the problem with real life is that gradients or rates are never simple like this (straight lines are replaced by curves). What happens when you have a more complicated water function and thus a more complicated gradient? This is where the synthesis of Greek mathematics, medieval mathematicians and predecessors of Newton and Leibniz culminated in a stroke of brilliance. The more complicated equations were solved by using the triangle we superimposed in Figure 9b and then shrinking and multiplying it in a clever way, which will now be explained. The diagram below shows perhaps a more realistic water function. Imagine our tap had some ice in it – it is an outside tap – and it snowed the day before and the tap was turned on first thing in the morning. The flow rate seemed fine at about the time it was turned on (say t=0) but between t= 0 and t=1 it started to splutter and then at t=1 the rate went up and started to increase variably. Thus, the gradient or rate of flow is not a simple constant of 1 litre per minute anymore, and instead of a straight line we now have a variable curve on our graph because values are both increasing *and* changing rates of increase. So calculations become much more difficult.

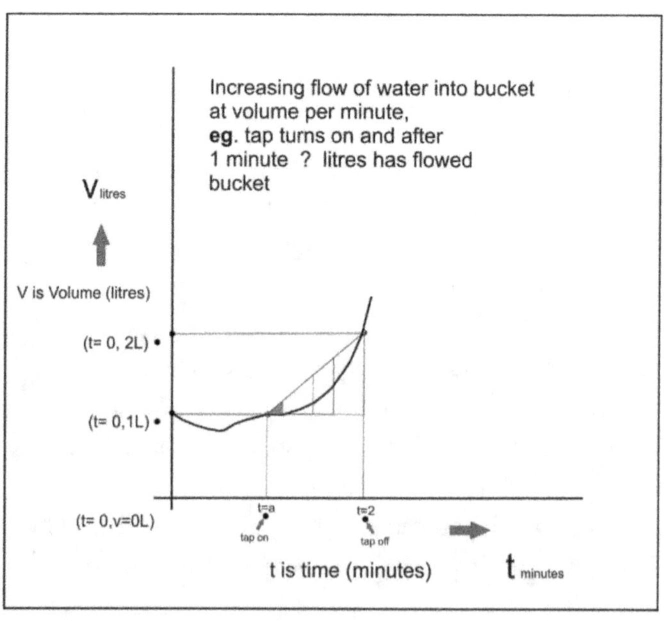

Figure 9c

However, both Newton and Leibniz would have faced the difficult problem that whereas before we had a straight line where the rate of increase could be easily calculated by dividing rise by run, now there was a curve, so how could the 'rise over run' calculation be made? Well, first a line was still able to be drawn between, say, t=1 and t=2, but if one drew this line at every point the line would vary through a multitude of lines, so how could this be solved? They decided to use a straight line called a tangent. If one has a circle and you draw a line just outside the circle but the straight line touches the circle at one point, that is called a tangent. Now, you can draw a tangent touching the bulge in a curve also, but they both realised independently that the tangent at each point joining the curved section was the key. They then made a 'moving triangle' with this changing tangent as its sloping side (hypotenuse), showing rise and run. Just as the first triangle that we saw above represented the slope or the rate of change of the function, we can also imagine a triangle 'sliding' up and down the curve. The important question then becomes – how to discover the 'generalised slope' of this sliding triangle at a particular point. It seemed logical to let the tangent with this triangle slide down the curve to the lower point, but to do this we would need to shrink the triangle to a very tiny triangle so that the curve was almost straight and not curved at the end of this 'slide' to where the lower point was. This meant that you effectively had a generalised tiny triangle where we could calculate rise over run at any point. They called this calculating the gradient or slope by generalising it to its *limit* as the sides approached lengths of zero. If they could work out rise over run at one point, they could work it out at any point and so they took the next step of generalising what that gradient would be using the function itself which here we have called V = f(t). This produced what is called differentiation of the function V = f(t) which has notation V = f'(t) (an example is given below). Thus, differentiating a function is just getting the gradient of any point in a new function V=f'(t) which automatically gives you the gradient at any point (t, f(t)) ie. you plug in time t=a and so inserting a in this new function V = f'(t) ie. V = f'(a) will give you the rate of change at t=a like a calculator would calculate, say, a percentage of a if you plugged in a percentage of a. We all are familiar with a 'percentage function' eg. to get 10% of a we multiply a by 10/100, ie. 1/10. Thus, in maths notation we would put f(a) = 1/10 x a or f(a) = *a*/10. If we graphed this, we would get a gentle straight line and find that rise over run is very gentle, ie. 1 unit rise for every 10 units run, ie. thus differentiating f(a) = *a*/10. means getting

$f'(a) = 1/10$. Now we could have used the 'sliding triangle' method but the triangle would not vary as it slid down the line, so we did not need to use any more complicated technique because the percentage function $f(x) = x/10$ or $f(t) = t/10$ is very simple. We would not normally use time t as a variable with getting a percentage but there is no reason why $f(t) = t/10$ is not a proper form of the percentage function.

Now unlike the percentage function, we have not specified what $V=f(t)$ is but only what we do to differentiate it. So this was called calculating the gradient or slope of the water function to its limit. Now assume this water function $V= f(t)$ obeyed the equation $V=t^2$. That is, if 2 minutes elapsed, the volume of water would be 4 litres (2x2) poured into the bucket. If 3 minutes elapsed, the volume of water would be 9 litres (3x3), and so we can see the rate of flow of f(t) increasing. The method to find a general form of function derived from f (t), namely, f '(t) is called differentiation. Now it was found that just as an equation like this water function of $V=t^2$ existed this method produced the derived function, $f'(t) = 2t$ for the rate of change (just like for $f(a) = a/10$ we found the rate of change was simply $f'(a) = 1/10$.) So $V=t^2$ differentiated is $f'(t) = 2t$, just like $f(a) = a/10$ differentiated is $f'(a) = 1/10$). I will leave out how they derived $f'(t) = 2t$ from $f(t) = t^2$ for simplicity's sake, which is called differentiating by first principles. Recall also that 2t is just shorthand for 2xt. Now it was also found that differentiating any function with a power in the form $f(t) = t^n$ or $f(x) = x^n$ when differentiated to $f'(t)$ is nt^{n-1} (or $f'(x) = nx^{n-1}$) which is known as the 'Power Rule'.

Leibniz's notation (preferred over Newton's) for f '(t) was $dV/dt = 2t$ (called differentiating V with respect to t or if you have a function with x and y instead of t and V, this would mean $dy/dx = 2x$ for $y= f(x) = x^2$). For example, this means that if we know the flow rate and the rate at which it is changing, we can calculate the volume of water that falls in the bucket between time t=1 and t=2. How do we do this using our knowledge of the rate $dV/dt = 2t$? This process is called integration, which we discuss below, and can be done by changing the vertical axis on our graph from just V (volume) to V l/m (flow in litres per minute).

This method was found to produce standard reliable differentiations for a great range of functions. For example, if applying the power rule to $V= t^3$

then the slope or gradient is $dV/dt = 3t^2$. Differentiation turned out to be extremely useful and not only made new calculations and new equations possible but in physics a huge range of equations could be created because physics always deals with changing quantities. Great advances in astronomy were made possible and indeed calculus has been used in all the space missions to calculate trajectories and destinations. Thus, Joseph Fourier was able to use differentiation to produce equations for the rate of change of heat passing through a steel girder, or James Maxwell was able to produce equations for the rate of change of electromagnetic or light waves through a medium.

There remains one major component to explain, and that is what is known as integration. Integration is the opposite of differentiation in that it adopts the reverse process. For example, if we were to integrate $dV/dt = 2t$ using Leibniz notation[291] then this would mean $\int dV/dt\ dt = \int 2t\ dt = t^2$ (for mathematical readers, we have assumed the constant is zero), which just reverses the differentiation. However, apart from it merely being the reverse of differentiation, we can gain a feel for what it is (and why it is important) by returning to our bucket. Firstly, like with differentiation, there are many standard results for integrating and indeed the fact that it reverses differentiation gave mathematicians a good methodology to find these standard results. Now how then do we integrate using $\int dV/dt\ dt = \int 2t\ dt = t^2$ to find the volume of water falling into the bucket between time t=1 and time t=2? This is denoted like this $\int_1^2 2t\ dt$: which means integrate from t=1 to t=2 using the differentiated function 2t and it means you do this:

$$\int_1^2 2t\ dt = t^2 \Big|_1^2 = 2^2 - 1^2 = 4 - 1 = 3$$

(3 litres of water fell in the bucket for 1 minute) (1)

Secondly, another method of integration, known as the Newton-Leibniz formula but better known as the fundamental theorem of calculus, is now stated. Recall our equation for the water function as being V=f(t). First, assume there is another unknown equation function Q or Q(t) that tells us the volume of water that has fallen in the bucket. With the constant flow rate in Figures 9a and 9b, working out what Q(t) is would have been easy – it would simply be t x 1 litre, eg. 3 minutes x 1 litre = 3 litres of water after the tap is on for 3 minutes. But if the rate of water was changing during those 3 minutes, we do not know what the Q(t) equation is and so we need calculus to work that out.

Assume therefore that this unknown equation is just Q(t), which is f(t) differentiated. Now recall that if we said that $f(t) = t^2$ then Q(t) would be 2t, but instead of using this approach we are using a different approach. First, we take the time from t=a to t=b then use integration, ie. the Newton-Leibniz formula which is $\int_a^b Q(t)dt = f(b) - f(a)$ to calculate the volume of water that has entered the bucket from time a to time b. Note that although $\int_a^b Q(t)dt$ looks daunting in appearance, it just equates to f(b) – f(a) which is simple to calculate. Note also that the rate was changing as shown in Figure 9c, so we cannot simply use V= t x 1 litre.

Now the amazing aspect of this equation is that even though we do not know what the Q(t) function is, we can use our graph in 9c to get the answer providing we know what f(t) is. In other words, it is a way of bypassing the differentiation calculation ((1) above) when we want to find the volume of water that fell into the bucket.

Assume that a=1 and b=2 in Figure 9c. Then we can immediately say $\int_1^2 Q(t)dt = f(2) - f(1)$. Thus, the volume of water flowing into the bucket between t=1 and t=2 is = f(2) – f(1). But what does this mean? Recall that $V=f(t) = t^2$ and so f(2) –f(1) is just $t_1^2 - t_2^2$ or $2^2 - 1^2 = 4 - 1 = 3$ .Thus, unlike the graph in Figure 9a, where only 1 litre of water filled the bucket during the time from t=1 to t=2, here we know that despite the complexity of a changing rate of flow between t=1 to t=2, we have calculated that 3 litres of water flowed into the bucket. For instance, at time t=1, the flow rate for our water function was 2t = 2 litres per minute, whereas at t=2 the flow rate was 2t = 4 litres per minute. Now at, say, 1.3 minutes, the flow rate was 2t=1.3 = 2.6 litres per minute. In fact, there would have been an infinite array of flow rates between t= 1 minute and t= 2 minutes, but the Newton-Leibniz formula effectively does an infinite calculation in one line to arrive at the volume of water. What is fascinating is that this $\int_1^2 Q(t)dt$ not only represents the volume of water that flowed into the bucket from time t=1 to t=2, it also represents the area under the curve which becomes useful for further applications such as calculating areas of geometric figures – even in 3 dimensions. This area under the curve is

shown in Figure 9d. Note that, unlike the previous graphs, the vertical axis is now V/min (volume of water flow per minute).

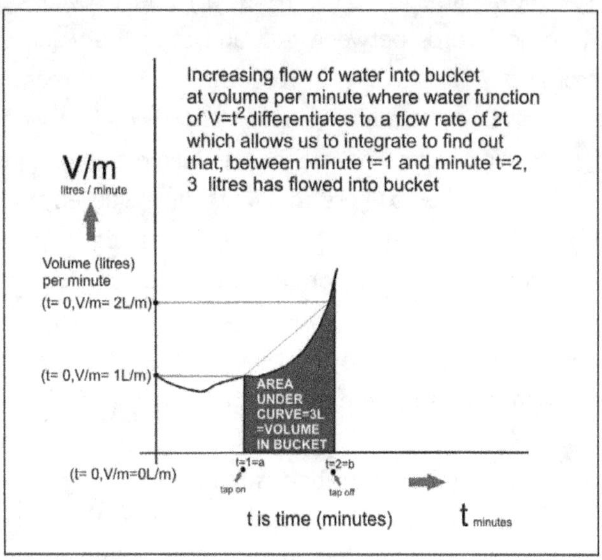

Figure 9d

Why does this integration represent both the volume of water poured between time t=1 and time t=2 and also the area under the curve? To see why this is so, we must consider a narrow strip next to the area under the curve as seen in Figure 9e:

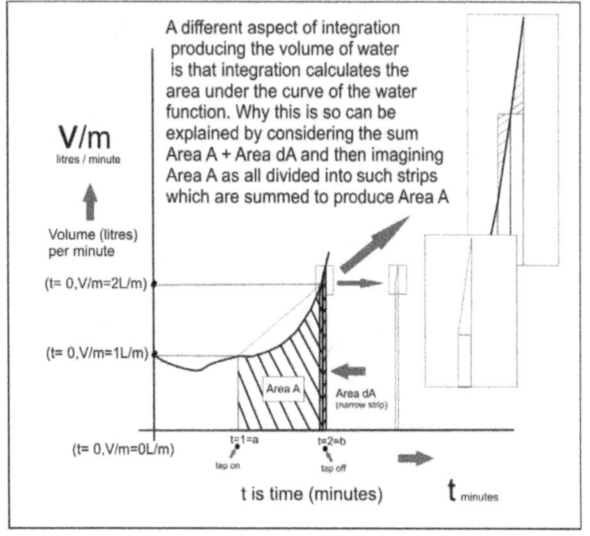

Figure 9e

Area A is the area under the curve and a narrow strip Area dA has been added next to it. Now note that Area dA is really made up of a narrow rectangle starting at time t+dt (or t= 2+dt) and it occurs at Volume = 2L+dL. In other words between t=2 and t=dt the volume of water increased from V= 2 litres to V = 2+dV ie. a tiny little increase in volume of water. Now this tiny increase in volume is approximately the area of the tiny strip, if you think about it, because the area of a rectangle is just the one side multiplied by another or here the width dt multiplied by the height dV. There is a discrepancy at the top which is the spiked-shaped tip of the strip on which the reader could comment, 'this may produce a tiny error when calculating the area of the strip (=dt x dV)'. This is true but consider the insets to the right of the strip and now imagine we place similar strips inside Area A right across to time t=1, as shown in Figure 9f.

Then one can see that if a calculation is made of the area of each strip, there are two ways to calculate this area if you consider the far right inset and the two shaded areas. Call these two areas the 'add spike' area and the lower one the 'minus spike' area. So if we calculate the add spike area, it will be approximately dt x 2 + ½ dt x dV but the minus spike area will be dt x 2 – ½ dt x dV. The immediate thought might be to say – well, we can then alternate between adding add spike areas to minus spike areas and they will cancel out with some adjustments, but there is an even better way which really achieves the same thing. When we add all the strips, we are

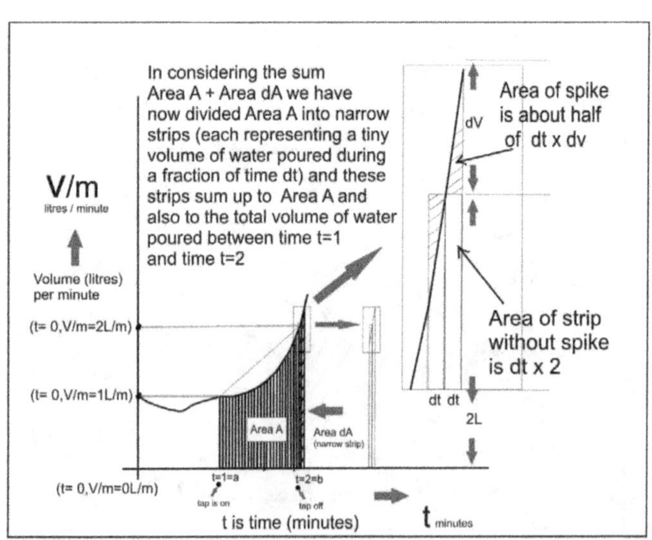

Figure 9f

missing the spikes all between t=1 and t=2, and these strips will resemble a long sawtooth just underneath the curve and so we are missing some area represented by the missing spikes. But now redraw those same strips so that the sawtooth is just above the curve. Now we would have an excess of area represented by the protruding spikes just above the curve. Then consider that if we multiply the strips to infinity, both the 'above the curve' sawtooth outline and the 'below the curve saw tooth outline' will, in the limit of this great shrinkage, become the curve itself. So the result is that we can say the addition of all the strip areas equals the area under the curve in that limit and as each strip's area (with the spike or less the spike) will be dt x dV for each strip. Thus, if we can add all these infinite strips of dt x dV between t=1 and t=2 then this will be the total volume of water poured between t=1 and t=2 (as we saw dt x dV is the volume of water that poured at the instant of time dt). Now it so happens that integration calculates the area that we want and this is what we have done by using the Newton-Leibniz formula. It is interesting (and both Newton and Leibniz would have been aware of this) that the Greeks used a similar thought process to calculate the area of a circle. They worked out the area of, say, a hexagon inscribed inside a circle and then outside the circle and reasoned that the area fell between the two values. They then looked at greater and greater sided polygons and saw that as the sides of these polygons approached infinity, the precise area of the circle could be found.

LEIBNIZ USES CALCULUS TO DERIVE π

The invention of calculus supercharged Isaac Newton's work in solving the great problems of physics. One instance displays Newton's genius. A problem relating to what type of curve allowed a bead to slide down it the fastest was published in a journal by a great mathematician (Bernoulli). A solution was published anonymously but Bernoulli recognised it as Newton because he "recognized the lion by his claw".[292] Apparently, Newton had come home from a long day working at the Mint and had solved the problem in a few hours – see Chapter 1 for his roles as both Warden and Master of the Mint.

Apart from the genius of Newton, there was one occasion when Newton praised Leibniz after he derived the series for π using a new method. In a letter to the Royal Society, Newton wrote: "Leibniz's method for obtaining convergent series is certainly very elegant, and it would have sufficiently revealed the genius of its author, even if he had written nothing else."[293]

Another mathematics 'giant' of the times, Christian Huygens, in a letter to Leibniz in 1674, wrote: "You have discovered a very remarkable property of the circle, which will forever be famous among geometers."[294]

Mathematician David Acheson and past president of the Mathematical Association has written an excellent summary of Leibniz's key steps in deriving the π series which, in essence, is set out below. He describes Leibniz's method as "one of the most extraordinary results in the whole of mathematics, linking π with odd numbers."[295] (The meaning of this will become clear.) However, to understand Leibniz's method, we must extend our 'crash course' in mathematics a little further whilst trying to 'keep it simple'. Even if the reader has not fully grasped the basics of calculus in this chapter, it is still hoped that some of the wonder in what Leibniz did will be apparent.

As the assumption has been made that the reader has no mathematical knowledge, we must explain what π is. π is a special number (related to circles – see Figure 9g below) which is generally described as a decimal – often just 3.142, but in fact the decimals can be extended infinitely if you wish. It is also known as an irrational number – which means it cannot be represented as p/q (ie. there aren't any integers p and q that when put into a fraction can result precisely in π). The number comes from the ratio of the circumference of a circle measured against its radius, as shown in the diagram below.

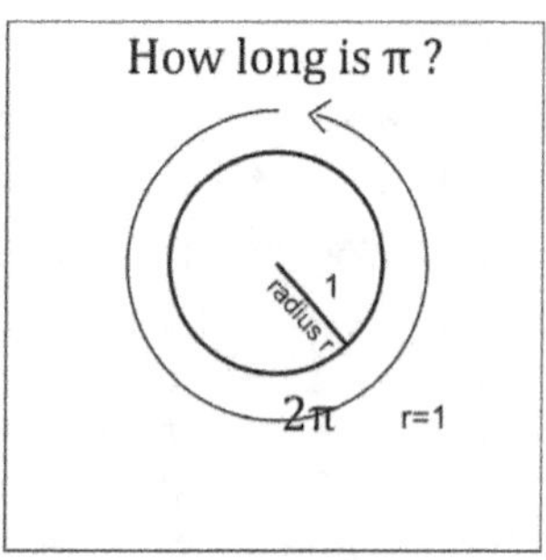

Figure 9g

The circumference = 2πr, so in Figure 9g, r=1 means the circumference = 2πr = 2π. Therefore, if the radius were r=3, the circumference would be 2πr = 2π x 3 = 6π. However, it is very useful to keep r=1 for simplicity. Thus, in the diagram, π would represent half the circumference of the circle.

Now, π can not only be used to measure things like the circumference of a circle but also to measure the internal angle of any rotation inside the circle. When people say, "Her car did a full 360°" this can also be said, "She did a full 2π rotation!" which is shown in Figure 9h.

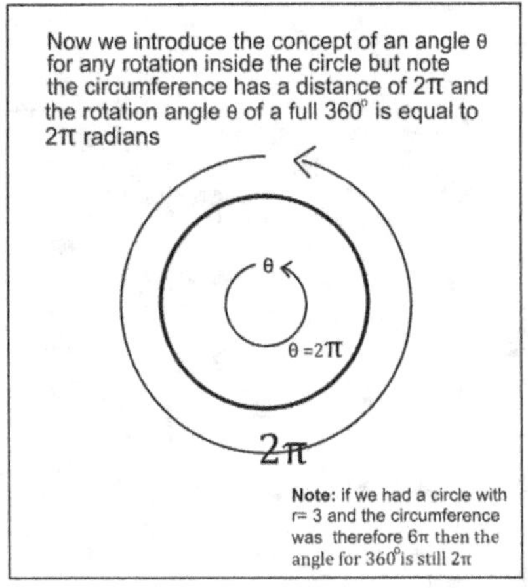

Figure 9h

Now, Leibniz actually derived π/4, which is virtually the same because whatever π/4 equals you just multiply it by 4 and you have π. We need one more thing to understand the Leibniz proof and this is some trigonometry.

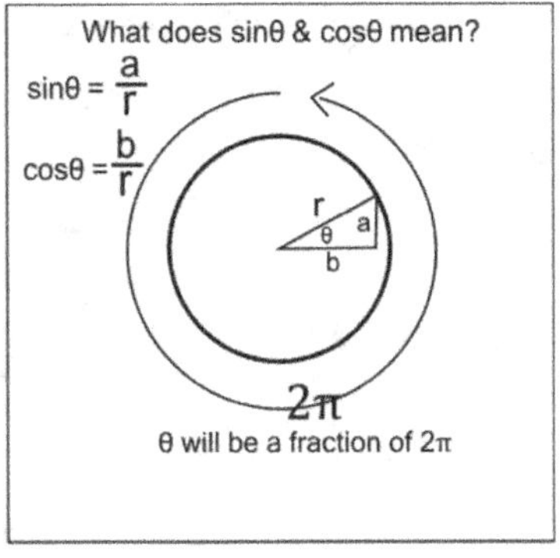

Figure 9i

All that sinθ means is that you take the up side of the triangle, ie. length a, and divide it by the slope side of the triangle (called the hypotenuse), which is length r or a/r. So if r=1 then $\sin\theta = a$ and similarly $\cos\theta = b/r = b$ (cos uses the bottom length of the triangle).

Here is both a summary of Leibniz's derivation highlighting key steps outlined by David Acheson[296] and a little more maths to help the reader understand it:

1. Having regard to the trigonometry which is explained above, Leibniz used the special properties of a right-handed isosceles triangle as shown in Figure 9j.

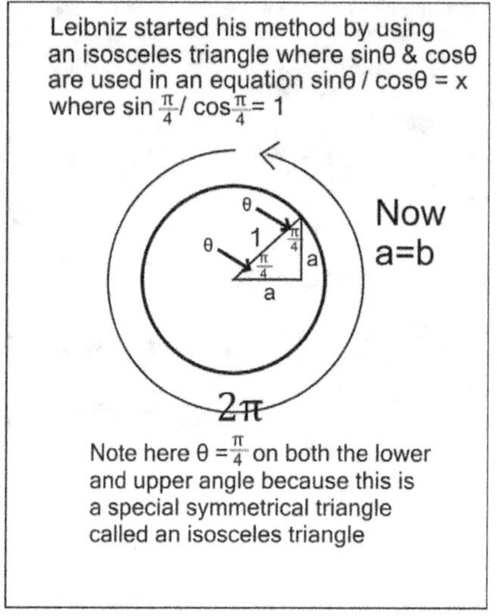

Leibniz started his method by using
an isosceles triangle where sinθ & cosθ
are used in an equation sinθ / cosθ = x
where sin π/4 / cos π/4 = 1
θ
θ
1
π/4
π/4
a
a
Now
a=b
2π
Note here θ = π/4 on both the lower
and upper angle because this is
a special symmetrical triangle
called an isosceles triangle

Figure 9j

2. If we consider an unusual graph – not the usual x and y graph that is taught at school, but another graph using x and θ, namely, $x = \sin\theta/\cos\theta$ – then Figure 9k shows the curve that plots this function. Recall that our earlier graph for the bucket used t as the horizontal variable and V as the vertical variable.[297] Here, the horizontal variable is θ and the vertical variable is x. Now note that when x=1 then $\theta = \pi/4$.

3. Recall our water function $V=t^2$ where differentiating the function resulted in $dV/dt = 2t$ (the rate the water flow increased). Now imagine if we had a little more complicated water function, say, t^2 / t^3,

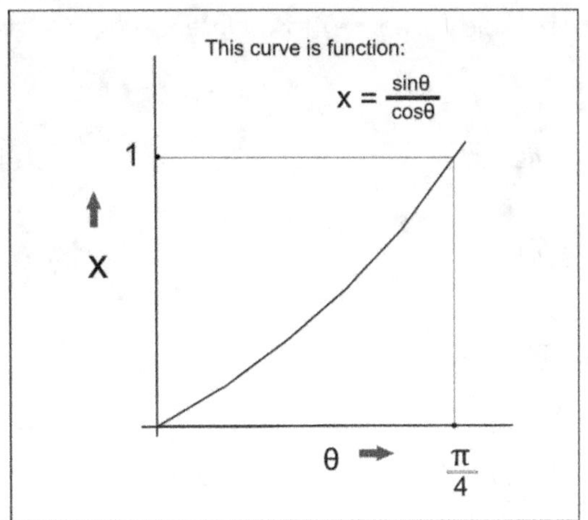

Figure 9k

how do you differentiate that? Leibniz had a rule for differentiating this – assume that $u=t^2$ and $v=t^3$ then his rule was:

$$\frac{d}{dt}\left[\frac{u}{v}\right] = \frac{\left[\frac{vdu}{dt} - \frac{udv}{dt}\right]}{v^2}$$

This just meant that:

$$\frac{d}{dt}\left[\frac{t^2}{t^3}\right] = \frac{\left[t^3 2t \;\; - t^2 3t^2\right]}{(t^3)^2} = \frac{1}{t^2} = -t^{-2}$$

A good insight here is that we could have derived this same result using the power rule that we discussed earlier in the chapter. Recall that where $f(t)= t^n$ is differentiated to $f'(t)$, the result is nt^{n-1}. Now, applying this rule to t^2 / t^3 we get $t^2 / t^3 = t^{-1}$ and so here $n= -1$ in nt^{n-1} and so $nt^{n-1} = (-)\, t^{-1-1} = -t^{-2}$, which is the same result.

4. Now just as differentiating t^2 turned out to be $2t$, the functions $V= \sin t$ or $V= \cos t$ can be differentiated (we will not expose the reader to the 'first principles' workings behind this). Instead of calling the function $V= f(t)$, let us just call it $f_1(x) = \sin x$ and $f_2(x) = \cos x$ or $y_1 = \sin x$ and $y_2 = \cos x$.

5. It turns out that differentiating the first function $y_1 = \sin x$ results in $dy/dx = \cos x$ and differentiating the second function $y_2 = \cos x$ results in $dy/dx = -\sin x$.

6. Using Leibniz's rule (3. above) for differentiating x = *sinθ/cosθ* results in:

$$\frac{d}{dt}\left[\frac{u}{v}\right] = \frac{\left[\frac{vdu}{dt} - \frac{udv}{dt}\right]}{v^2}$$

which using 5. is

$$\frac{dx}{d\theta} = \frac{[\cos\theta.\cos\theta - \sin\theta.(-sin\ \theta)\,]}{(\cos\theta)^2}$$

7. Here, Leibniz used an elegant manoeuvre. He switched back to x = *sinθ/cosθ* and rearranged the equation so it appeared now as follows:

$$\frac{dx}{d\theta} = \frac{[\cos\theta.\cos\theta - \sin\theta.(-sin\ \theta)\,]}{(\cos\theta)^2} = \frac{[cos\ \theta)^2 + (sin\ \theta)^2]}{(\cos\theta)^2} =$$

$$= \frac{[(cos\ \theta)\ \ ^2]}{(\cos\theta)^2} + \frac{[(sin\ \theta)\ \ ^2]}{(cos\ \theta)^2}$$

$$= 1 + \ \left[\frac{\sin\theta}{\cos\theta}\right]^2$$

$$= 1 + x^2$$

Then invert $\frac{dx}{d\theta}$ to $\frac{d\theta}{dx} = \frac{1}{1+X^2}$

8. Now, at this stage, Leibniz used a series which was known to expand from a similar equation to a series, namely:

$$= \frac{1}{1+X} = 1 - x + x^2 - x^3$$

(which was known to work for x between – 1 and +1 – again, deriving this series on first principles is not shown here for simplicity).

So Leibniz substituted x^2 for x in this equation and this resulted in:

$$\frac{d\theta}{dx} = \frac{1}{1+X^2} = 1 - x^2 + x^4 - x^6 \ldots$$

9. Leibniz then used calculus to integrate this, ie.

$= \int \frac{1}{1+X^2}\,dx = \int \frac{d\theta}{dx}\,dx$ [recall this reverses differentiation] so

$$\theta = 1 - \frac{x^3}{3} + \frac{x^5}{5} - \frac{x^7}{7} + \cdots$$

[Note: when you integrate, you add a constant such as +C because we could have many curves x = *sin*θ/*cos*θ +1 or x = *sin*θ/*cos*θ +2 or x = *sin*θ/*cos*θ + C and if differentiated they all equal 1/1+X^2 because a constant such as 1, 2 or C when differentiated equals 0. Now here, if you look at the θ – x graph, it intersects the origin (0,0) and so we know there is no constant (or C = 0), ie. the function is merely x = *sin*θ/*cos*θ. Now, you may query as to why

$$\theta = 1 - \frac{x^3}{3} + \frac{x^5}{5} - \frac{x^7}{7} + \cdots]$$

has the 'constant' 1 in it, which seems contradictory. The 1 in this equation is not the universal variable constant C. It is part of the series that you get when you use

$$\frac{1}{1+X^2} = 1 - x^2 + x^4 - x^6$$

If x = *sin*θ/*cos*θ intersected, say, x = 7 then the constant would be 7 and the equation would be

$$\theta = 1 - \frac{x^3}{3} + \frac{x^5}{5} - \frac{x^7}{7} + \cdots +7)$$

Now, recall in 2. above (the graph) we saw that when x = 1 then θ = π/4. Leibniz then substituted 1 for x in the above and this produced his famous equation:

$$\theta = \frac{\pi}{4} = 1 - \frac{1^3}{3} + \frac{1^5}{5} - \frac{1^7}{7} + \cdots = = 1 - \frac{1}{3} + \frac{1}{5} - \frac{1}{7} + \cdots$$

Chapter Ten Plate

Branes floating in 3D hyperspace are seen as fundamental particles of reality pursuant to M-Theory (string theory) which will be explained in Chapter 26. P-Branes are even smaller than atoms or quarks. A DO-Brane is a point and a member of a family of Branes. Some physicists use Leibniz's monads in a similar way as fundamental particles to encode reality.

Source: Author: Trace Theeman

TEN

LEIBNIZ'S REALITY

In the age of Leibniz, great thinkers such as Galileo, Descartes and Locke had seen empiricism as the key to reality. That is, observe what our senses see, hear, etc., record and analyse. Objects of our reality have shapes, dimensions or qualities that can be measured by observation. Extension or actualisation of matter and movement could be measured and analysed by mathematics, as indeed both Newton and Leibniz had done. Effectively, this is the 'physics' of the mechanistic Clockwork Universe which was replacing Plato's theory of the (eternal and perfect) forms or Aristotelian theory of final causes. Previously, according to Plato, we could discover truths by analysis of what *form* an object will take in the real world by our idea of that form, yet somehow our ideas come from outside this world – the metaphysical world.[298] In fact, Plato believed the world was an illusion and indeed Plato's philosophy contributed to Descartes's famous scepticism, 'How do I know that the world exists?' which he started to resolve with 'I think therefore I am'. But Aristotle criticised Plato's idea of a form, objecting that an object's essence (or *substance*) is not because it resembles our idea of the form but its essence results from its purpose or goal, eg. an ashtray can take many forms but it has one purpose which defines the object.[299] Aristotle had proposed four types of causes for objects in reality:

1. Material cause – this is the material or *substance* that makes up the object
2. Formal cause – the design of the object
3. Efficient cause – the action that produced the object
4. Final cause – the purpose of the object or its goal of existing.[300]

Important for Leibniz was the metaphysical concept of the underlying essences of reality – the *substances* or what he called monads, discussed in Chapter 8. Both monads and Leibniz's logical calculus formed the basis for 'Leibniz's reality'. Also important was the distinction between Aristotle's final causes or goals being replaced by the mechanistic philosophy of the Scientific Revolution. Leibniz held that mechanists were right in supporting efficient mechanical causality but wrong in denying that mechanical causality replaces purpose – there was truth in both mechanism and final causes.[301] This is why Leibniz rejected Locke's reality – Locke being the founder of British empiricism.[302]

However, what of metaphysics? Almost a century after Leibniz, another great philosopher, David Hume, became one of the great sceptics of metaphysics to rank with Descartes in this scepticism. He came up with such insights as we do not *know* that the sun will rise tomorrow – we only 'know' this from repeated experience, not reason. This conclusion is an induction, not a deduction. Science had to heed this principle because it has now been demonstrated in spectacular ways. For example, scientists previously accepted the proposition 'All swans are white', which was based on empirical scientific method and observation. However, after black swans were discovered in Australia, this 'scientific fact' was found to be false. Hume defined his scepticism with his famous 'Hume's Fork' which was:

1. "Do these ideas concern matters of fact, in which case, do they rest on observation and experience? And
2. Do they concern relations between ideas, as for example in mathematics and logic?

If the answer to both questions is 'No' then, he says, commit those ideas to the flames, for they can contain nothing but sophistry and illusion."[303]

It seems modern science has proven Hume wrong. For example, the discovery of quarks occurred after the theoretical work of Murray Gellman and George Sweig on the mathematical symmetries of subatomic particles predicted quarks would exist. The Higgs Boson was confirmed by experiment many years after the Higgs theory that a Higgs particle must account for mass. Perhaps the most spectacular 'theory to actuality' was the development of the anthropic principle, which predicted the discovery of carbon-12 merely because the fact that humans existed implied that carbon-12 must exist (see Chapter 5 – recall astronomer Fred Hoyle's scientific prediction using the 'anthropic principle'). Thus, 'theoretical' physics – 'metaphysics' – still has an important role to play despite the mechanistic empiricism of Galilieo, Descartes and Locke becoming dominant over the Greek theory of forms. Although Leibniz is seen as '100 leagues from empiricism', according to philosopher Julian Marias[304], he seemed to accept the role of mechanism and indeed made outstanding contributions to mechanistic science.

Yet, as mentioned, he saw truth in both mechanism and final causes by seeing the fabric of mechanism with the underlying reality of substance as a pre-ordered harmonious universe which he saw as pre-programmed with what he termed 'pre-established harmony'. Copleston describes Leibniz's leading idea: "the universe as a harmonious system in which there is at the same time unity and multiplicity, co ordination and differentiation of parts."[305] Leibniz wrote: "Since, however, there is really no wisdom or appetite in nature, the beautiful order arises from the fact that Nature is the clock of God (horologium Dei)" and this idea of the harmonious cosmos was also found in Newton (as we saw in Chapter 5), Kepler and philosopher Nicholas of Cusa.[306]

A harmonious fine-tuned universe as seen by both Newton and modern scientists (see Chapter 5) is quite viable, but Leibniz's pre-established harmony – the ultimate theory of determinism – has unnerved some philosophers. For example, Philosopher Austin Farrer explains: "We are tempted to conclude that Leibniz has introduced the *deus ex machina* with the fatal facility of his age. 'Where a little meditation on the characters in the play would furnish a natural *dénouement*, he swings divine intervention on to the scene by wires from the ceiling. It is easy for us to reconstruct for him the end of the piece without recourse to stage machines'. Is it? I fear it is not. There is really no avoiding the pre-established harmony."[307]

However, there may indeed be a way of avoiding the pre-established harmony. Professor Copleston refers to possible incompatibility in his esoteric philosophy and his popular philosophy, presumably because of the burning issues of religion and the then serious consequences of writing science or philosophy that was deemed heretical.[308] For example, this incompatibility might explain how to reconcile his ostensible concept of freedom, which seemed inconsistent with pre-established harmony and, in fact, Professor Copleston suggests that some people may be inclined to think that "Leibniz had his tongue in his cheek when he spoke as though contingency is simply relative to… knowledge [because] …they regard unpredictability as being essential to the notion of free choice. Leibniz spoke of choices and decisions as being *a priori* certain yet free. These two characteristics are incompatible, and *Leibniz, as a man of outstanding ability, must have seen that they were incompatible*. Therefore, we must take it that his real mind was revealed in his private papers and not in his published writings"[309] [my italics]. Thus, in the light of the tumultuous age of Leibniz, with religious wars and persecutions, perhaps the pre-established harmony *can* be avoided, although some scholars will surely be aghast at such a suggestion. However, Leibniz may also have been 'tongue in cheek' by seeking to appease the strict religious critics of his age (ie. by postulating monads as made with pre-established harmony to preserve God's determinism). Be that as it may, it would seem that any such concept is greatly altered in the light of Heisenberg's Uncertainty Principle (which was unavailable to Leibniz) and indeed was such a shock to great thinkers such as Einstein that he remarked, "God does not play dice with the universe." Leibniz may also have made a similar remark had he lived to learn the principle. Indeed, it is quite plausible to propose that God, not wanting to create a universe of puppets, would embed the Heisenberg Uncertainty Principle to create true free will and banish the principle of determinism forever – a form of fine tuning to allow free will.

Philosophers like Hume were followed by the logical positivists who stated that everything empirical is 'reality' but everything metaphysical or abstracted is not real and should be dismissed. However, a problem emerged with this philosophy, namely, that drawing the line between reality and abstraction became difficult. The abstraction of mathematics is seen as real by physicists yet beyond that more abstraction is not real. The concept of emergence in science, eg. consciousness or life, is now beginning to look like a higher level of abstraction beyond the materialism of our perceived reality. Thus, we are seeing layers of abstraction, and in this regard a great insight into reality which will be repeated throughout this book is the physics saying: 'the map is not the territory' or in other words: 'the abstraction is not the reality'.[310]

The synthesis of Leibniz's reality of monads, pre-established harmony and worldview of a harmonious universe is governed by both his mathematical thought and logical calculus. Leibniz's famous insight on logic was contained in a 1696 letter to Gabriel Wagner referring to logic as the 'art of reason as the art of the intellect'.[311] Leibniz's influence upon the modern masters of logic such as Boole and Frege is evident. George Boole's widow wrote that her husband saw Leibniz anticipating his logic and felt "as if Leibniz had come and shaken hands with him across the centuries."[312] Logician Ernst Schröder saw Leibniz's conception of logical calculus as perfected by Boole.[313] In 1857, the secretary of the Royal Prussian Academy of Science, Friedrich Trendelenburg, wished to honour their first president (Leibniz) by translating some of Leibniz's works on logic. Trendelenburg observed that there was no logical relation between sign and intuition, but science has provided the opportunity to "bring the composition of the signs into immediate contact with contents of the concept."[314] Trendelenburg referred to both *Begriffsshrift* ('Conceptual Notation') and a 'characteristic language of concepts' or *lingua characterica universalis,* or Leibniz's term *characteristica universalis* (or another of Leibniz's terms was *alphabet of human thoughts*).[315] *Begriffsshrift* was in fact used by master logician Frege as the title of his landmark work of the same name who commented that *characteristica universalis* was too ambitious for Leibniz alone in that phase of the history of logic.[316]

There are six main principles that underpin Leibniz's principles of logic. Some of these might sound obvious but sometimes in logic or mathematics an obvious statement can provide profound truths not apparent from the obvious statement itself. For example, symmetry may sound obvious, eg. if you rotate a circle or flip its mirror image, you get a symmetry or even back to where you started (its 'identity' in maths language), but as stated earlier, symmetry has produced major discoveries (quarks – Gellman – Sweig) and indeed quantum mechanics is governed by symmetries. These six principles are:

1. ***The Principle of Contradiction***: 'in virtue of which we judge that which involves a contradiction to be false, and that which is opposed or contradictory to the false to be true'.[317]

2. ***The Principle of Sufficient Reason***: 'by virtue of which we consider that we can find no true or existent fact, no true assertion, without there being a sufficient reason why it is thus and not otherwise, although most of the time these reasons cannot be known to us.'[318]

3. ***The Predicate-in-Subject Principle***: this, in short, is the principle that the name of the subject implies its features (predicates) – it is reminiscent of Plato's theory of forms because we see the thing in our reality and the idea of its form appears in our consciousness (somehow) and this implies an innate predicate implied by the subject – Leibniz really takes this further, stating that our knowledge includes the idea based on reason.[319]

4. ***The Identity of Indiscernibles***: 'If a has all its properties in common with b, then a and b are one and the same. Hence, if a and b are not identical, then there must be some difference between them.'[320]

5. ***Necessary vs contingent truths, analytical vs synthetic statements,*** *a priori* ***vs*** *a posteriori* ***statements***: These are technical philosophical terms not confined to Leibniz but Leibniz made heavy use of them. Of particular note is that a necessary truth is true in all possible worlds (a concept Leibniz introduced), but a contingent truth is dependent on it both being actualised and being logically possible. If state A and state B are possible then Leibniz states they are compossible, but not all states are compossible – even to God. For instance, Leibniz states: "The universe is only a collection of compossibles, and the actual universe is the collection of existent possibles... And as there are different combinations of possibles, some better than others, there are many possible universes, each collection of compossibles making one of them."[321]

6. ***Principle of the Best***: "The actual world is the best of all possible worlds. 'Best' means 'simplest in hypotheses, richest in phenomena.'". The best world is an optimal solution to two simultaneous requirements: it

> contains as much reality (perfection) as possible, while being maximally simple and therefore intelligible.[322] This also includes Leibniz's principle of perfection, which is to attain the best. Metaphysical perfection is for God (a necessary truth), whereas moral perfection is for mere mortals [a contingent truth].

And so combining Leibniz's system of monads, pre-established harmony, mathematics and logical system into *Leibniz's reality,* was his concept of reality opposed to *Newton's reality*? The difference is exemplified in the heated dispute between Leibniz and Newton which went through the go-between Samuel Clarke over their competing concepts of space and time. For Newton, there is the absolute concept of space and time – neither space nor time is relative (which Einstein spectacularly disproved). This was necessary for Newton as a solid unmovable medium to display the physics of gravity, mass and motion. Leibniz saw space as a system of relations between objects in the universe as opposed to Clarke's concept of space being more like a container in which objects could exist and move. Objects are actualised like 'geometric extension made concrete'.[323] So Leibniz rejected absolute space as the 'idol of some modern Englishmen'.[324] He saw the notion of a finite universe moving forward in empty space as 'fantastic and chimerical', for there would be no observable change whatsoever – no discernible change between one position and another, nor any sufficient reason for God to have chosen one position over another.[325] Leibniz held a similar view on time because in answer to the question 'Why would God not have created the universe one minute or one billion years later?' Leibniz would respond "…there would be no sufficient reason for God's creating the world at instant X rather than at instant Y [and therefore] one also proves that there are no instants apart from things."[326] This is surprisingly modern of Leibniz because Stephen Hawking famously said that there was no time before the Big Bang, ie. outside the existence within the universe because that *is* existence. Professor Copleston comments: "But though Leibniz doubtless succeeded in his object of drawing attention to the paradoxical character of the views on space and time put forward by Newton and Clarke, it does not follow that his own theory is, I do not say adequate, since the last word on space and time has scarcely been said even in the post-Einstein era, but self-consistent. On the one hand, monads are not points in space, and they have no real relative situation extending beyond the phenomenal order. 'There is no spatial or absolute distance or propinquity of monads.' ...Space, therefore, belongs to the phenomenal order... But Leibniz does not appear to have worked out successfully the relation between the subjective and objective elements in space and time."[327] Kant argued for Newton's concept of space as absolute by stating that if the universe were created with just a glove, the question would arise whether it was a right-handed or left-handed glove and thus the universe can only be oriented one way; otherwise, it would be a mirror image.[328] In other words, it cannot be reduced to spatial relations between parts which may be inconsistent with Leibniz on one view. On the other hand, chirality – right-handedness or left-handedness – is a standard concept in mathematics, ie. in symmetry and indeed in biology, where the chirality of DNA and amino acids is vital for life. Thus, I do not see this as inconsistent with Leibniz's view of there being spatial relations. Indeed, if Leibniz is a founder of topology, as Euler believed, then the concept of symmetry or chirality is vital in topology. Topology involves spatial relations between points regardless of the distances between them. This seems to be what Leibniz is stating about space. In fact, this goes to the heart of a famous Greek paradox – Zeno's paradox (discussed later in this chapter) – which some philosophers have sought to solve by doing away with distances between points in favour of spatial relations – see Grünbaum's *Philosophical Problems of Space and Time*.[329]

Modern scientists see value in Leibniz's monadology, including Sochichi Uchii of Kyoto University, emeritus, who explains: "...A monad is, according to Leibniz, a simple substance with the primitive force, and this force governs its state transition. What we usually regard as 'world' is phenomena produced by the activities of monads. Many people may think that this whole idea is crazy, but as John Archibald Wheeler said: 'crazy ideas are worth pursuing'. He goes on to explain Leibniz's grand vision of reality: "The spheres of phenomena and reality must be strictly separated. Physics or dynamics studies motions which are supposed to underlie phenomena, but the laws of dynamics must be grounded in metaphysical principles." Leibniz referred to this as 'ontological grounding', ie. "the physical world requires some further 'grounding'... to postulate some further entity or principle as the condition of this very world as we experience it".[330] Uchii refers to Leibniz's topology and in particular, "The problem of space and geometry... [has been]... masterfully treated by Vincenzo De Risi [2007].[331] Leibniz's work on Analysis Situs (an analysis of situations) deals with this question. In a nutshell, bodies coexisting in the phenomenal world are situated or related with each other, and this situation is the phenomenal expression of the groups of monads corresponding to them. Thus, space results as a totality of such relations… [or] more precisely space in general can be defined by considering all possible situations, and the actual space can be defined in terms of the actual situation of the world."[332]

Uchii believes that Leibniz's monadology is a good model to develop the 'holy grail' of physics or unification of physics, which I take to mean both the grand unified theory – combining the strong and weak forces, the electromagnetic forces and the forces of gravity to give the 'theory of everything'. Topological principles and symmetry figure in this view (as indeed they do in quantum mechanics, eg. gauge symmetry – quarks, etc.). The two important reasons Uchii provides to support this are that monads provide:

(1) the basis of space and time, together with

(2) the basis of the law of state-transition of the phenomenal world.

(And he adds, "We know neither of course.")[333]

Uchii explains how monads, ie. simple *substances* (they are really the 'atoms' of the underlying reality), can provide such a model: "…in *Specimen Dynamicum*, [Leibniz] said that this primitive force (in a monad) is transformed into the derivative force in the phenomenal world. This statement can be simply translated as: the derivative force is a coded appearance of the primitive force. 'Coded' because the primitive force belongs to reality (the sphere of monads), but the derivative force belongs to the phenomena, how the reality is represented to intelligent creatures like humans. In order to connect two entirely different spheres, we need coding. Further, notice that any monad's states (perceptions) themselves represent the whole reality, the sphere of monads. Thus we have to assume at least two different systems of coding, one for states (perceptions, which occur in reality), another for phenomena. These must be different because Leibniz consistently says that there is no space or no time for reality whereas phenomena (including motion) occur in space and time."[334]

Developing monads further, Uchii relies on the timeless nature of monads to postulate that monads in 'reality' ground metric time in the phenomenal world via coding which preserves order of states. It is as if the underlying reality is topological – no metrics – just order and other information embedded in a continuum of monads which undergoes a transformation or mapping into our phenomenal world (ie. a metric non-topological world). However, in this transformation – underlying reality → phenomenal world – the criterion (coding) for congruence of distance cannot give congruence of time.[335] Presumably, this coding might be similar to the information encoded in Einstein's tensors in the space-time of his General Theory of Relativity.

I should pause before completing Uchii's main concepts and comment that I think the word 'congruence' is unfortunate because there are no metrics in the underlying reality that can be congruent to distance in our phenomenal world, and that is why Uchii uses the term 'criterion for congruence'. In other sections of his paper, he refers to the topological term – *homomorphisms* – rather than congruent, which is preferable. *Homomorphisms* relate to a transformation of elements from one space to another space where all the relations between elements of both spaces are preserved. Thus, if there is a continuum of monads, the relations between the monads are preserved in such a way that the encoding of monads establishes relations which upon transformations into our phenomenal world are preserved. The problem with this is the dubious pre-established harmony which led to Leibniz's famous statement – 'monads do not have windows'. The only resolution to this is that whilst there can be relations between monads in Uchii's scheme, there can be no interactions because interactions only occur in the phenomenal world, which has the time and the motion to 'cloth' interactions.

Uchii continues to explain the main concepts of his 'modern monadology'. "…Leibniz, in *Initia Rerum,* dwells on the analogy between space and time. And the crucial point emerges when he describes a path of a movable object:

A 'path' [via] is the continuous and successive locus of a movable thing (*Initia Rerum,* Loemker 1969, 668)."

Uchii refers to drawing a line on a blackboard and says: "Although your motion of drawing does not exist anymore, its path exists on the blackboard. This is the crucial connection of time and space, via motion". He further quotes Leibniz:

> ***"But we know as co existing, not merely those things we perceive together, but also those which we perceive successively, provided only that, during the transition… from the perception of one to the other, the former is not destroyed and the latter generated" (Loemker 1969, 671, [Uchii's] italics).***

And Uchii's comment includes: "…we take time for seeing the whole path by successive perceptions… [and]… it is clear that the metric of time (of your motion) is somehow connected to the length of the line."[336]

Leibniz gave a simple analogy for monads underlying our reality to create objects. He referred to mathematics and said that monads could be analogised as points which have no length yet together they make up a line.[337] Earlier in this chapter, I referred to Zeno's paradox and how topology involves spatial relations between points regardless of the distances between them. I also referred to the concept of Grünbaum solving Zeno's paradox by eliminating distance. What was Zeno's paradox? *The Oxford Dictionary of Philosophy* relates one of the Zeno paradoxes as follows:

> ***"Suppose a runner needs to travel from a start S to a finish F. To do this he must first travel to the midpoint, M and thence to F but if N is the midpoint of SM he must first travel to N and so on ad infinitum… but it is impossible to accomplish an infinite number of tasks in a finite time. Therefore the runner cannot complete (or start) his journey."***[338]

(It is also framed as a race between Achilles and a tortoise and thus called the Achilles Paradox.)

Leibniz's monads have no length and thus could produce the time, space and motion required to resolve the paradox using the concepts of both Uchii and Grünbaum,[339] so it seems that monads may be immutable after all.

A monad has also/been likened to an 'Aleph'[340] and has been written about in at least two famous literary works known as *The Aleph*.[341] In one of these,

Jorge Luis Borges's short story *The Aleph*, it was described and perhaps may give us a romantic image of a monad:

> *"...I saw a small iridescent sphere of almost unbearable brilliance. At first I thought it was revolving; then I realized that this movement was an illusion created by the dizzying world it bounded... all space was there, actual and undiminished. Each thing (a mirror's face, let us say) was infinite things, since I distinctly saw it from every angle of the universe. I saw the teeming sea; I saw daybreak and nightfall... I saw bunches of grapes, snow, tobacco, lodes of metal, steam... I saw a woman in Inverness whom I shall never forget; I saw her tangled hair, her tall figure, I saw cancer in her breast... I saw horses with flowing manes on a shore in the Caspian Sea at dawn... I saw all the ants on the planet... and I felt dizzy and wept, for my eyes had seen that secret and conjectured object whose name is uncommon to all men but which no man has looked upon – the unimaginable universe."*[342]

Chapter Eleven Plate

Leibniz's lengthy theodicy involves a type of theological multiverse to explain evil and suffering because in creating the world God would have chosen the best of all possible worlds. His treatise Theodicy, *which he wrote in 1710 towards the end of his life, was the culmination of all his scientific, mathematical and philosophical work and was written at the behest of Queen Charlotte of Prussia.*

ELEVEN

LEIBNIZ'S BEST OF ALL WORLDS THEODICY

Leibniz's fertile mind was active between 1677 and June 1690 when he wrote 522 pieces of over 3,000 pages, including groundbreaking maths and science, but now he was urged by Queen Charlotte of Prussia to turn his mind to one of the great questions of the ages... why God allowed evil and suffering. Professor M.R. Antognazza sets out the key parts of Leibniz's *Theodicy*:[343]

> *"Of all the infinitely many worlds which are possible, God created (or more precisely 'actualised') the best."*[344]

> *"Now, God being God – namely, the most perfect Being encompassing all unlimited perfections, including absolute power (omnipotence), absolute knowledge (omniscience), and absolute goodness (omnibenevolence) – he cannot but choose the best. An omniscient being can never fail to know what is best. Being also absolutely good, he can never fail to choose it. Being also omnipotent, he cannot fail to bring it about. In brief, God being God, he is morally (although not logically or metaphysically) necessitated to choose the best. As Leibniz writes in the* Theodicy, *'if we could understand the structure*

and economy of the universe, we would find that it is made and governed as the wisest and most virtuous would wish , since God could not fail to do thus. This necessity nevertheless is merely moral'... Of course, this does not yet answer the question of what criteria a possibility must satisfy in order to qualify as the 'best'... Minimally, however, one can attempt to give some content to the notion of 'best' by saying that it must mean something like 'as good as possible', given the circumstances.' The question becomes what counts as 'good'... Thus, the best possible world is the world in which there is the greatest possible combination of beings, with all their perfections (that is positive qualities), in all their varieties." (GP VI, 236).[345]

However, Professor Antognazza recognizes a major flaw in Leibniz's *Theodicy*:

"Assuming (as Leibniz does) that there are infinitely many possible worlds, and that the actual world, as many others, is also itself infinite, how can there be a 'best' one? For there to be infinitely many possible worlds means that no matter how many we can think there are, there are more... And... no matter how great is the number of beings we think there are in each of these infinite worlds, there are more... Leibniz's strategy for repelling this (for his purposes, quite devastating) objection, is to appeal to the principle of sufficient reason. The very fact that there is a world shows that it must be possible for there to be a best of all possible worlds. Otherwise God would have lacked sufficient reason for choosing this one over all others..." (Theodicy, *GP VI 232).*[346]

Note that Leibniz's principle of sufficient reason is a type of logical cause and effect principle, as Professor Antognazza explains: "In each of these possible worlds, 'nothing is without a reason' ...everything has a sufficient reason, why it should be thus and not otherwise, even though most often these reasons cannot be known to us." (Monadology, § 32;GPVI)[347]

Philosophers have also objected to the 'best' of all possible worlds as being feasible by stating that all that is required is a 'better' world. Indeed, Voltaire found it easy to ridicule the 'best of all possible worlds' theodicy in *Candide* (1759), but it has been observed that Voltaire may not have appreciated the intricacy of Leibniz's argument (especially in view of Leibniz's status in history as a brilliant mathematician, scientist and thinker). Indeed, it seems that Newton himself subscribed to a similar view in speaking of God's powers: "I can conceive all my owne powers (knowledge, activating matter,

& C.) without assigning them any limits." Newton followed this to examine God's powers, holding that God could create anything not physically impossible and indeed every feature of this world was a conscious choice by God "to create one out of a number of possible (ie. non-contradictory) worlds."[348]

This argument is explained again by Professor Antognazza:

> *"Surely this actual world would be even better if, say, one extra fawn had escaped death in the Thuringian forest fire of 1016 CE? It is at this point that the full force of Leibniz's net of theses regarding possible worlds; universal harmony, purely extrinsic denominations, identity of indiscernibles, principle of sufficient reason, and so on, comes into play. There is no fact, however seemingly minor, that does not affect the entire universe, since each fact is the result of an infinite chain of reasons which explain why it is thus and not otherwise. Had it been otherwise, the infinite chain of reasons branching in all directions would have been different."*[349]

Leaving aside the major flaw until later, and despite Voltaire's ridicule, it turns out that Leibniz's fawn insight may have had an element of truth beyond what anyone thought after mathematician/meteorologist Edward Lorentz discovered a 'fawn' in the form of a 'butterfly', namely, in the Butterfly Effect from chaos theory, as will be explained in the next chapter.

Can the flaw in Leibniz's *Theodicy* be overcome? Leibniz made clear that utopian worlds were possible but stated they would be inferior to ours and added: "I cannot show you this in detail. For can I know and can I present infinities to you and compare them together? But you must judge with me *ab effectu* [from the effect] since God has chosen this world as it is".[350]

Infinity is the pivotal concept to analyse. Mathematician Ian Stewart, author of *Infinity, A Very Short Introduction*, has considered the theological aspects of infinity:

> *"Some version of the infinite occurs in many religions. I'll focus solely on Christianity to keep the topic within bounds. As Philoponus illustrates, the philosophy and mathematics of infinity became intimately entwined with early Christian beliefs… the notion that God has no limits became entrenched; it was pretty much the definition of the deity."*[351] *However, "Early theologians*

> *seem not to have considered God to be literally infinite. Around 200 A.D. in De Principiis (on first principles), Origen, the first Christian theologian of repute, maintained that God's power is finite. The reason is that perfection can't have blurred edges. Its limits must be sharp. Latin perfectus means 'complete'. If God's power were infinite, it would be incomplete, hence imperfect."*[352]

Today, the more standard usage of 'infinite' or 'perfect' by Christian theologians means a God as having attributes *not limited in any way* as this passage from the *New Advent Catholic Encyclopaedia* (online) demonstrates: "When we say that God is infinite, we mean that He is unlimited in every kind of perfection or that every conceivable perfection belongs to Him in the highest conceivable way. In a different sense we sometimes speak, for instance, of infinite time or space, meaning thereby time of such indefinite duration or space of such indefinite extension that we cannot assign any fixed limit to one or the other. *Care should be taken not to confound these two essentially different meanings of the term*"[353] [my italics]. In effect, this encyclopaedia warns that the term 'infinite' should not be conflated with the mathematical /scientific meaning of 'infinite'.

So if we are to avoid conflation, which meaning is to be adopted for theodicy? It is suggested that the impartial observer who is to be satisfied by the theodicy would be better served by an objective 'dogma free' usage of the term 'infinite' in the mathematical/scientific sense. For example, the 'impartial inquirer' may well be an atheist unmoved by preferred usages of religions. The theist would wish to satisfy all inquirers. The objection that, in the metaphysical sense, God is not of the scientific or mathematical world is overcome by stating that the more logically consistent usage of 'infinite' satisfies not only mathematics and science but rigorous philosophical and logical methodology. Therefore, it is in all of the philosophical, mathematical and scientific senses that 'infinite' is used in analysing Leibniz's ideas. Philosopher St Thomas Aquinas, who was fond of observing what God cannot do, referred to three categories of possibilities in relation to God's power, explaining the third as follows:

> *"Thirdly, in mathematics the word possible is used relative to the sort of potentiality exemplified in geometry… things impossible in the third way God cannot do…" and later Aquinas says: "For Socrates not to have run when he has run is called incidentally impossible…"*[354]

Or in another place Aquinas said that God cannot: "...make a man have no soul, or make the sum of the angles of a triangle not be two right angles. He cannot undo the past, commit sins, make another God, or make Himself not exist." (Summa contra Gentiles – 2 – Book II.)

It should be noted that Aquinas would have meant 'God cannot change the formula for a triangle without changing the axiom upon which it is based'. (The axiom for the triangle formula assumed by Aquinas is based on Euclidean geometry.) Note that in non-Euclidean geometry, this two-angle theorem would be wrong. This is an important point because when we are viewing different worlds ('trans worlds') for a theodicy, it could appear, at first sight, that God has the power to merely change things that appear logically impossible, like this triangle formula, when in fact for the triangle formula to change in a 'trans world', God would need to change the axiom as well. This is a similar concept to the mathematical multiverse hypothesis we saw in Chapter 5 where modern theories postulate a change in world and a change in the laws of that world. Aquinas also said: "In the theories of sciences we always base ourselves on something already known, whether we are proving propositions or discovering definitions. But we can't go on this way forever; that would spell the death of all science, its proofs, and its definitions, since you can't bridge the infinite."[355] Thus, it is suggested that Aquinas's approach accords with mathematical and scientific meanings of 'infinite' and indeed these last italicised words lend support to the approach.

It follows that God is not 'infinite' and, more importantly, any exercise of His power that involves infinite tasks could not be actualised into an object of that power. Why not? Presumably creating objects requires time; otherwise, cause could not follow effect. Stephen Hawking has said that before the universe existed there was no time, suggesting that time is not a necessity for cause and effect to occur. He goes further and postulates that the universe has no boundary (called the Hartle-Hawking proposal) and thus can create itself. In Chapter 23, we will consider whether this proposal could be an exception to the cause and effect principle despite being beyond our understanding. However, with our present physics and the concept of 'time's arrow', it is a reasonable assumption to keep time as necessary for action like tasks. Even during a finite time, if an exercise of power needed completion of infinite tasks to actualise an object then the object could not be actualised because the last task would never be completed. If one objected that the runner in Zeno's paradox (discussed at the end of the last chapter) could make infinite steps over a finite distance (the infinite fractions contained in the continuum of that distance) and lead to 'completion', this does not refute that impossibility because unlike Zeno's runner reaching his destination (which we know), we do ***not*** know whether the object can be actualised. Furthermore, the knowledge that runners (like Zeno) do complete their journeys and all the ways that philosophers have postulated to overcome Zeno's paradox merely create a possibility that God could theoretically complete an infinite set of tasks *somehow*.

This still leaves the original reasonable impossibility that God cannot complete an infinite set of tasks. If so, that is a potential basis for a theodicy (being any reasonable defence of God's creation), yet God's omnipotence is preserved because not being able to complete an impossible task is not inconsistent with being omnipotent. This adopts a reasonable definition akin to St Thomas Aquinas's points about what God cannot do despite being omnipotent – in essence, omnipotence means being capable of doing things that are possible and not absurd. Professor Antognazza explains Leibniz's theodicy further in this regard: "In order to get our head around this whole business of possible worlds, it may be helpful for us to think of the worlds in which 'I' am in Paris right now and not stuck in London (as I actually am). However, according to Leibniz, these imaginary counterparts of me enjoying coffee under the Eiffel Tower are unfortunately not me at all but other individuals. Had God actualized one of the worlds in which someone *superficially* similar to me is having a good time in Paris right now, I would simply not have existed (except , of course, as a thought in God's mind). Hence, Leibniz berates those complaining that God did not put them in better circumstances… [because otherwise]… they would not have existed *at all* in the actual world."[356] This reflects the new thinking of the Scientific Revolution escaping the mystical world where God was conceived to have magician-like powers. Clearly, Leibniz draws on the mathematics of reality in any world that 'A' cannot be 'Not A' and so on (recall Leibniz's six main logical principles referred to in Chapter 10). Now God's creative power can be considered more closely within the more realistic framework of 'logically possible power' and not 'magician-like power.'

Philosophers have studied infinite tasks and have called them 'super tasks', which allows us to consider what objects can or cannot be created with regard to whether an object's creation requires a super task or not. It is convenient to call an object requiring a super task (ie. infinite tasks) a 'super object' and knowledge having a prerequisite of completion of super tasks 'super knowledge'. For example, it might require infinite iterative investigative checks to ascertain the super knowledge. *Some* super knowledge could be acquired by means other than infinite iterative investigative checks. For example, assume that to create an object its precise volume is required (to create an object its features would be required and volume is one of such features). The precise volume of an object could be argued to be super knowledge as calculation of volume would have required an infinite iterative series of steps (ie, adding of each slice of the object) and then summation of the infinite slices, but there could be no final 'step' to add the last slice. However, calculus provides an alternative means to acquire that super knowledge. Whether there are alternative means to the super task of calculating the infinite interactive combinations of states between the ranges of volumes of a cell and its energy spectrum is a more challenging question.

Crucially, this must reflect on God's creative knowledge[357] of what objects can be created. Knowledge of what can and cannot be actualised must therefore depend on whether for any proposed objects of creation an infinite number of tasks may be required to actualise an object, because if an infinite number of tasks is required to actualise a theoretical super object it cannot be actualised. If so, the super object's *method of creation* (not its theoretical existence) cannot be the subject of creative knowledge.

Why are tasks important in terms of creation of worlds and objects in trans worlds? Philosopher Max Black has studied the topic of 'super tasks' and used thought experiments using 'infinity machines' which perform infinite calculations in a finite time but has concluded that the paradox remains, as explained by this entry: "Max Black (1950) argued that it is nevertheless impossible to complete the Zeno task, since there is no final step in the infinite sequence. The existence of the final step was similarly demanded on *a priori* terms by Gwiazda (2012)."[358]

The study of 'super tasks' in philosophy has provided some new insights.

The *Stanford Encyclopaedia of Philosophy*'s entry for Zeno's paradox explains further: "…a number of philosophers – most notably Grünbaum (1967) – took up the task of showing how modern mathematics could solve all of Zeno's paradoxes… What they realised was that a purely mathematical solution was not sufficient: the paradoxes not only question abstract mathematics, *but the nature of physical reality*. So what they sought was an argument not only that Zeno posed no threat to the mathematics of infinity but also that that mathematics correctly described objects, time and space… While no one really knows where this research will ultimately lead, it is quite possible that space and time will turn out, at the most fundamental level, to be quite unlike [typical mathematical viewpoints of the] continuum…"[359] [my italics]

Thus, we have seen that God's creative knowledge does not include super knowledge of how to create super objects except a subset of those super objects whose method of creation could only be obtained by alternative means. Therefore, it can be seen that knowledge of the infinite possible designs of objects is not logically possible to know as no 'final design' can complete the super task. Creation faced an 'infinite wall' of infinite possible designs and insoluble equations (such as those that require infinite numerical iterations to solve). I call this the 'Infinite Wall Principle', explained in more detail in Chapter 13. Thus, God cannot have creative knowledge of all designs, let alone the 'infinity prone' calculations inherent in features of those objects. Some calculations can be achieved through calculus or infinite series to give finite solution-based knowledge but we can induce that many are impossible. If so, the question remains as to how to gauge the importance of this Infinite Wall Principle to creative power. This is a question that will be dealt with in Chapter 13 but first the outcome of the Scientific Revolution will be examined in the next chapter.

What then is the significance of theodicy in our 'post-modern' world? Firstly, from a clinical viewpoint, theodicy is seen as important in palliative care. In a 2013 technical paper entitled 'Theodicy and End-of-Life Care', health professionals are given guidance on how to comfort patients holding religious or spiritual beliefs.[360] Secondly, it would seem important for those who have lost beliefs and a purpose in life due to the apparent inexplicability of evil and suffering in our world today. Moreover, according to a ~7000 patient study reported in *New Scientist* (Teal Burrell, 28 January

2017), a sense of purpose correlated to big drops in death rates over a fourteen-year period. Author Teal Burrell stated, "Could it be that *purpose* is just another term for religious faith?"

If theodicy is important to people's lives then so is the Infinite Wall Principle for a credible theodicy. In Chapter 13 we will see how the Infinite Wall Principle strikes at the heart of reality, but first we must understand how the 'Clockwork Universe' reality broke down, as will now be explained in the next chapter.

A GENERAL MAP OF THE
WITH ALL THE NEW DISCOVERIES and MARGINAL DELINEATIONS, CONTAININ
WESTERN HEMISPHERE OR THE NEW WORLD
NIVERSAL SCALE
A CHART OF THE WORLD

D, OR TERRAQUEOUS GLOBE
The SOUTHERN HEMISPHERE
An Astronomical
EASTERN HEMISPHERE or THE OLD WORLD
NORTHERN ICY OCEAN
ATLANTIC OCEAN
AFRICA
INDIAN SEA
INDIAN OCEAN
NEW HOLLAND or TERRA AUSTRALIS
EASTERN SEA
NORTH PACIFIC OCEAN
SOUTH PACIFIC OCEAN
SOUTHERN ICY OCEAN
The ANALEMMA
The VICISSITUDE of SEASONS Explained

Chapter Twelve Plate

A general map of the world by Samuel Dunn (1794) with overlay of watch mechanism

*Source: Author: Michelle Vileneuve – watch mechanism circular inset from original https://creativecommons.org/licenses/by-sa/3.0/deed.en**

** or such deed or licence instrument as may be applicable to the licence*

TWELVE

THE DEMISE OF THE CLOCKWORK UNIVERSE

In the Preface, I quoted science writer Edward Dolnick's explanation of the Clockwork Universe and calculus:

> *"...at some point in the 1600s, a new idea came into the world. The notion was that the natural world not only follows rough-and-ready patterns but also exact, formal, mathematical laws. Though it looked haphazard and sometimes chaotic, the universe was in fact an intricate and perfectly regulated clockwork... God was a mathematician, 17th-century scientists firmly believed. He had written His laws in a mathematical code. Their task was to find the key."*[361]

Dolnick also described the outcome of Newton's and Leibniz's mathematical breakthroughs: "Because calculus was the ideal tool to study the natural world, the debate spilled over from mathematics to science and then from science to theology. What was the nature of the universe? What was the nature of God, who had designed the universe?"[362]

In fact, as the effects of the Scientific Revolution took hold, the great mathematician Laplace then claimed the universe was scientifically deterministic (later challenged, as explained below in relation to Heisenberg's Uncertainty Principle.) Today, a key question remains: what is the nature of reality?

Before Leibniz's theodicy, the Bible records one of the earliest defences of God in the narrative of Job when he rebukes God for his misfortune. God answers: "Where were you when I laid the foundations of the earth? … Who determined its measurements… or who stretched the line upon it…?" Job 38.4, 38.5, and in another place: "Do you know the balancing of the clouds…?" Job 37.16. In essence, Job was not a professional creator and so his challenge to God was misconceived.

Philosophers have carved their own theodicies from these 'clouds of reality', but advances in mathematics and science have now shown that "clouds are not spheres, mountains are not cones, coastlines are not circles," Benoit Mandelbrot.[363] We will see later that the turbulence of clouds, air, liquids, etc. may reframe our concept of reality. Also, a great insight has emerged from quantum physics that reality necessarily involves the *observer* as a key component of reality. The instrument of observation in this component of reality is our consciousness, which Dr Pim van Lommel explains:

> ***"Everything exists only in our consciousness and everything outside it, such as 'true objective reality' is unknowable… This view was shared by the philosopher Immanuel Kant, who argued that we can only know reality that we can experience in our consciousness. Perception is possible thanks to our power of reason (an aspect of consciousness) because our consciousness shapes reality as it appears to us."***[364]

Nevertheless, Immanuel Kant held that one can discover significant truths about reality by using pure reason or *a priori* reasoning (ie. using reason without empirical observation or experiment). Then, *a posteriori* or empirical analysis shuttles back and forth with *a priori* reasoning. Kant then held that anything beyond our experiences was unknown – we cannot make conclusions about things not experienced. He held that phenomena we experience cannot be used to make conclusions about the universe as a whole. Today, advances in science have reversed the credibility of such a philosophy and indeed Kant would be most surprised that quantum physics now requires an observer to construct quantum reality.

So what caused the demise of the Clockwork Universe paradigm? There were three paradigm shifts:

(i) Einstein's Theory of Relativity;
(ii) Heisenberg's Uncertainty Principle;
(iii) Chaos Theory.

(i) Einstein's Theory of Relativity

Einstein's Theory of Relativity showed that space was curved and time was not absolute, causing the first crack to appear in the 'clockwork paradigm' because Newton's laws of gravitation and motion, uncurved space, etc. had been a foundation of the Clockwork Universe. So Einstein's reality replaced Newton's reality. We will see later how 'Hawking's reality' replaced Einstein's reality through his insight that the laws of general relativity broke down at points of infinite density known as singularities (to the dismay of admirers of Einstein's Theory of Relativity).[365]

(ii) Heisenberg's Uncertainty Principle

It was Heisenberg's Uncertainty Principle that also fundamentally changed our view of reality. Stephen Hawking explains in his chapter on the Uncertainty Principle in *A Brief History of Time*: "The doctrine of scientific determinism was strongly resisted by many people, who felt that it infringed God's freedom to intervene in the world, but it remained the standard assumption of science until the early years of this century..."[366]

In essence, Heisenberg's Uncertainty Principle stated that it was not possible to know a particle's position and momentum at the same time. This also relates to Schrödinger's Wave Equation, which describes all possible alternative behaviours of the particle combined into a wave. A precise solution for both where the particle is and its momentum is not possible in quantum mechanics. The Wave Function is said to collapse when the particle's position is known – as explained by this extract from *Scientific American* regarding the probabilistic suite of alternative particle positions:

"...For at the moment of measurement, the wave function describing the superposition of alternatives appears to collapse into one member of the superposition, thereby interrupting the smooth evolution of the wave function and introducing discontinuity. A single measurement outcome emerges, banishing all the other possibilities from classically described reality," Peter Byrne, *Scientific American* (2008).[367] The reference to Heisenberg's Uncertainty Principle is specifically the indeterminacy of deriving precise values for position and momentum of the particle. If you have one, you cannot have the other.

This led to the wider principle, referred to by Hawking, that scientific determinism was now dead. As Hawking explained: "We now know that Laplace's hopes of determinism cannot be realised, at least in the terms he had in mind. The uncertainty principle of quantum mechanics implies that certain pairs of quantities, such as position and velocity of a particle, cannot both be predicted with complete accuracy... These quantum theories are deterministic in the sense that they give laws for the evolution of the wave with time [through Schrödinger's Wave Equation]... The unpredictable, random element comes in only when we try to interpret the wave in terms of the positions and velocities of particles."[368]

Thus, in 2017, we can understand Hawking when he said: "Since the Big Bang itself the universe has been governed by laws of nature. Laws have determined everything from the formation of entire galaxies to the stable orbits of planets around stars. This is a cosmic clockwork of determinism."[369]

Thus, we can see that, in one sense, determinism still survives in 'macro physics' – in classical physics. However, in philosophy, the fact that the brain relies on quantum processes in its electrical activity seems to have banished determinism in human beings – we have true free will (although philosophers continue to argue depending on the context).

(iii) Chaos Theory

Mathematician Ian Stewart relates the story of how the great mathematician Henri Poincaré took up the challenge in 1887 of the King of Sweden to solve what is called the 'Three-Body Problem'. The behaviour of two planets was well determined by Newton's laws but the behaviour of three planets appeared chaotic and anomalous despite being governed by 'smooth' deterministic equations. The paper Poincaré submitted introduced a revelation: "Simple, deterministic equations can have complex, apparently random solutions. We now call such behaviour 'chaos' – the full phrase should be 'deterministic chaos'…"[370] Stewart illustrates this discovery with an example of a whisk beating an egg and trying to predict where the particles of mixed-up egg end up:

> *"…Whatever prediction we make, particles indistinguishably close to one another must follow very different paths, ending up somewhere quite different in the bowl… Now we see the real flaw in Laplace's reasoning. In order to predict where the egg will go, we need to know exactly where it starts – accurate to thousands of decimal places. The slightest error in the predicted starting position will quickly translate into a big error in the predicted motion… Poincaré's discovery boils down to this – the dynamics of the three-body system mixes things up much like an egg whisk. Particles that start very close together end up far apart. The motion is deterministic, but this only carries practical consequences if you can measure the positions exactly – and you can't. So deterministic is not the same as predictability. And Laplace's 'vast intellect' has to be able to do more than just submit its data to analysis. It has to be able to obtain the data in the first place."*[371]

Laplace had postulated scientific determinism such that a 'vast intellect' could know everything about the universe, given enough data to submit to analysis. As Laplace said, "…for such an intellect nothing could be uncertain."[372]

Assuming this 'vast intellect' to be God, is God limited by logical impossibility or empirical impossibility from acquiring such data? Chaos is demonstrated by the butterfly effect, discovered by Edward Lorentz (shown in the metaphor that a butterfly flapping its wings can cause great changes to the weather, or tiny changes to mathematical modelling of weather conditions can have major changes to the modelling).

One puzzling aspect is the term 'deterministic' chaos, which is explained by Ian Stewart: "Chaos is apparent randomness with a purely deterministic cause… Chaos inhabits the twilight zone between regularity and randomness… The discussion is made more difficult by a philosophical problem: does true randomness really exist?"[373] If one blends the Heisenberg Uncertainty of particles in the quantum world with the indeterminacy accompanying any initial conditions then this 'uncertainty-chaos blend' appears truly random. The only way to rescue determinism might be to reject Heisenberg's Uncertainty Principle by replacing it with another hypothesis, eg. the Many Worlds hypothesis or another theory known as the hidden variable hypothesis.[374] Leaving aside Heisenberg Uncertainty, in the next chapter we will see that chaos, despite not being *truly* random, may have a *de facto* randomness quality because of an existing mathematical insolubility related to a common form of chaos – turbulence. In the next chapter, we will consider a puzzling question to further push the boundaries of reality – whether turbulence's inherent data is unknowable even if the starting conditions were known.

In summary, reality is now seen as probabilistic and chaotic. Biological systems may introduce a further dynamical aspect of reality related to chaotic systems – stochastic processes. The issue of whether chaos is deterministic or truly random and indeterministic will also be addressed in the next chapter.

Thus, both Newton's Reality and Leibniz's Reality relied upon the deterministic Clockwork Universe when it now appears that the chaotic universe is more indeterminate, probabilistic and stochastic. Newton talked about the chaos in the universe and may have intuitively pondered the possible mechanisms but did not live to see the mathematics of chaos with attractors, fractals, stochastic cycles and the implications they have for 'Newton's reality'.

This Groupe Exhibiting the Future Enjoyment of Reserved Knowledge wherein an Arch Angel is explaining a Solar System to the admiring Newton Gallileo Copernicus & Bacon is taken from the Picture of the State of
Retribution with the most Respectful Affectionate Regard is Dedicated to the Society for the Encouragement of Arts Manufactures & Commerce by James Barry R.A. Professor of Painting to the Royal Academy -

THIRTEEN

IS A PERFECT WORLD POSSIBLE?

This chapter does not have the density of Newton's Principia *or* Hawking's *Black Hole papers but has some challenging biological concepts explained by easier mathematics than the* Newton-Leibniz *calculus.* However, *to lighten the reader's burden, each concept section has a short summary, allowing the reader to skip any dense sections and just rely on summaries before reaching the intriguing conclusion of this analysis.*

Chapter Thirteen Plate

An archangel revealing the physical nature of the universe to a group of natural philosophers and mathematicians. Etching by James Barry, 1795, after his painting. It appears to have these great thinkers listening attentively: Francis Bacon; Nicolaus Copernicus; Galileo Galilei; Isaac Newton; Thales; René Descartes; Archimedes; Robert Grosseteste; Roger Bacon; James Barry

*Source: Author: Gallery: https://wellcomeimages.org/indexplus/image/L0022460.html https://creativecommons.org/licenses/by-sa/3.0/deed.en**

**or such deed or licence instrument as may be applicable to the licence*

Let us try and test Leibniz's *Theodicy* in the light of today's mathematics and science. Voltaire made fun of Leibniz's theory that God had made the optimal best of worlds probably because of the apparent worldly chaos with its evil and suffering. Voltaire probably had in mind that God had the power to make things perfect and laughed at the present state of our world. Voltaire was profoundly affected by the Lisbon earthquake in 1755. However, could God have made things perfect as a matter of pure logic, mathematics, biology and physics? The crucial question becomes 'Can God create a perfect human being or even a perfect universe?' To create life as we know it, God must have conceived of a body design with its parts and functions governed by equations describing the mathematical relations for the dimensions, materials, qualities and 'bio-functions' of those parts. Mathematical relations would have needed conversion for different units to measure such things, as will be seen in this chapter.

BIOLOGICAL CIRCUITS NEED SOLUTIONS FOR LIFE TO FUNCTION

A cell with its cell wall, nucleus, DNA, cell machinery, etc. can be viewed as 'bio-hardware' running 'bio-software' for its processes or 'bio-systems'. Cells make proteins which are the building blocks of the body and also include globular proteins such as enzymes which catalyse biochemical reactions. Enzymes also synthesise amino acids in readiness for the cell's ribosomes to synthesise proteins. Possible concentrations or numbers of amino acids, enzymes, ribosomes and proteins are important quantities to know if one were to design a human being. Quantities readily translate to a mathematical equation and are important, but essential factors or features of proteins would also include shape, folding, binding sites, specificity, affinity, pH, temperature, UV light, pressure, catalytic activity, thermodynamics, chirality, molecular weight, atomic number and protein variants. So each of these features could also translate to a mathematical equation. However, to keep things simple, quantities such as those described above will be our initial focus for an equation to demonstrate the Infinite Wall Principle that was outlined at the end of Chapter 11.

Assuming that there is a given biological function – bio-function f – in a human being or organism then there will be a number of proteins that form the bio-hardware to house that function. Harvard professor of computer science and applied mathematics, Leslie Valiant, has conveniently introduced some mathematics in relation to evolution but it can be used for our purposes also:

> *"...Biological organisms are governed by protein expression networks... The protein expression networks upon which our biology depends are known to have more than 20,000 genes and the outputs they produce depend in a highly complicated way on the innumerably many possible input combinations. These circuits define how the concentration levels of the many proteins in our cells are controlled in terms of each other. We can seek to describe them mathematically. For example, the amount produced of our seventh protein may depend on the concentrations of three others – say, the third, twenty-first and seventy-third. The dependence is something specific, perhaps $f_7 = 1.7x_3 + 3.4x_{21} + 0.5x_{73}$ or more likely something else. But in any case it is some particular dependency $f_7(x_1, \ldots x_{20,000})$ on all the available proteins and possibly on some additional parameters, such as temperature. Valiant –* **Probably Approximately Correct.**"[375]

In this example, Professor Valiant effectively has an equation where the function f is the sum of the different proportions of concentrations of 3 proteins which is written mathematically as $f(x_3, x_{21}, x_{73}) = 1.7x_3 + 3.4x_{21} + 0.5x_{73}$. A more familiar form might be $f = ax + by + cz$ where x, y and z are the more familiar variables. Unfortunately, if we have a function f equating to 3 variables x, y and z, then this is challenging to show in a graph because it requires 4-dimensional representation. A more easily graphed function is $f(x,y) = 2x + 3y$ or say $z = 2x+3y$ which is a linear equation (ie. producing a straight line in a normal 3-dimensional x, y and z graph.) It is convenient to simplify the 3-protein equation to a 2-protein equation matching our function f into what we will call our Protein A – Protein B Equation ie. $f(x,y) = ax + by$ or $z = ax + by$ where x is the concentration of protein A and y is the concentration of protein B and a is the proportion of protein A and b is the proportion of protein B in some system of a living body and z could be the concentration of enzymes in that system indicative of its state at any time. Now, the concentration of protein A and B could be p and q parts per million litres or p and q parts per trillion litres or tinier. Similarly, the proportions of a (of A) and b (of B) could be any of an infinite set of numbers, small or large.

Now, $z = ax + by$ is classified as an indeterminate equation because it cannot be solved with 4 unknowns (ie. you cannot determine x, y, a or b) – even if the concentrations of protein A and B were known, it would still be indeterminate as there would still remain 2 unknowns, a and b. Note that even if you eliminate, let's call them 'crazy solutions', the problem is insurmountable (eg. you may know that A cannot equal zero or that B cannot be greater than 3 times a, etc.).

There is another layer of indeterminacy. Just like Zeno's distance to the finishing line, there is an infinite set of fractions and irrational numbers between Zeno and the line – known as the continuum. Thus, an infinite wall of numbers prevents the analytic calculation of a unique solution. However, if an equation is unsolvable analytically, it may still be solvable numerically, eg. successive iterations by a computer using trial and error such as stochastic modelling, choosing discrete inputs (ie. rounded numbers) or uniform distributions, curve fitting or other methods.

In conclusion, an indeterminate equation is unsolvable analytically but *possibly* solved numerically. In computing algorithms, even a numerical method might be solved using uniform distributions but unsolvable using non-uniform distributions.[376] Thus, remembering the analogy in Chapter 5, where author Paul Davies refers to a divine designer having a 'creation machine' of thirty knobs, we can extend this thought model to a 'biological setting'. If so, the creation machine will meet unsolvable biological design problems because not all design equations can be solved analytically. The right settings of the knobs are unknown or unascertainable. This means design problems must be solved numerically such as by evolution where trillions of trial and error iterations test possible solutions that are otherwise unsolvable.

SUMMARY

Biological circuits need solutions for life to function, but solutions may not be known or even exist because, even if there are infinite solutions, not knowing the correct solution may result in death to the organism. Unless there is a way around this 'infinite wall' or super task – that is, a way of knowing solutions – there can be no perfect biological synchronisation of circuits co-existing with interacting biological circuits needed for life to function perfectly without disease, dysfunction or death.

HOW THE HIPPASUS DICHOTOMY MAKES REALITY INHERENTLY INSOLUBLE

Drawing further insights from Greek mathematics, Hippasus shone a light on an important anomaly in Euclid's geometry. In his book *Infinity*, mathematician Ian Stewart explains the anomaly which concerns the dichotomy between rational and irrational numbers (which we will call the Hippasus Dichotomy for convenience) by use of some geometry:

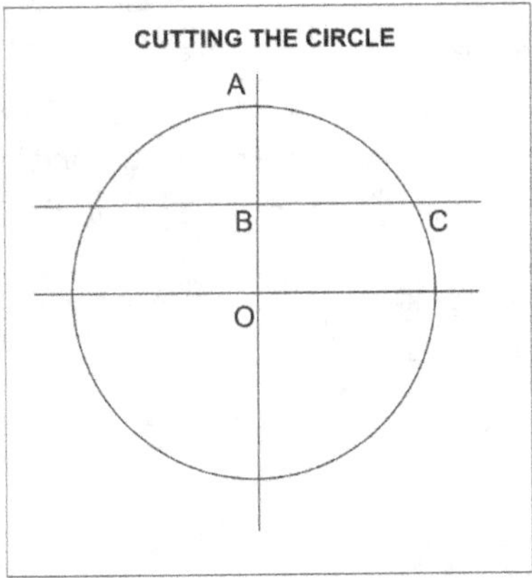

Figure 13a: Geometry demonstrating the Hippasus Dichotomy

(Summarising Ian Stewart's example shown in Figure 13a above): If a circle is divided by two perpendicular lines dividing it into four equal quadrants then call the origin 'O' and the top intersection 'A'. Let the radius of the circle equal 2. Then OA equals 2. Now draw another line perpendicular to OA (above and parallel to the horizontal line) so that it bisects OA at intersection B exactly in half – meaning that OB = 1 and BA = 1. Call the intersection of this last line through the right side of the circle – intersection C (see Figure 13a) . Pythagoras's theorem states that $OB^2 + BC^2 = OC^2$ or $1^2 + BC^2 = 2^2$ which implies that BC^2 is $\sqrt{3}$. $\sqrt{3}$ is an irrational number and so does not exist (as explained in the quote from Tim Gowers below). Ian Stewart explains: "So in the world of rational geometry, in which the only meaningful lengths are rational, the point C *does not exist*… Euclid's *Elements* tacitly assumes this kind of thing can't happen"[377] [my italics]. Why is this anomalous? "Between any two distinct rational numbers there lies at least one more rational number; indeed, infinitely many." Thus, Stewart described the line through point C by saying, "It 'squeezes' through a point-sized gap in the circumference of the circle."[378] We are more interested in the philosophical significance of the Hippasus Dichotomy discussed below when we analyse the totality of an interacting bio-system. The essential problem with this anomaly is that irrational numbers such as √2 or √3 create an *a priori* insolubility quality. Mathematician Timothy Gowers explains:

> ***"The square root of 2 is about 1.4142356 Where is infinity in a simple statement like the above, which says merely that one smallish number is roughly equal to another? The answer lies in the phrase 'the square root of 2' in which it is implicitly assumed that 2 has a square root… this phrase forces us to ask what sort of object the square root of 2 is. And that is where infinity comes in: the square root of 2 is an infinite decimal." Gowers.***[379]

In essence, √2 does not exist because it is just a mathematical construct as it has no exact value – only an approximation (another mathematical construct is i= √-1).

Gowers continues to state that there is a theoretical impossibility (ie. a 'logical impossibility') in multiplying two (unequal) irrationals ie. infinite decimals (or adding them). Thus, just as physicists discovered a schism in reality – the wave particle duality (see Chapter 21) – so too a mathematical duality is revealed which we have called the Hippasus Dichotomy. We will see later that this also creates another impossibility or 'infinite wall', namely, in synchronising equations needed to design the totality of systems and circuits of a human being.

However, before considering the biological circuits, some important concepts need to be understood which come from a surprising connection – weather and turbulence. Mathematician/meteorologist Edward Lorenz used equations to model weather systems and discovered that solutions to the equations were irregular – chaotic – and that there was great sensitivity in the chaotic dynamics of his models to tiny errors. He found that minute changes in the initial conditions could result in huge changes to the behaviour of the system (colloquially called the Butterfly Effect – after the quote "Does the flap of a butterfly's wings in Brazil set off a tornado in Texas?")[380] The Butterfly Effect can also be seen in biology if you compare biological cycles, systems and circuits to weather cycles and systems in terms of the mathematics of chaos theory. Ian Stewart also refers to the Lorenz weather modelling and compares the continuum of an ideal fluid model to the real discrete atoms as a 'philosophical curiosity', which demonstrates the conundrum we have called the Hippasus Dichotomy.[381] He comments: "The equations of classical mechanics replace the discrete physical reality by a smooth ideal liquid,"[382] echoing the reason Max Planck quantised reality with the Planck length.

Ian Stewart describes turbulence as "One of the great unsolved problems of physics... turbulence is a manifestation of chaos in the deterministic equations of fluid flow."[383] In Professor J.M. McDonough's *Introductory Lectures on Turbulence*,[384] he explains: "The understanding of turbulent behaviour in flowing fluids is one of the most intriguing, frustrating – and important – problems in all of classical physics. It is a fact that most fluid flows are turbulent, and at the same time fluids occur, and in many cases represent the dominant physics, on all macroscopic scales throughout the known universe – from the interior of biological cells, to circulatory and respiratory systems of living creatures... to geophysical and astrophysical phenomena... the 'problem of turbulence' remains to this day the last unsolved problem of classical mathematical physics." In this regard, recall the physics saying mentioned in Chapter 10, Leibniz's Reality: 'The map is not the territory'.

Ian Stewart also states: "[The Navier Stokes equations for turbulence] ... describe a highly idealised fluid, one that is infinitely divisible and homogenous. But a real fluid is composed of atoms (take your pick among competing levels of detail, from tiny hard balls to quantum swirls of probability)... If a real fluid were to obey the Navier Stokes Equations... it would have to shred its own atoms. So conceivably turbulence is a macroscopic effect of atom structure".[385] This 'atom shredding' insight could equally apply to any particle of size equal to or greater than the Planck length, thus making smooth perfect behaviour of interactions between different systems/frames impossible without 'quantum shredding' (shredding the Planck length to cut into the continuum – 'solution finding').

In conclusion, turbulence pervades biological and physical systems and based on our present state of knowledge is insoluble analytically – it can be modelled but empirical evidence is critical for design that involves turbulence – thus, wind tunnels are needed for aircraft design despite computer modelling (which needs the empirical data). Perhaps an analytical method (ie. *alternative means* in the language of theodicy) such as calculus could be found to bypass this insolubility but at present turbulence remains insoluble, which is sufficient for theodicy. (The Hippasus Dichotomy, as shown in Stewart's 'cutting of the circle' example, may be relevant to the definitional essence of turbulence, namely, chaotic vortices and that 'cutting these vortices' means *a priori* insolubility even if the initial starting conditions were known – meaning chaos despite being not *truly* random would have a *de facto* randomness quality because its features would be analytically unknowable.)

SUMMARY

The Hippasus Dichotomy has been defined here as a dichotomy between rational and irrational numbers. Discrete numbers can rely on rational numbers, but the continuum of irrational numbers between discrete numbers better reflects reality but in doing so an inherent insolubility comes with it. Planck's quantum length to measure particles as quanta is an example of discrete phenomena which if removed would leave an insoluble continuum of infinitesimal lengths prone to insolubilities, super tasks and paradoxes like Zeno's paradox (also called Achilles paradox). The startling insight is that the square root of a number may be a much-needed solution to a problem yet a square root does not exist (ie. has no exact value – which is problematic). There may be mathematical ways around any particular problem but importantly in the context of Leibniz's theodicy there may equally be no way around it which is a sufficient factor for a theodicy to work. This impacts on the nature of reality. Einstein saw the problem when he said: "As far as the laws of mathematics refer to reality, they are not certain; and as far as they are certain, they do not refer to reality."[386] In this regard, a physics saying comes to mind: 'The map is not the territory'.[387] That is, we construct a mathematical map but it is not the reality. Yet maths and the language of logic, science, etc. are all we have, so there will be mismatches and misjudgements which distort our concept of reality on many fronts. Werner Heisenberg also saw the problem when he said: "The problems with language here are really serious. We wish to speak in some way about the structure of atoms... but we cannot speak about atoms in ordinary language."[388] As explained in Chapter 12, Heisenberg saw our concept of reality melt beyond this inherent mathematical uncertainty when he postulated another uncertainty in reality – Heisenberg's Uncertainty Principle.

SYNCHRONISATION OF BIOLOGICAL CIRCUITS IS AFFECTED BY CHAOS, IRRATIONALS AND INFINITIES

So applying Leibniz's logical principles together with the above principles we are in a position to conduct a thought experiment on creation. If God were to design the perfect human being then God would imagine the perfect living body with its functions and systems all interacting together in perfect synchronisation. These functions and systems would be defined by mathematical relations, ie. mathematical functions – let us call them bio equations. If so, there would be interactions such that the output of those bio equations would be fed directly or indirectly as the input to other interacting systems such that the body synchronised. These synchronised 'settings' or 'states' would produce a code or codes representing the 'solution' or 'solutions' to this vast synchronisation of bio equations. The body has thirteen main systems such as circulatory system, digestive system, nervous system and each has its own sub systems.[389] Generally, a sub system of a system R_a would produce a state S_a of the body, and a second sub system of system R_b would also interact to alter that state to state $S_{ab,}$ and a third sub system might interact to alter that state, and so on. A code or codes would then define operating parameters that the 'bio hardware' making up the perfect body needs to satisfy to produce the 'perfect' synchronised performance.

To illustrate the synchronisation problem, it is useful to refer to a real biological model. Modelling of cellular processes with respect to cellular replenishment of cartilage has been carried out involving metabolic pathways by setting up a matrix of over fifty external and internal metabolites (substances used by the cell) with their accompanying reaction rates (called fluxes). Further analysis of this model and an added thought experiment on human biological circuits are set out in the Schedule at the end of this book for reference. As shown in the Schedule, it turns out that these biological circuits have infinite possible solutions – an infinite wall – and, because a biological design solution is a super task and cannot be completed, a perfect human being is impossible to make. The reference in this biological circuit (being just one of a multitude of circuits in the body) and to the infinite possible solutions displays how infinite solutions do not provide a solution at all unless you are able to eliminate the unworkable solutions (ie. the 'crazy' solutions) and have a workable number of 'good' or optimum solutions. Biologists try and eliminate these crazy solutions to get to the optimum set of solutions, but this intuitive or numerical process is not perfect. Empirical test results can provide knowledge of what to eliminate in this regard. Without empirical test results, the elimination of unwanted solutions must be performed intuitively or numerically but if this requires a super task of eliminating infinite solutions, an analytical solution will be impossible as there can be no last elimination when applying the Infinite Wall Principle.

The example of the biological model and an added thought experiment set out in the Schedule indicates that the magnitude of iterations, including the continuum of infinite values tested by possibly thousands of bio equations, is so great that even the possibility for analytic 'short cuts' (calculus, summation of series, numerical methods) without access to empirical evidence diminishes greatly. Of course, a perfect code may be found in a finite time. On the other hand, the impossibility of finding a perfect code remains an alternative, which is sufficient for theodicy. However, if a perfect code had been found analytically in the metaphysical realm (or multiple prior universes – see under 'A New Cosmic Theology?' towards the end of this chapter) prior to creation then it would be implausible to postulate that God withheld it for some 'greater good' reason – ie. God would have had no *sufficient reason* to withhold it – to use Leibniz's term (italics). Surely, even for 'perfect humans', enough hardship would still exist for the world to remain the 'Vale of Soul-making' (as the poet John Keats described it). Using Ockham's Razor, it is more plausible to postulate the simpler explanation that the imperfect human being exists because no 'perfect' solution is possible.

The above analysis of the biological circuits displays inherent insolubilities arising from mathematics itself. However, there is an even more obvious potential logical 'showstopper' for the perfect human circuit which is broadly an extension of Leibniz's own logical calculus. One of the earliest analyses of a circuit-like problem was presented by the great mathematician Leonard Euler. In the early 18th century, there were seven bridges in the town of Königsberg (present-day Kaliningrad). The problem posed to Euler was whether it was possible to begin at a starting point and cross each bridge only once and return to the starting point. Euler proved it was not possible. If crossing the bridges once fulfilled a function that required synchronisation with another function in the nearby village, the synchronisation would fail if this required a bridge to be crossed twice before returning to start. Circuits are wired to execute functions but where two circuits interact, the logic of one circuit may not interact with the other – a synchronisation failure.

The Bridges of Königsberg problem represented a simple circuit, but designing a human array of connected circuits each having functions to carry out to synchronise with other circuits is a steep uplift in complexity. It is clear that 'logic gates' would appear, producing some interaction failures, and so not all designs will be able to carry out all functions and synchronise when required. Even allowing for more and more bypasses to these failed interaction points then these layers add more and more complexity, opening possibilities for more complex failures of synchronisation when a new bypass is placed to solve each specific synchronisation failure point.

How important would these synchronisation failure points be in designing the master biological circuit plan for the human body? Noting first that chaos can affect circuits and that Lorentz showed small changes can make big outcomes, it is important to look at the master program software that would iterate with evolution to solve these synchronisation failures. DNA seems to be the master iteration machine for a perfect master biological circuit, but the precision required to prevent synchronisation failure during the trillions of iterations of evolution is of a very high order, as seen in this example. The right-handed chirality in DNA and the asymmetry of the weak nuclear forces can affect the evolutionary behaviour of DNA based on work by physicist Dilip Kondepudi. Ian Stewart explains the asymmetry of DNA: "One consequence of this asymmetry is that the energy of a molecule and that of its mirror image are not quite the same. Until recently this was thought to be unimportant because the difference is vanishingly small – to be precise one part in a hundred quintillion… physicist Dilip Kondepudi showed that if nature is biased in favour of the lower energy version of some biologically significant molecule (even by this tiny amount), then within a mere hundred thousand years a massive 98 per cent of those molecules will be of the lower energy variety. The difference is amplified by the reproductive processes of life."[390] Clearly, the slightest change in DNA – think chaos or synchronisation failure or even 'quantum shredding' – can cause genetic defects and consequently imperfections resulting from insolubility of circuit design solutions which may be amplified greatly in the human body. In short, the infinite wall we have described can, by mere mathematical impossibility, have devastating effects on the organism. Yet the miracle is that we are here!

SUMMARY

Biological circuits are governed by the mathematics of circuits made famous by Euler's analysis of the Bridges of Königsberg problem which meant that circuits are subject to a logical calculus similar to Leibniz's logical calculus but applied to the permutations and combinations in circuits. In addition, circuits can be subject to chaos and interactions can involve irrationals, infinities and super tasks. Thus, an infinite wall inherently blocks perfect solutions to many desired modes of living that a biological circuit needs for synchronisation and proper functioning. Tiny changes in the genetic code in DNA which must evolve such circuits can produce massive changes because such changes can be affected by chaos, irrationals and infinities. Again, there may be 'short circuits' around any particular 'impassable/bottleneck circuit gates' but importantly in the context of Leibniz's theodicy there *may not be* short circuits, which is a sufficient basis for theodicy to work.

Just as Leibniz commented on infinity – "I cannot show you this in detail. For can I know and can I present infinities to you and compare them together?" – so too we cannot show that there is a limit or infinite wall beyond which the failed synchronisations do not decrease (or cannot be eliminated), and therefore the remainder of Leibniz's comment must also apply: "But you must judge with me *ab effectu* [from the effect] since God has chosen this world as it is."[391]

On the basis of our present mathematical and scientific knowledge, it would seem that God would need 'magician-like' methods to bypass the 'super task' conundrum of the effects of chaos and infinities upon circuit failures and genetic defects which prevents any analytic solution for the creation of super tasks unless there were alternative means available. In today's weird world of quantum mechanics, it might seem that magician-like forces already exist – Einstein called one of them 'spooky action at a distance' – and so it is tempting to reinstate the medieval way of thinking about a magician-like God as opposed to the more cosmic God under consideration here. The critical difference is that even the power to create quantum or magician-like forces will not create knowledge that requires an infinite iteration or super task unless there are alternative means. If there were no alternative means then this likelihood would, if true, mean that without a solution there would be a logical impossibility to create a 'perfect human being' using analytical design. Thus, it seems a reasonable possibility therefore that, at creation, in the absence of a perfect solution, gaps in 'perfect solutions' appeared, prompting the creation of a human being capable of using the ranges of possible solutions through evolutionary processes, bio rhythms, homeostasis or stochastic processes to plug the gaps (the reader should not conflate these 'gaps' with the concept of 'God of the gaps', which refers to the separate controversial topic of intelligent design not addressed here. Recall in Chapter 5 the suggestion that perhaps a new principle – the anthrobiopic principle – should be introduced to discover what laws, constants, processes and mechanisms generated life. The Infinite Wall Principle might then permit as possible that there is no intelligent design in evolution because the features of life are autogenous. Life itself may arise from those laws, constants, processes and mechanisms which are possibly 'bio fine-tuned' to be autogenous via clever hidden mechanisms in unknown fabrics of reality and at unknown times).

It would seem that the flaw in Leibniz's original theodicy can be solved by a simple variation of the theodicy, being that God created the best (or optimal) worlds by choosing from all possible theoretical worlds that were possible to survey and then created a possible world within the logical envelope permitted by Leibniz's logical calculus and the functions (and resulting operating states) that are possible to design into bio systems, biological circuits and genetic systems. Could God have created a perfect universe free of disaster, danger, disability, disease, etc? Leibniz's logical restrictions on what is possible or compossible can be applied. Philosopher Swinburne has laid down a core principle for theodicies which he terms 'The Logical Straitjacket', which he explains as follows: "God will seek to provide all the good things and none of the bad things… But he cannot – for reasons of logic. For, as simple non-religious examples will make evident, some good states are logically incompatible with each other."[392] For example, if two armies wage war, only one can win. We can't have A and not A without distorting language, and indeed Leibniz adopted a similar theme with his logical calculus. Thus, we can now see that both a 'Biological Straitjacket' and 'Physical Straitjacket' have emerged in relation to Leibniz's theodicy for the following reasons:

Physical systems cannot produce perfect outcomes

Just like the bio equations interacting to synchronise all critical states of the human body, so too a symphony of equations for all the elements and 'zoo' of quantum particles would need to interact to achieve intended 'laws and constants of the universe' (at least those contributing to the anthropic principle[393], such as the laws and constants that make human life possible, eg. those allowing the formation of carbon-12 inside the centre of a red giant star at a temperature of 100 million degrees). These interactions – perhaps not as complex as biological interactions – would still be susceptible to a similar type of insolubility to that seen in the super tasks that appear to be involved in creation.

Physical systems display the Infinite Wall conundrum

This type of insolubility is reflected in the following insight from physicist Carlo Rovelli: "The very way in which the equations of the Standard Model make predictions about the world is absurdly convoluted. Used directly, these equations lead to nonsensical predictions where each calculated quantity turns out to be infinitely large. To obtain meaningful results it is necessary to imagine that the parameters entering into them are themselves infinitely large, in order to counterbalance the absurd results and make them reasonable. This convoluted and baroque procedure is given the technical term 'renormalisation'"[394]

Physical systems cannot produce perfect predictions

Despite the 'logical possibility' of Heisenberg's Uncertainty Principle being superseded (by such concepts as the Hidden Variables hypothesis), the outstanding success of quantum mechanics indicates it as the best theory to describe reality. This means that God would have been constrained by a physical straitjacket to create such a feature as a physical 'fine tuning measure'[395] as part of the universe because of the reasonable possibility that creative insolubilities existed as outlined above, ie. there was a creative necessity compelling a creator to 'lubricate' reality with a 'solution flexible' or 'solution finder' wave equation. (The qualification to this is that the universe could have been designed without Heisenberg's Uncertainty but even without it, the presence of chaos, irrationals and infinities still leaves a substantial 'uncertainty' mechanism in any reality.) Another instance of 'fine-tuning' could have been the quantum nature of our microscopic universe, which avoids insolubilities of the continuum of infinitesimals and irrationals. With the infinite wall 'worldview', God's survey of 'all possible worlds' envisaged by Leibniz may possibly be explained by the generation of megaverses or the multiverse as a creation solution finder. Thus, the multiverse or multiple universe hypotheses, now so much in vogue as an alternative to divine 'fine-tuning', becomes a divine instrument of fine tuning. (See later in this chapter under 'A New Cosmic Theology?'.)

The uncertainty inherent in chaos and turbulence

The discussion of Chaos Theory and in particular the inherent insolubility of turbulence and its consequent deficient epistemological features creates a logical impossibility of 'hard' omniscience[396] for any part of the universe affected by turbulence (and indeed Chaos's Butterfly Effect means that even tiny effects can have big unknown effects). This Achilles heel of omniscience is compounded by Heisenberg's uncertainty, wave particle duality and quantum entanglement ('spooky action at a distance'[397]).

The Infinite Background problem

There is a further philosophical impossibility that is akin to the Infinite Wall Principle – in fact, it has been called the 'Infinite Background Problem' (also allied to the 'Frame Problem') raised by philosopher Hubert Dreyfus in relation to Artificial Intelligence (AI) with reference to the philosophy of Heidegger.[398] (AI in this context is suggested as analogous to considering what a divine intelligence can logically do or not do.) Dreyfus outlined the problem facing an AI-based computer in analysing incoming environmental data: "[Each receptor or *element* receiving such data must be interpreted]... according to different rules and which rule to apply depends on the context... But if each context can be recognised only in terms of features selected as relevant and interpreted in a broader context, the AI worker is faced with an infinite regress of contexts."[399] Philosopher Michael Wheeler, in commenting on Dreyfus's description, states: "An infinite regress would be bad enough, but may not be the worst of it. As Horgan and Tienson (1994) point out, the context sensitivity of cognition cannot be achieved by a system first retrieving an inner structure... [ie. retrieving the key knowledge]... and then deciding whether or not it is relevant, as that would take us back to square one. But then how can the system assign relevance until the structure has been retrieved? The result is a kind of cognitive paralysis."[400] Taking the infinite background problem to its implications, the set of contexts, ie. background data, is infinite and so is unknowable as it is a super task. This creates an impossibility problem for a perfect event control system of reality – our perfect universe free of disaster, danger – because pre-determining events requires an infinite database of knowledge to avoid each potential disaster or bad event.

This means there is an inherent impossibility problem (ie. an impossible super task) with a universe or a simulated universe. This problem – inherent in reality – is the presence of irrational numbers and infinite possible values of data needed to design features for a perfect human being or a perfect universe. Trial and error iteration via processes like evolution may deliver some of such features instead of outright design. Strangely, experts on evolutionary algorithms state some features will still be impossible even if evolved (iterated) for infinity.[401] It would seem that the Lisbon earthquake which prompted Voltaire to write his novel *Candide*, satirising Leibniz's 'best of all worlds' theodicy (discussed in Chapter 11), would be explicable in view of these factors. God's knowledge would not include unknowable gaps in design or background data required to create the universe free of both defects in life or prone to bad events. Even Newton saw the impossibility of true omniscience when he rejected Descartes's theory of the motion of objects: "and if one follows the Cartesian doctrine, not even God himself could define the past position of any moving body accurately and geometrically [since the original place] no longer exists in nature."[402]

Such a philosophy may have saved Charles Darwin's loss of faith upon observing the cruel behaviour of a wasp laying its eggs into a live cricket, as he explained: "I cannot persuade myself that a beneficent and omnipotent God would have designedly created the Ichneumonidae with the express intention of their feeding within... living bodies."[403] The outcome of the Infinite Wall Principle is 'undesigned' evolution, which is unforeseeable to an unknown extent. Thus, to create a perfect physical universe without natural evil involves a Physical Straitjacket with super tasks and their logical impossibilities. There may be mathematical ways or 'short circuits' around some anomalies, but importantly in the context of Leibniz's theodicy there may not be a 'fix' to avoid the impossibility which is a sufficient basis for theodicy to work. Thus, in terms of theodicy, the credible possibility remains that these obstacles prevented creation of the 'perfect' universe free from disaster, danger, disability, disease, etc. which is sufficient for theodicy.

Leibniz's extended theodicy (the Infinite Wall Principle) makes sense of the world

It seems that extending Leibniz's principles creates a theology that does make sense of the world. Creation requires both cosmic and biological evolution to solve otherwise insoluble problems, such as creating a biology that maths could not solve theoretically. Some problem tasks cannot even be solved numerically, rendering them impossible for evolution.[404] The vast universe is then explained as a necessary iterative 'sandbox' where almost infinite possibilities of permutations and combinations of matter become output of this stochastic creation process, both for the building blocks of matter and life itself (recall Chapter 5 and nuclear physicist Dr N.M. Clarke's astonishment relating to the evolution of carbon-12, or Fred Hoyle's prediction that a certain peculiar resonance must exist; otherwise, life would not exist). Human affairs cannot be free of suffering, evil or unfairness not only due to the side effects of the evolution of the universe (physical and biological) but the Infinite Background Problem (akin to the Infinite Wall Principle) producing infinite regresses which inevitably involve anomalies, ethical dilemmas, culture wars and other clashes of human affairs.

The Infinite Wall Principle (Leibniz's extended theodicy) seems to catch the most powerful potential challenges to God which the 'new atheists'[405] use to 'debunk' God – the greatest challenge being the presence of natural evils such as disability, disease, disaster and despair – and the problem of evil and suffering. Evil is often explained by the need for free will, but natural evil is not as easily explained. Remaining obstacles would be nuanced objections such as Dostoevsky's famous challenge in *The Brothers Karamozov*: when one brother challenged another by asking him to imagine that he was "building the edifice of human destiny with the object of making people happy in the finale" but previously asked, "if everyone must suffer, in order to buy eternal harmony with their suffering... why should [children] have to suffer[?]..." and indeed his final question, "would you agree to be the architect on such conditions?" was answered, "No, I would not agree," by his brother.[406] Dostoevsky's insight may mean that a personal God may not have created the world, but this does not weaken Leibniz's theodicy as a defence of the God he was more likely to believe in – the first mover – the Cosmic God.

When we sum up the synthesis of thought from Newton's Cosmic

Universe, fine-tuning, Leibniz's 'best of all worlds' theodicy and the new concepts involving chaos, ubiquitous insolubility and uncertainty presented in this chapter, an intriguing and radically new worldview emerges. Even Einstein showed disbelief about the Uncertainty Principle – God would not play dice with the universe, he said. He was uncomfortable with uncertainty and chaos. Nor did Einstein believe that God would need the universe to be fine-tuned and, in fact, he was uncomfortable about inserting the cosmological constant (Λ) into his famous field equation in his theory of General Relativity, namely, when $G=8\pi\gamma T$ became $G=8\pi\gamma T + \Lambda g$ and later described this as his 'greatest blunder'.[407] Author David Bodanis summarises Einstein's views regarding inserting Λ: "[Einstein] didn't believe that any deity or force of nature would have started the universe in accord with ultra simple principles, then awkwardly thrown in such a correction." [408]Yet with our understanding of chaos, the Butterfly Effect, irrationals, infinitesimals, infinities, quantum shredding and super tasks – the Infinite Wall Principle – makes it easy to see that this counterintuitive insertion of Λ might be expected with many other insertions that are needed for cosmic fine-tuning in perfecting Leibniz's theodicy. Einstein eventually discarded Λ when it appeared the discovery of the expansion of the universe made Λ unnecessary.[409] Now Λ has re-emerged as it appears the universe does in fact need extremely tiny fine-tuning – presumably it would still be uncomfortable for Einstein.These days, Λ is more likely to be introduced in reference to dark matter, according to mathematical physicist Roger Penrose.[410]

A New Cosmic Theology?

In a 2018 paper, Roger Penrose and others proposed an interesting hypothesis which originated from his visualising the eventual evaporation of supermassive black holes.[411] It seems that the CMB map (derived from the WMAP and Planck satellites) – the model image for the Cosmic Background Radiation emanating from the Big Bang – shows anomalous points, coined Hawking points. On this basis, Penrose and his co-authors have proposed the Conformal Cyclic Cosmology (CCC) hypothesis which proposes that our future becomes the Big Bang for the next aeon and our Big Bang was preceded by previous Big Bangs – a perpetual cycle of universes forward and backward in time. In Chapter 5, I referred to the 'fantastic probabilities' of the origin of life, to quote prominent biologist Eugene Koonin. I produced my own calculation of what these odds might be, namely, 6.27×10^{180} universe life times to even get a minimal form of life (assuming no evolutionary infinity intervenes). The implication is that it would take more than trillions of universe lifetimes for the knowledge inherent in the clever mechanisms of life to originate. The Infinite Wall Principle states, by pure logic, that God may not have such knowledge and must obtain it empirically due to the infinite walls described. If so, the tantalising question becomes 'What is the repository of knowledge – "memory" – between each aeon in the cosmic cycle of multiple universes?' It would seem that only an entity applying mind, memory and goals to empirical knowledge building methods (such as evolution) surviving between aeons fits the description – God. Hawking famously said that with no time before the Big Bang, no God is needed. Yet this model of the cosmos would see God needing timelessness to store evolutionary 'design' knowledge trillions of universes evolve.

The synthesis of concepts in this chapter connects chaos, turbulence, the Hippasus Dichotomy (irrationals/infinities), quantum shredding and the Infinite Wall Principle – and for want of a common term we can call all these phenomena 'universal chaos'. The inherent insolubility in physical systems such as the packing of particles and interaction of turbulent forces produces mathematical roadblocks to a 'perfect reality'.

The overarching insight is that a universe cannot be created or generated without universal chaos if it is to achieve meaningful life. Unbeknown to Einstein, Newton or Leibniz was the existence of this universal chaos (although it must have consciously or subconsciously dawned on them). Newton spoke of chaos – in the general sense of the word, namely, the chaos of the cosmos – and was aware of the three-body problem which involves the chaos (and its new mathematics) that was discovered later in the 20th century. (The fact that the three-body problem displays chaos with an element of randomness yet has now been solved is beyond the scope of this book, but for further reading on the problem, readers can find two good references noted here.[412]) What Newton did not realise (at least as far as we can discern from his writings) was that the difficulty in solving the three-body problem was caused by chaos despite his neat laws of motion governing their movement (in fact, there are no 'nice equations' or geometry governing their motion[413]). Universal chaos imposes a creative limit to any 'constructed reality' – far reaching showstoppers to 'precision' or 'perfect creation'. The Heisenberg Uncertainty Principle discussed earlier may be a feature of creation rather than an *a priori* or analytical necessity (ie. necessary in principle) and so may not contribute to this creative limit (its contribution may be that it guarantees free events and true free will). Moreover, this limitation (of universal chaos) contributes to the need for fine tuning because it will distort our reality, just as the limitation of the speed of light distorted Newton's reality.

What distortions of Newton's reality then follow from this creative limit? It has many implications but the vastness of the universe, which has long puzzled cosmologists, is now explained in the context of a creator facing a creative limit. Only the trillion trial and error iterations of cosmic events iterating away via the enormous permutations and combinations of physical and biological particles can surpass this limit. Cosmic evolution, such as the formation of new elements inside dying stars or life forming at the mouth of volcanic vents in oceans, can break this creative limit, climb the infinite wall and deliver the ultimate solution – the building blocks of life such as carbon-12, the periodic table of elements, self-replicating ribozymes, organic molecules and of course DNA (arguably the super-iterator). So the calculated accidents of these iterating building blocks slot into an almost infinite iterative newsreel of events, pausing only for intermission at the origin of first life before renewing their tireless evolutionary computation towards the final (not yet perfect) frame of this newsreel – human life. Leibniz and Newton might marvel at this new mechanism in the cosmic clockwork.

NOTES TO PART I – MASTERS OF THE ENLIGHTENMENT

1. PM Rattansi, *Isaac Newton and Gravity*, Priory Press Ltd, 1974, p13
2. Ibid, p13
3. R. Iliffe, *Priest of Nature – The Religious Worlds of Isaac Newton*, Oxford University Press, 2017, p46
4. Ibid, p244
5. I.Newton quote from N. Guiccardini, *Isaac Newton and Natural Philosopy*, Reaktion Books, 2018, p43
6. N. Guiccardini, *Isaac Newton and Natural Philosophy*, Reaktion Books, 2018, p49
7. Iliffe, Newton, *A Very Short Introduction*, Oxford University Press, 2007, p18
8. N. Guiccardini *op. cit.* pp49–50
9. Iliffe, Newton, *A Very Short Introduction*, Oxford University Press, 2007, p42
10. Ibid, p43
11. Ibid, p41
12. N. Guiccardini *op. cit.* p81 refers to his election in 1672 as a 'member' of the Royal Society whilst Rattansi *op. cit.* on p41 refers to him as a 'Fellow' of the Royal Society
13. Newton quote from P.M. Rattansi, *op. cit.* p41
14. Newton quote from P.M. Rattansi, *op. cit.* p41,42

15. N. Guiccardini *op. cit.* p54
16. Iliffe, Newton, *A Very Short Introduction*, *op. cit.* p45
17. N. Guiccardini *op. cit.* p57
18. N. Guiccardini *op. cit.* p59
19. Newton's formula for lens was $pq=f^2$ where f is the focal length of the lens – *Oxford Dictionary of Physics, Seventh Edition*, 2015, Oxford University Press, p382
20. J. Uglow, *Charles II and the Restoration*, Faber & Faber, 2009, p219, 221
21. Ibid, p235
22. Iliffe, Newton, *A Very Short Introduction*, *op. cit.* p48
23. P.M. Rattansi, *op. cit.* p89
24. Iliffe, Newton, *A Very Short Introduction*, *op. cit.* p52–53
25. W. Dunham, *The Mathematical Universe*, John Wiley & Sons, 1994, p130
26. W.Dunham, *The Mathematical Universe* (ibid), pp131–132
27. Ibid, p114
28. P.M. Rattansi, *op. cit.* p30, and N. Guiccardini *op. cit.* p234 confirming 1713 as the year of distribution of the key document used in adjudicating the dispute, the *Commercium epistolicum* (see p214 N. Guiccardini)
29. W.Dunham, *The Mathematical Universe* (ibid), p134
30. Quote from R. Westfall quoted in Edward Dolnick – *The Clockwork Universe* – *op. cit.* p232.
31. L. M. Principe, *The Scientific Revolution, A Very Short Introduction*, Oxford University Press, 2011, p10
32. J. Baggott, *Mass – The Quest to understand matter from Greek atoms to Quantum fields*, Oxford University Press, 2017, p20
33. W. Isaacson, *Leonard Da Vinci*, Simon & Schuster, 2017, p149
34. Quote from W. Isaacson, *Leonard Da Vinci op. cit.* p149
35. S. Law, *Philosophy*, Dorling Kindersley, 2007, p251
36. Quote from J. Uglow, *Charles II and the Restoration op. cit.* p9
37. Quote from a translation by Eugene O'Connor included in T. Butler-Bowdon, *50 Philosophy Classics*, Nicholas Brealey Publishing, 2013, p100, ref. p103
38. *Oxford Dictionary of World History*, Oxford University Press, 2015, p31
39. J. Baggott, *Mass op. cit.*p20
40. L M Principe, *The Scientific Revolution, A Very Short Introduction*, *op. cit.* p11
41. *Oxford Dictionary of World History*, *op. cit.* p400
42. *Longman Illustrated Encyclopaedia of World History*, Ivy Leaf, 1991, pp541–542
43. Ibid, p601
44. L. M. Principe, *The Scientific Revolution, A Very Short Introduction*, *op. cit.* p18
45. Quote from L. M. Principe, *The Scientific Revolution, A Very Short Introduction*, *op. cit.* p19
46. Ibid, p19
47. R. Iliffe, *Priest of Nature – The Religious Worlds of Isaac Newton*, *op. cit.* p160
48. Quote of Archbishop Tillotson regarding the mid 17th century from J. Uglow, *Charles II and the Restoration*, *op. cit.* p9

49. S. Law, *Philosophy, op. cit.* p 275
50. Ibid, p275
51. *Oxford Dictionary of World History, op. cit.* p296
52. R. Iliffe, *Priest of Nature – The Religious Worlds of Isaac Newton, op. cit.* p395
53. *Oxford Dictionary of World History, op. cit.* p208
54. Ibid, p128
55. Ibid, p167
56. J. Uglow, *Charles II and the Restoration, op. cit.* p1
57. *Longman Illustrated Encyclopaedia of World History, op. cit.* p408
58. *Oxford Dictionary of World History, op. cit.* p41
59. R. Iliffe, *Priest of Nature – The Religious Worlds of Isaac Newton, op. cit.* pp132–148
60. Ibid, p399
61. A. Keay, *The Last Royal Rebel, The Life and Death of James, Duke of Monmouth,* Bloomsbury Publishing plc, 2017, p297–301
62. Iliffe, *Newton, A Very Short Introduction, op. cit.* p103
63. Ibid, pp103–104
64. A. Keay, *The Last Royal Rebel, op. cit.* p382
65. P.M. Rattansi, *op. cit.* p78
66. A. Keay, *The Last Royal Rebel, op. cit.* p303
67. Iliffe, *Newton, A Very Short Introduction, op. cit.* p105
68. I. Stewart, *Significant Figures*, Profile Books, 2017, p63
69. Iliffe, *Newton, A Very Short Introduction, op. cit.* p112
70. Ibid, p112
71. I. Stewart, *Significant Figures, op. cit.* p63
72. Ibid, p64
73. Iliffe, *Newton, A Very Short Introduction, op. cit.* p112–113
74. J.Perez-Pariente, *National Geographic History* (May/June 2018 Issue), *The Alchemy Paradox*, p50
75. J. Uglow, *Charles II and the Restoration, op. cit.* p223
76. J.Perez-Pariente, *National Geographic History* (May/June 2018 Issue) *The Alchemy Paradox,* p54, and J. Uglow, *Charles II and the Restoration, op. cit.* p223
77. L M Principe, *The Scientific Revolution, A Very Short Introduction*, Oxford *op. cit.* p80
78. Ibid, p89
79. N. Guiccardini *op. cit.* pp15–16
80. N. Guiccardini *op. cit.* pp17–19
81. R. Iliffe, *Priest of Nature – The Religious Worlds of Isaac Newton, op. cit.* p14
82. Ibid. pp22–23
83. L. M. Principe, *The Scientific Revolution, A Very Short Introduction,* Oxford *op. cit.* p66
84. N. Guiccardini *op. cit.* p21
85. L. M. Principe, *The Scientific Revolution, A Very Short Introduction,* Oxford *op. cit.* p35

86. Ibid, pp64–65 Newton began his *Principia* (see note 6) in 1684 which proposed a gravitational theory inconsistent with Descartes's mechanical theory of gravity
87. Newton's famous work *The Principia* or *Philosophiae Naturalis Principia Mathematica* (The Mathematical Principles of Natural Philosophy) was commenced in 1684 and completed in 1687 (see note 7) – see GB Ferngren – *Science & Religion – A Historical Introduction*, John Hopkins University Press, 2017, p125 – Chapter 9, *Isaac Newton*, by SD Snobelen
88. *The Principia* was completed and published in 1687 – see Iliffe, *Newton, A Very Short Introduction*, Oxford University Press, 2007, p69, p98
89. Boyer, *A History of Mathematics*, p 398, quote from P. Watson, *Ideas, A History from Fire to Freud*, Weidenfeld & Nicolson, 2005, p653, note 30
90. Quote from G.B. Ferngren – *Science & Religion – A Historical Introduction* Chapter 15, *Cosmogonies*, by R.L. Numbers & P.J. Susalla, p221
91. Ibid, p222
92. Quote from G.B. Ferngren – R.L. Numbers & P.J. Susalla, *op. cit.* p222
93. Ibid, p222
94. Ibid, p222
95. Quote from L. M. Principe, *The Scientific Revolution, A Very Short Introduction*, Oxford University Press, 2011, p90
96. *Oxford Dictionary of Physics*, *op. cit.* see 'light', p314
97. See Chapter 21 for a history of quantum mechanics
98. See Chapter 12 under '(iii) *Chaos Theory*
99. Oxford Dictionary of Physics, *op. cit.* see 'quantum chaos' p459
100. Quote from Iliffe, Newton, A Very Short Introduction, *op. cit.* p83–84
101. Iliffe, *Newton, A Very Short Introduction, op. cit.* p84
102. Ibid, p86
103. Quote from Iliffe, *Newton, A Very Short Introduction, op. cit.* p86
104. Ibid, pp86–87
105. Ibid, p88–89
106. Ibid, p93
107. P.M. Rattansi, *op. cit.* p70
108. Ibid, p73
109. Iliffe, *Newton, A Very Short Introduction, op. cit.* p86; *Oxford Dictionary of Physics*, Seventh Edition, 2015, Oxford University Press, p383
110. Ibid, p383
111. R. Arianrhod, *Seduced by Logic, Emilie du Châtelet, Mary Somerville and the Newtonian Revolution*, University of Queensland Press, 2011, pp59–61
112. P.M. Rattansi, *op. cit.* p74
113. Quote from P.M. Rattansi, *op. cit.* p74–75
114. P. Watson in his book, *Ideas, A History from Fire to Freud*, Weidenfeld & Nicolson, 2005, p 653 observes that scientist J.D. Bernal observed that despite Copernicus the theory was not 'in any way explained'
115. J. Hannam, *God's Philosophers, How the Medieval World Laid the Foundations of*

Modern Science, Icon Books, 2010, pp34–35

116. P. Watson, *Ideas, A History from Fire to Freud*, op. cit. pp670–671
117. Iliffe, *Newton, A Very Short Introduction, op. cit.* p94–95
118. Arp and over thirty contributors, *1001 Ideas that changed the way we think*, Pier 9, Murdoch Books, 2017, p160
119. Ibid, p181
120. Ibid, p205
121. S. Law, *Philosophy, op. cit.* p24
122. J. Baggott, *Mass, op. cit.* pp45–46
123. R. Iliffe, *Priest of Nature – The Religious Worlds of Isaac Newton op. cit.* p93
124. R. Iliffe, *Priest of Nature – The Religious Worlds of Isaac Newton op. cit.* p94
125. R. Iliffe, *Priest of Nature – The Religious Worlds of Isaac Newton op. cit.* p99
126. R. Iliffe, *Priest of Nature – The Religious Worlds of Isaac Newton op. cit.* p96
127. R. Iliffe, *Priest of Nature – The Religious Worlds of Isaac Newton op. cit.* p97
128. N. Guiccardini *op. cit.* pp165–167 for discussion of Newton's concepts of absolute time, space, velocity and absolute resting frame. The author noted that his contemporaries "especially Huygens, Leibniz and George Berkeley" were sceptical (p167)
129. *Oxford Dictionary of Physics, op. cit.* p550
130. Quote of Isaac Newton from J. Baggott, *Mass, op. cit.* p41
131. *Oxford Dictionary of Physics, op. cit.* p383
132. Quote from Ernest Mach (1938–1916), *The Science of Mechanics* quoted from Note 10, p268 Baggott, *Mass, op. cit.* (*Oxford Dictionary of Physics, op. cit.* p331)
133. *Oxford Dictionary of Physics, op. cit.* p331
134. Physicist Peter Higgs shared the 2013 Nobel Prize for his gauge theory where mass emerges from a broken symmetry of gauge symmetry. His papers (1964 & 1966) were published about fifty years before it was verified by the Large Hadron Collider in 2012 – see *Oxford Dictionary of Physics, op. cit.* p251
135. R. Iliffe, *Priest of Nature – The Religious Worlds of Isaac Newton, op. cit.* p207
136. Iliffe, *Newton, A Very Short Introduction, op. cit.* p124
137. Iliffe, *Newton, A Very Short Introduction, op. cit.* p98
138. María Colín-García , Alicia Negrón-Mendoza, Sergio Ramos-Bernal, Fernando Ortega-Gutiérrez, José Luis García, Julio Valdivia-Silva, Alejandro Heredia, *Comets and the Origin of Life: Pioneer of Organics for Chemical Evolution* Available from: https://www.researchgate.net/publication/234080819_Comets_and_the_Origin_of_Life_Pioneer_of_Organics_for_Chemical_Evolution (as at Aug 24 2018).
139. P. Watson, *Ideas, A History from Fire to Freud, op. cit.* p653
140. Ibid. p653
141. Ibid. pp647–648
142. P.M. Rattansi, *op. cit.* pp27–36
143. L. M. Principe, *The Scientific Revolution, A Very Short Introduction, op. cit.* p87
144. Ibid, p88

145. P.M. Rattansi, *op. cit.* p44
146. *Oxford Dictionary of Physics, op. cit.* p356
147. N. Guiccardini, *op. cit.* pp96–97
148. Ibid, pp93–95
149. Ibid, p101
150. See O. Gal & R. Chen-Morris, *Baroque Science*, University of Chicago Press, 2013, pp180–184, where the authors refer to 'massaging' of figures to cope with errors and Newton's reference in making equations equate to stating certain results 'very nearly (*quam proxime*)' equate or approximate a postulated result – once the implication of chaos theory is understood, it will be seen that approximating and iterating towards an optimal equation is a necessary part of science
151. Iliffe, *Newton, A Very Short Introduction, op. cit.* p93
152. I. Stewart, *Calculating the Cosmos*, Profile Books, 2017, p48
153. Quoted from Edward Dolnick – *The Clockwork Universe* – Harper Perennial, 2011, p234
154. See note 33
155. R. Iliffe, *Priest of Nature – The Religious Worlds of Isaac Newton, op. cit.* p206, including preceding quote of Newton
156. R. Iliffe, *Priest of Nature – The Religious Worlds of Isaac Newton*, ibid, p291
157. R. Iliffe, *Priest of Nature – The Religious Worlds of Isaac Newton*, ibid, p311
158. R. Iliffe, *Priest of Nature – The Religious Worlds of Isaac Newton*, ibid, p311 with reference to Note 31 "CUL Add. Ms. 3996 fol.128r ; Newton to Briggs, 12 September 1682, and Newton to Bentley, 10 December 1692 and 17 January 1692/93, in Correspondence, 2:233–34; and I. Newton, Principia, 381" p 453, R. Iliffe, *Priest of Nature*, ibid.
159. R. Iliffe, *Priest of Nature – The Religious Worlds of Isaac Newton*, ibid. pp311–312
160. The author Jessica Riskin refers to the great four meanings of Newtonianism in the *Oxford Illustrated Companion to the History of Science* – Editor in Chief – J.L. Heilbron, Oxford University Press, 2008, published by Tess Press, imprint of Black Dog and Leventhal Publishers, p212
161. G.F. Lewis & L.A. Barnes, *A Fortunate Universe – Life in a Finely Tuned Cosmos*, Cambridge University Press, 2016, p xiv
162. Paul Davies, *The Mind of God*, Penguin Books, 1992
163. Paul Davies, *The Goldilocks Enigma*, ibid. p139, p146
164. Paul Davies *The Goldilocks Enigma – Why is the Universe just right for life?* First Mariner Books, 2006, p2
165. Wikipedia entry for *Cosmological Constant* as at 4 September 2015
166. Paul Davies, *The Goldilocks Enigma*, ibid. p149
167. Dr N.M. Clarke, *Nucleosynthesis, Life, Bent Chains and the Anthropic Principle* – see http://www.np.ph.bham.ac.uk/history/nucleosynthesis
168. Dr N.M. Clarke, *Nucleosynthesis, Life, Bent Chains and the Anthropic Principle*, ibid.
169. Marcus Chown, *Life's Subatomic Secret – New Scientist* – 22 October 2016, p34

170. M. Chown, p34, ibid.

171. Dr N.M. Clarke, *Nucleosynthesis, Life, Bent Chains and the Anthropic Principle*, ibid.

172. Quote contained in G.F. Lewis & L.A. Barnes, *A Fortunate Universe – Life in a Finely Tuned Cosmos*, ibid, p20, quoted from J.Barrow & F.Tipler, *The Anthropic Cosmological Principle*, 1986, p21

173. G.F. Lewis & L.A. Barnes, *A Fortunate Universe – Life in a Finely Tuned Cosmos*, ibid, p268

174. R. Penrose, *The Road to Reality, A Complete Guide to the Law of the Universe*, Vintage, 2005, p759

175. G.F. Lewis & L.A. Barnes, *A Fortunate Universe – Life in a Finely Tuned Cosmos*, ibid, p16

176. The precise many worlds hypothesis proposed by Everett is not settled – see http://plato.stanford.edu/entries/ qm-everett/

177. Quoted from G.F. Lewis & L.A. Barnes, *A Fortunate Universe – Life in a Finely Tuned Cosmos*, ibid, p20

178. *The fractal universe*, Landscape Theory Part 3 – Stanford News, 10 September, 2018

179. *Oxford Dictionary of Physics, op. cit.* p87, 636

180. G.F. Lewis & L.A. Barnes, *A Fortunate Universe – Life in a Finely Tuned Cosmos*, ibid, pp347–348

181. Swinburne, 1996, cited from Taliaferro and Harrison, *The Routledge Companion to Theism* Taylor & Francis, 2013, p250

182. Herman Philipse, *God in the Age of Science? A critique of religious reason*, Oxford University Press, 2014.

183. Herman Philipse, *God in the Age of Science? A critique of religious reason*, ibid, pp270–272

184. Herman Philipse, *God in the Age of Science? A critique of religious reason*, ibid, pp272–273

185. G.F. Lewis & L.A. Barnes, *A Fortunate Universe – Life in a Finely Tuned Cosmos*, ibid, p282–286, quote on p286

186. That is, whether the theoretical possibility of a 'multilaw' multiverse (not the 'unilaw' megaverse or Everett's many worlds) increases the probability, given that there is no indication of multiverses (as opposed to Linde's 'unilaw' megaverse) as existing except as mathematical constructs.

187. www.philpapers.org The Fallacy of Fine Tuning, Victor J. Stenger

188. Victor Stenger, *The Problem with the Cosmological Constant, Reality Check*, www.csicop.org 2011, Skeptical Briefs, Volume 21. 1 Spring 2011

189. M.Tegmark, *Our Mathematical Universe, My quest for the ultimate nature of reality*, Penguin Books, 2014, p141

190. N. Lane *The Vital Question* Profile Books, 2015, pp43–45

191. N. Lane, 'Life: Inevitable or fluke?' *New Scientist*, 23 June 2012

192. Even though considered a leading theory, this article refers to it as the worst of origin of life theories – see Harold S Bernhardt *The RNA world hypothesis:*

the worst theory of the early evolution of life (except for all the others) – Biology Direct 20127:23. Published online, 13 July 2012 doi: 10.1186/1745-6150-7-23

193. Leslie Orgel, *The Implausibility of Metabolic Cycles on the Prebiotic Earth* PLOS Biology, January 2008, Volume 6(1) :e18 (2008)
194. Quote from D. Axe, *Undeniable*, Harper Collins, 2016, pp227–228
195. This reflects the morphing from a 'tiny' prokaryote microbe to a 'large' eukaryote complex cell as described by biologist Nick Lane: "Prokaryotes are typically about 15,000 smaller than eukaryotes..." – N. Lane *The Vital Question*, ibid, p158
196. N. Lane, *Life: Inevitable or fluke?* ibid, p36
197. L. Valiant, *Probably Approximately Correct*, Basic Books, 2013
198. L. Valiant, *Probably Approximately Correct*, ibid, pp106–108
199. L. Valiant, *Probably Approximately Correct*, ibid, pp112–113
200. Samir Okasha, *Philosophy of Science – A very short introduction*, 2002, Oxford University Press, p52 under topic 'Can Science Explain Everything?'
201. See *The Origin of Chirality in the Molecules of Life – A revision of awareness to the current theories and perspectives of this unsolved problem*, Guijarro & Yus, RSC Publishing, 2009
202. A. Flew, *There is a God: How the world's most famous atheist changed his mind*, Harper One, 2007, pp90–91
203. Quoted from A. Flew, *There is a God: How the world's most famous atheist changed his mind*, ibid, pp131–132
204. G.B. Ferngren, *Science & Religion – A Historical Introduction*, John Hopkins University Press, 2017, pp364–365
205. G.B. Ferngren, *Science & Religion – A Historical Introduction*, ibid, pp364–365 – reference to Worthing is Worthing, Mark, *God, Creation and Contemporary Physics*, Minneapolis: Fortress Press, 1996
206. R. Arianrhod, *Seduced by Logic, Emilie du Châtelet, Mary Somerville and the Newtonian Revolution*, *op. cit.* p110
207. M.R. Antognazza, *Leibniz*, ibid, pp1,8–9
208. *On the Principle of Individuation (De principio individui)*, Look, Brandon C., 'Gottfried Wilhelm Leibniz', The Stanford Encyclopedia of Philosophy, ibid.
209. Look, Brandon C. 'Gottfried Wilhelm Leibniz', *The Stanford Encyclopaedia of Philosophy*
210. B. Magee, *The Great Philosophers*, Oxford University Press, 2009, p99
211. B.Russell, ibid, p531
212. Look, Brandon C. "Gottfried Wilhelm Leibniz", *The Stanford Encyclopaedia of Philosophy*, ibid.
213. Ibid.
214. Ibid.
215. W.Dunham, *The Mathematical Universe op. cit.* pp143–144 including quote from Leibniz
216. Quoted by W. Dunham, ibid, from J. Hoffman, *Leibniz in Paris*, p15

217. W.Dunham, *The Mathematical Universe*, ibid, p144
218. W. Dunham, *The Calculus Gallery, Masterpieces from Newton to Lebesque*, Princeton University Press, 2018, p21
219. M.R Antognazza, *Leibniz*, ibid, p4
220. Look, Brandon C. 'Gottfried Wilhelm Leibniz', *The Stanford Encyclopaedia of Philosophy*, ibid.
221. Iibid.
222. B. Magee, ibid, p98
223. M.R. Antognazza, *Leibniz*, ibid, p10, it appeared in the *Acta eruditorum*, ibid, p26
224. B.Russell, ibid, p532
225. Isaac Newton, *The Principia – Mathematical Principles of Natural Philosophy* – The Authoritative Translation by I. Bernard Cohen and Anne Whitman, University of California Press, 1999, pp646–647
226. N. Guiccardini *op. cit.* p195
227. Iliffe, Newton, *A Very Short Introduction, op. cit.* p120 and N. Guiccardini *op. cit.* p195
228. N. Guiccardini *op. cit.* p195
229. N. Guiccardini *op. cit.* p196
230. De Analysi per aequationes numero terminorem infinitas (On analysis by Means of Equations with an Infinite Number of Terms)
231. Iliffe, *Newton, A Very Short Introduction, op. cit.* p120
232. Leibniz discovered the series for π described as his 'first greatest achievement' by maths author Ranjan Roy quoted in *The Math Book*, Clifford A. Pickover, Sterling Publishing, 2009, p110 – it is also interesting to note that like Newton and Leibniz independently inventing calculus, another two mathematicians independently invented the same series for π that Leibniz invented (p110, Pickover – ibid)
233. Newton's method is a numerical method for solving a difficult function, say, $f(x) = 0$, where you guess and test each guess, ie. you guess successive points at, say, x_1, x_2, x_3 and use $f'(x_1), f'(x_2), f'(x_3)\ldots$ and then $x_2 = x_1 - f(x_1)/f'(x_1)$ to test each guess where x_1 will lead from its tangent on $f(x_1)$ to the x value on the x axis to (hopefully) show $f(x_1) = 0$ to give a better approximation than your guess. This can be applied successively – see Pickover, ibid, p154 and *Oxford Concise Dictionary of Mathematics*, Oxford University Press, 2015, pp327–328
234. N. Guiccardini *op. cit.* p197 & Iliffe, *Newton, A Very Short Introduction, op. cit.* p122
235. Iliffe, Newton, *A Very Short Introduction, op. cit.* p124
236. M.R. Antognazza, *Leibniz*, ibid, p11
237. M.R. Antognazza, *Leibniz*, ibid, pp6–7 and B.Russell, ibid, p532
238. M.R. Antognazza, *Leibniz*, ibid, pp8–9
239. M.R. Antognazza, *Leibniz*, ibid, p12
240. Will and Ariel Durant, *The Age of Voltaire – The Story of civilisation Part IX*,

Simon & Schuster, 1965, p584

241. Longman *Illustrated Encyclopedia of World History, op. cit.* p357
242. Ibid, p357
243. Will and Ariel Durant, *The Age of Voltaire*, ibid, p 397
244. *Longman Illustrated Encyclopedia of World History,* ibid, pp921–922
245. M.R. Antognazza, *Leibniz*, ibid, p2
246. Author, Johann Christoph Gottshed, per: Will and Ariel Durant, *The Age of Voltaire*, ibid, p400
247. *Oxford Dictionary of World History, op. cit.* p283 and M.R. Antognazza, *Leibniz*, ibid, p8
248. M.R. Antognazza, *Leibniz*, ibid, p9
249. M.R. Antognazza, *Leibniz*, ibid, pp11–12
250. M.R. Antognazza, *Leibniz*, ibid, p10
251. Leibniz letter quoted from M.R. Antognazza, *Leibniz*, ibid, p36, in which she cites the source as Guerrier, 206–8
252. McDonough, Jeffrey K. 'Leibniz's Philosophy of Physics', *The Stanford Encyclopaedia of Philosophy* (Spring 2014 Edition), Edward N. Zalta (ed.), URL = <https://plato.stanford.edu/archives/spr2014/entries/leibniz-physics/>.
253. Iliffe, Newton, *A Very Short Introduction, op. cit.* p120
254. S. Law, *Philosophy op. cit.* p40
255. I. Berlin, *The Proper Study of Mankind,* Vintage, 2013, p390
256. S. Law, *Philosophy op. cit.* pp282–283
257. J. Marías, *History of Philosophy,* Dover Publications, 1967 (English Translation) p243
258. Indeed Professor Copelston – former professor of history of philosophy at London University – so classified them – see F. Copelston, *The Rationalists, Descartes to Leibniz*, Bloomsbury Publishing plc, 1958
259. S. Law, *Philosophy op. cit.* p279
260. The Latin *a priori* is mostly used in philosophy to mean analytic with *a posteriori* to mean empirical facts
261. See 'Copenhagen Interpretation', *Oxford Dictionary of Physics, op. cit,* pp103–104
262. 1999, distributed by Warner Bros.
263. R. Arianrhod, *Seduced by Logic, Emilie du Châtelet, Mary Somerville and the Newtonian Revolution*, p34
264. R. Arianrhod, ibid, p22
265. R. Arianrhod, ,ibid, p33
266. R. Scruton, *A Short History of Modern Philosophy,* ibid, p67
267. I. Berlin, *The Proper Study of Mankind*, iIbid, p359
268. I. Berlin, *The Proper Study of Mankind*, ibid, p359
269. R. Arianrhod, ibid, p104
270. F. Coppelston, *The Rationalists, Descartes to Leibniz*, Bloomsbury Publishing plc 1958, p298
271. R. Arianrhod, ibid, pp105–106

272. F. Coppelston, ibid, pp297–298
273. F. Coppelston, ibid, p298
274. F. Coppelston, ibid, p298
275. A force acting on an object produces (or consumes) work that is transferred to the object and manifests itself as increase (or decrease) of its energy (which is an increase of kinetic or potential energy). The practical difference between the two quantities is that a small force can increase the kinetic energy by a large amount if it acts for a long distance or period of time in the sense that kinetic energy does not directly measure the strength of the force, but also takes into account the duration and distance the object has travelled.
276. McDonough, Jeffrey K., "Leibniz's Philosophy of Physics", *The Stanford Encyclopaedia of Philosophy* (Spring 2014 Edition), Edward N. Zalta (ed.), URL = <https://plato.stanford.edu/archives/spr2014/entries/leibniz-physics/>.
277. Newton's advances based on the inverse-square law is seen as one of the great four meanings of Newtonianism by Jessica Riskin in the *Oxford Illustrated Companion to the History of Science* – Editor in Chief – J.L. Heilbron, Oxford University Press, 2008, published by Tess Press, imprint of Black Dog and Leventhal Publishers, p212
278. *Theoria motus abstracti* – see McDonough, Jeffrey K. 'Leibniz's Philosophy of Physics', *The Stanford Encyclopedia of Philosophy* (Spring 2014 Edition), Edward N. Zalta (ed.), URL = <https://plato.stanford.edu/archives/spr2014/entries/leibniz-physics/>
279. *The Oxford Handbook of Leibniz*, edited by M.R. Antognazza, Oxford University Press, 2018, pp3–4 (Life and Works by M.R. Antognazza).
280. Encyclopedia Britannica Online (as at 22 October 2018)
281. Encyclopedia Britannica, Online, ibid.
282. Encyclopedia Britannica, Online, ibid.
283. Encyclopedia Britannica, Online, ibid.
284. *The Oxford Handbook of Leibniz*, ibid, p xviii
285. *The Oxford Handbook of Leibniz*, ibid, p 4
286. Quoted from *The Stanford Encyclopaedia of Philosophy*, ibid.
287. Quoted from *The Stanford Encyclopaedia of Philosophy*, ibid.
288. B. Magee, *op. cit.* p109
289. B. Magee, *op. cit.* p109
290. B. Magee, *op. cit.* pp109–110
291. Leibniz notation was preferred over Newton's only due to the evolution of mathematical usage in Europe.
292. S.G. Krantz, *Differential Equations Demystified*, Macgraw Hill Books, 2005, p188
293. Quoted from R. Roy, The Discovery of the Series Formula by Leibniz, Gregory and Nilakantha, *Mathematics Magazine*, Vol 63, No 5 (Dec 1990) pp 295–296 – Leibniz was not the first to derive π according to this article using series. However, it appears his method to do so was unique. The

derivation is important in mathematics as series were starting to be used in general mathematics as opposed to geometry in the second half of the 17th century when Leibniz and Newton were in their prime. Gregory also was trying to derive infinite series for any function and later discovered the famous Taylor Series – see pp291–292 of this article.

294. Quoted from D. Acheson, *The Calculus Story, A mathematical adventure*, Oxford University Press, 2017, p106
295. D. Acheson, *The Calculus Story*, ibid, Chapter 17, p106
296. D. Acheson, *The Calculus Story*, ibid, Chapter 17, p106 et. seq.
297. The horizontal variable is called the abscissa and the vertical variable is called the ordinate.
298. J. Marías, *History of Philosophy, op. cit.* pp43–46
299. S. Law, *Philosophy op. cit.* p249
300. *Oxford Dictionary of Philosophy*, Oxford University Press, 2016, p76
301. F. Coppelston, ibid, p269
302. F. Coppelston, ibid, p269
303. B. Magee, *op. cit.* p153
304. J. Marías, *History of Philosophy, op. cit.* p243
305. F . Coppelston, ibid, p266
306. F. Coppelston, ibid, p267, including preceding Leibniz quote.
307. Austin Farrer, Fellow of Trinity College, Oxford, *Introduction to Theodicy – The Project Gutenberg ebook of Theodicy*, 2005, www.pgdp.net
308. F. Coppelston, ibid, p286
309. F. Coppelston, ibid, p287
310. A saying of Polish American Alfred Korzybski
311. Peckhaus, Volker, 'Leibniz's Influence on 19th Century Logic', *The Stanford Encyclopaedia of Philosophy* (Spring 2014 Edition), Edward N. Zalta (ed.), <https://plato.stanford.edu/archives/spr2014/entries/leibniz-logic-influence/>
312. Peckhaus, Volker, 'Leibniz's Influence on 19th Century Logic', *The Stanford Encyclopaedia of Philosophy*, ibid.
313. Peckhaus, Volker, 'Leibniz's Influence on 19th Century Logic', *The Stanford Encyclopaedia of Philosophy*, ibid.
314. Peckhaus, Volker, 'Leibniz's Influence on 19th Century Logic', *The Stanford Encyclopaedia of Philosophy*, ibid.
315. Peckhaus, Volker, 'Leibniz's Influence on 19th Century Logic', *The Stanford Encyclopaedia of Philosophy*, ibid.
316. Peckhaus, Volker, 'Leibniz's Influence on 19th Century Logic', *The Stanford Encyclopaedia of Philosophy*, ibid.
317. R. Scruton, *A Short History of Modern Philosophy, op. cit.* p 70
318. R. Scruton, *A Short History of Modern Philosophy, op. cit.* p 70
319. See also Roger Scruton's summary of this principle – R. Scruton, *A Short History of Modern Philosophy, op. cit.* p71
320. R. Scruton, *A Short History of Modern Philosophy, op. cit.* p 72

321. F. Coppelston, ibid, pp278–279
322. R. Scruton, *A Short History of Modern Philosophy, op. cit.* p 73
323. McDonough, Jeffrey K. 'Leibniz's Philosophy of Physics', *The Stanford Encyclopaedia of Philosophy op. cit.*
324. F. Coppelston, ibid, p305
325. F. Coppelston, ibid, p305
326. F. Coppelston, ibid, p306
327. F. Coppelston, ibid, pp306–307
328. D. Papindeau, *Philosophy*, Shelter Harbour Press, 2009, p40
329. A. Grünbaum, *Philosophical Problems of Space and Time*, D. Reidel Publishing Company, Chapter 6, 'The Resolution of Zeno's Paradox of Extension for the Mathematical Continua of Space and Time'
330. M.R. Antognazza, *Leibniz*, ibid, p79
331. De Risi, Vincenzo, 2007, *Geometry and Monadalogy: Leibniz's Analysis Situs and Philosophy of Space*, Birkhauser, 2007
332. S. Uchii, Leibniz *Ultimate Theory,* 3 February 2017, Ref. 2Fphilsci-archive.pitt.edu /12787 – University of Pittsburg, p1
333. S. Uchii, Leibniz *Ultimate Theory*, ibid, p6
334. S. Uchii, Leibniz *Ultimate Theory*, ibid, pp2–3
335. S. Uchii, *Leibniz's Theory of Time*, 2015, 2Fphilsci-archive.pitt.edu /11448 – University of Pittsburg, p8
336. S. Uchii, *Leibniz's Theory of Time*, ibid, p9
337. M.R. Antognazza, *Leibniz*, ibid, p98
338. *Oxford Dictionary of Philosophy, op. cit*, p512
339. Grünbaum, *Philosophical Problems of Space and Time*, ibid.
340. An Aleph is the first letter of a number of writing systems including Hebrew, Aramaic, Arabic and other languages and in mathematics is a 'creature of infinity' (Aleph Null, $\aleph_0$, is also the size of the lowest infinite number, ie. number of natural numbers – technically known as cardinality of the natural numbers).
341. Jorges Luis Borges & Paulo Coelho's works
342. J.L.Borges, *The Aleph* (short story) quoted from V. Mihály, *Mythical Spaces – The Aleph as Seen by Borges and Coelho*. Acta Universitatis Sapientiae, Philologica, 4, 1(2012) 200–208
343. M.R. Antognazza, *Leibniz*, ibid. p10 & Russell, *op. cit.* p533
344. M.R. Antognazza, *Leibniz, A Very Short Introduction*, ibid, p65
345. M.R. Antognazza, *Leibniz, A Very Short Introduction*, ibid, p65
346. M.R Antognazza, *Leibniz, A Very Short Introduction*, ibid, p65
347. M.R. Antognazza, *Leibniz, A Very Short Introduction*, ibid, p41
348. R. Iliffe, *Priest of Nature – The Religious Worlds of Isaac Newton, op. cit.* p97, including preceding quote of Newton
349. M.R. Antognazza, *Leibniz, A Very Short Introduction*, ibid, pp67–68
350. T. Butler-Bowden, *50 Philosophy Classics,* Nicholas Brealey Publishing, 2013, p178

351. Ian Stewart, *Infinity, A Very Short Introduction* – Oxford University Press, 2017, p46
352. Ian Stewart, *Infinity, A Very Short Introduction* – Oxford University Press, 2017, p47
353. *The Nature and Attributes of God* (www.newadvent.org).
354. Quaestiones Disputatae de Potentia, 1.3, 7 from *Aquinas, Selected Philosophical Writings*, Oxford University Press, 1993, pp 244–246
355. St Thomas Aquinas, *Original Manuscript – Expositio super Librum Boethii de Trinitate* ed. B.Decker 2nd Ed. Leiden 1959 1255 to 1259 AD.
356. M.R. Antognazza , *Leibniz, A Very Short Introduction*, ibid, p68
357. Philosophers call knowledge akin to this 'middle' knowledge often in relation to what A might have decided to do or what might have happened if B did something else – called counterfactuals
358. Manchak, John and Roberts, Bryan W. 'Supertasks', *The Stanford Encyclopaedia of Philosophy* (Winter 2016 Edition), Edward N. Zalta (ed.), URL = <https://plato.stanford.edu/archives/win2016/entries/spacetime-supertasks/>.
359. Huggett, Nick, 'Zeno's Paradoxes', *The Stanford Encyclopedia of Philosophy* (Winter 2010 Edition), Edward N. Zalta (ed.), URL = <https://plato.stanford.edu/archives/win2010/entries/paradox-zeno/>
360. Simon Dein, John Swinton, and Syed Qamar Abbas, *Theodicy and End-of-Life Care* – J Soc Work End Life Palliat Care. 2013 Apr; 9(2–3): 191–208. Published online, 18 Jun 2013, doi: 10.1080/15524256.2013.794056
361. Edward Dolnick, *The Clockwork Universe*, Harper Perennial, 2011, p xvii to xviii
362. Edward Dolnick, *The Clockwork Universe*, ibid, p 260
363. *The Fractal Geometry of Nature*, 1
364. P. van Lommell, *Consciousness Beyond Life – the Science of the Near Death Experience*, 2007, Harper One, p305
365. Stephen Hawking, *A Brief History of Time*, Bantam Press, 1989 p54 & p66
366. Stephen Hawking, *A Brief History of Time*, ibid, p57
367. Peter Byrne, *The Many Worlds of Hugh Everett*, Scientific American, 21 October 21, 2008
368. Stephen Hawking, *A Brief History of Time*, ibid, pp182–183
369. National Geographic production aired on cable channels in 2017 – *Genius by Stephen Hawking* – an episode in the series entitled 'Why are we here?'
370. Ian Stewart, *The Beauty of Numbers in Nature*, Ivy Press, 2017, p174
371. Ian Stewart, *The Beauty of Numbers in Nature*, ibid, p175
372. Ian Stewart, *The Beauty of Numbers in Nature*, ibid, p173
373. Ian Stewart, *The Beauty of Numbers in Nature*, ibid, p176
374. The hidden variable hypothesis is to the effect that although it seems as if we do not know the particle's position and momentum until the observer observes wave collapse, there are in fact hidden variables science is yet to discover which will determine the unknown position and momentum.
375. Leslie Valiant, *Probably Approximately Correct*, Basic Books, 2013, pp19–20

376. Leslie Valiant, *Probably Approximately Correct*, ibid, pp106–107
377. Ian Stewart, *Infinity: A Very Short Introduction*, Oxford University Press, 2017, pp29–30
378. Ian Stewart, *Infinity: A Very Short Introduction*, ibid, p30
379. Timothy Gowers, *Mathematics: A Very Short Introduction*, Oxford University Press, 2002, pp56-57
380. Philip Merilees, 139th meeting of the American Association for the Advancement of Science in 1972 – See also p182–183, 148–149, Ian Stewart, *The Beauty of Numbers in Nature*, Ivy Press, 2017
381. Ian Stewart, *Does God Play Dice*, ibid, p129
382. Ibid.
383. Ian Stewart, *The Beauty of Numbers in Nature*, ibid, p178
384. Professor J.M. McDonough's 'Introductory Lectures on Turbulence' [Dept of Mech.Eng & Math – University of Kentucky]
385. Ian Stewart, *Does God Play Dice*, Penguin, 1990, p169
386. Quoted from F. Capra, *The Tao of Physics*, 1992, Flamingo – Harper Collins, p49
387. A saying of Alfred Korzybski – a Polish American Scholar
388. Quoted from F. Capra, *The Tao of Physics*, ibid, p53
389. These thirteen systems are the Circulatory System, Digestive System, Endocrine System, Excretory System, Immune System, Integumentary System, Lymphatic System, Muscular System, Nervous System, Reproductive System, Respiratory System, Skeletal System and the Urinary System, and they would have hundreds of subsystems.
390. Ian Stewart, *The Beauty of Numbers in Nature*, Ivy Press, 2017, p59
391. T. Butler-Bowden, *50 Philosophy Classics*, Nicholas Brealey Publishing, 2013, p178
392. Swinburne, *Providence and the Problem of Evil*, Oxford University Press, 1998, p14
393. Recall that in Chapter 5 the Anthropic Principle was explained.
394. Carlo Rovelli, *Seven Brief Lessons on Physics*, Penguin Random House, 2016, pp32–33
395. A physical 'fine-tuning measure' is not necessarily anthropic but could merely be a reality problem-solving measure to aid orderly operations of the laws of the universe.
396. Hard omniscience really entails knowing things that logically cannot be known, such as what lies at the end of an infinite series of terms or tasks – a more reasonable definition of omniscience is knowledge of what it is possible to know – another example might be that if a biological circuit required solving an equation that cannot be solved, it is not possible to know the solution.
397. Einstein's famous description of quantum non locality
398. Dreyfus draws from Heidegger's concept that to understand a being we must understand the 'world' that the being is in – the context – see Sub

Chapter 1 'The Problem of Being' – Chapter (unnumbered) entitled – Heidegger's Existential Philosophy J. Marías, *History of Philosophy op. cit.* pp427–428. The philosophy of Wittgenstein also has relevance because he saw the structure of reality of the world reflected by the structure of language – sentences – each sentence was like a small piece of the picture of reality and as John Searle comments: "Wittgenstein is always anxious to insist in the" [Philosophical Investigations] 'that language is indefinitely extendable'. B. Magee, *op. cit.* pp322–327 – quote on p327. Thus language, being capable of portraying the background reality – discussed here, namely, all the little pieces of reality, constitutes an infinite pool of potential 'reality data'.

399. Dreyfus, 1992, pp288–9 quoted from M. Wheeler, *Cognition in Context: Phenomenology, Situated Robotics and the Frame Problem* – International Journal of Philosophical Studies, 16(3), 323–49,2008
400. M. Wheeler, *Cognition in Context: Phenomenology, Situated Robotics and the Frame Problem* – International Journal of Philosophical Studies, 16(3), 323–49, 2008 (quote from section 2. Stalking the Frame Problem)
401. Leslie Valiant – *Probably Approximately Correct*, ibid, pp106–107
402. R. Iliffe, *Priest of Nature – The Religious Worlds of Isaac Newton, op. cit.* p99
403. Quoted from T. Stafford, *The Choice Engine*, New Scientist, 6 April 2019, p35
404. Recent work by Harvard professor Leslie Valiant has demonstrated that a general evolutionary algorithm for a range of functions required for life using non-uniform distributions of input data used by such algorithms will be unsolvable. (See Chapter 6, L. Valiant – *Probably Approximately Correct.*)
405. The new atheists are those who use science and reason as 'modern thinking' without God on the basis of the Scientific Revolution and religion's outdated notion that religion is not accessible to reason.
406. Fyodor Dostoevsky *The Brothers Karamazov*, Vintage Books, 2004 Edition, Random House, pp244–245
407. D.Bodanis, *Einstein's Greatest Mistake*, Abacus, 2017, p166
408. D.Bodanis, *Einstein's Greatest Mistake*, Abacus, 2017, p118
409. R. Penrose, *Fashion, Faith and Fantasy*, Princeton University Press, 2016, p5.
410. R. Penrose, *Fashion, Faith and Fantasy*, ibid, p5
411. D.An, K.A. Meissner, P. Nurowski, R. Penrose, *Apparent Evidence for Hawking Points in the CMB Sky* – arXiv: 1808,01740v3 [Astri-ph.co] 17 December 2018
412. Emeritus Mathematics Professor, Ian Stewart in his *The Great Mathematical Problems*, Profile Books, 2014, states: "...we now know that three-body dynamics is chaotic – so irregular that it has elements of randomness," p136, and Emeritus Mathematics Professor Paul J. Nahin in his book *An Imaginary Tale – The story of* $\sqrt{-1}$, Princeton University Press, 1998, states: "...the Finnish mathematician astronomer Karl F. Sundman (1873–1949) solved the three-body problem during the period 1907–19, and in 1991 a chinese *student*, Quindong Wang, solved the N-Body problem for any N. These solutions are in the form of infinite convergent series, however, which converge far too

slowly to be of any practical use." P110. (Professor Nahin continues to say that super computers surpass this mathematics anyway now, so numerical rather than analytical solutions are currently used for N-body calculations.)

413. I. Stewart, *The Great Mathematical Problems*, ibid, p136

PART TWO

MASTERS OF THE COSMOS

God does not play dice with the universe
Einstein

MARIE CURIE

CURIE

Chapter Fourteen Plate

Marie Curie, her husband Pierre (centre) and assistant in their laboratory in Paris

FOURTEEN

MARIE CURIE

It was early 1899. On the forest floor in Bohemia (now the Czech Republic) lay a pile of 'waste' black rock called Pitchblende which had already been mined for uranium. A wagon full of the black rock mixed with pine needles and dirt was now arriving at Marie Curie's makeshift laboratory – a shed in Rue l'Homond, Paris – amidst her excitement.[1] It was not the residue of uranium that excited her but her discovery of a mysterious glowing element which would prove to be a million times more radioactive than uranium.[2] She and her husband, Pierre, would now start one of the greatest feats of scientific endurance in history – to prise open the black rock and recover this mysterious glowing substance – radium. For this quest was to take them almost four years of painstaking, backbreaking processing of the waste ore in various vats and chemical plant. Not only would the recovery of radium herald the dawn of a new science – nuclear physics – but it would break a scientific paradigm, namely, that the atom was the smallest unit of matter. The mysterious rays which Marie coined as radioactivity gave her the insight that the atom had not yielded all its secrets – there was a subatomic reality remaining to be discovered. Her pioneering role in this new reality delivered her the title of the first and only woman in history to receive two Nobel Prizes.

Although she made her home in France, her native country was Poland, which at that time was under Russian rule. She was born on 7 November 1867 as Maria Salomea Sklodowska, the youngest of four children, but was nicknamed Manya.[3] Her parents were teachers. At first, they lived at 16 Freta Street, Warsaw, but her father, who taught maths and physics, was offered a job at a boys' high school which saw them move to an apartment in the western district of Warsaw. Her character would develop as determined with strong principles, yet she was extremely shy and very emotional[4], which would make for a tumultuous life. She would be tested with the greatest challenges through both her self-imposed quests of scientific endurance and personal tragedies. She found great prowess in maths, science and languages and had a natural curiosity for science – she would explore her father's study with its scientific instruments such as barometer and electroscope, which she was told were 'physics apparatus'. She was primed for science.

The first tragedy was the death of her sister Zosia, aged fourteen, from typhus, which then saw her mother's health worsen, having long suffered with tuberculosis. Her mother had been the principal of a girls' school near Freta Street, before the move to the western district prompted her to resign due to travel times, but her health had declined anyway. Her mother had then enrolled her at the age of ten into a private school for girls run by a Madame Sikorska.[5] Although the Russian curriculum was designed to stamp out Polish culture, Madame Sikorska and other teachers at her school still taught subjects with Polish culture to maintain a type of cultural resistance to Russian domination. This political oppression was thus centre stage in the mind of Manya (Marie), and indeed the intensity of this mindset was underscored by visits to the school from Russian inspectors, who would arrive unannounced to surprise students and test them (although a system of warning bells would chime as a warning for the girls to put away their Polish books before the inspector entered the classroom). Often, Manya, as the brightest student in the class, would be asked to respond to the inspector's question before her answer in perfect Russian to satisfy the inspector. Yet her emotional state would see her holding back tears and fear whilst she did so.[6] (Her father was also a Polish patriot whose Russian employer had cut his salary dramatically and demoted him some years before this, and so it was this oppressive atmosphere that moulded Manya and perhaps steeled her for the challenges of later life.)[7] Manya would later write: "Constantly held in suspicion and spied upon, the children knew that a single conversation in Polish, or an imprudent word, might seriously harm not only themselves, but also their families."[8]

However, when she was fourteen, Manya's father increased her challenges by enrolling her in a government-run school despite the risk of the Russian teachers treating her badly. Perhaps it was the fact that only government schools could award graduation certificates which would enhance Manya's career as a gifted student. Manya's patriotic spirit was galvanised at the school by a strong friendship forged with the librarian's daughter – fellow Pole Kazia Przyboroska. Every day, they would unceremoniously spit on a statue erected in honour of those loyal to the Russian Tsar, and when the Tsar was assassinated, they were reprimanded for dancing in the street.[9] It was a time that evoked extreme passions in the wake of injustices inflicted upon Poland by their oppressors. Overcoming the oppressive atmosphere at school, she received the gold medal by coming first in her class in 1883 – she was fifteen.[10] However, after she achieved this, she suffered a type of 'nervous collapse', which has been suggested as depression.[11]

Her father then arranged for her to have a year off from her studies and fortunately she was able to stay with relatives who were landowners with impressive estates, giving her a year spent in picturesque settings. She stayed with two uncles on her mother's side at Zwola, south of Warsaw. She wrote: "Ah, how gay life is at Zwola! There are always a great many people, and a freedom, equality and independence such as you can hardly imagine." After that, she stayed north of Warsaw at her Uncle Xavier's horse stud where she learnt to ride.[12]

Fresh from her year off, she had returned to Warsaw as a high school graduate and then worked as a tutor and governess. She continued her studies at a type of 'underground educational institution' known as the 'flying university', which defied the Russian educational system. Her older sister Bronya was determined to go into medicine, but that required going to Paris. Manya and Bronya then made a pact that Bronya would go to Paris first, to begin medicine, whilst Manya remained in Poland to support her by co-contributing (with her father) to Bronya's medical studies. This saw Manya take a well-paid job 100 kms north of Warsaw. She described leaving Warsaw: "That going away remains one of the most vivid memories of my youth… My heart was heavy as I climbed into the railway car. It was to carry me for several hours, away from those I loved."[13] She had been hired as tutor to the two eldest daughters of the Zorawski family, who lived in a small manor house situated on the same property as the sugar beet factory where Mr Zorawski worked as estate manager. Although Manya was keen to progress her studies in the peace and quiet of the countryside, she also pursued the patriotic activity of supporting the education of the poor and uneducated farmers and their families. This activity arose from an ideology called 'positivism', conceived by a French mathematician Auguste Comte (1798 to 1857), philosopher and social theorist[14], whose concepts came to dominate 19th -century philosophy.[15] Manya supported this philosophy (which will be examined in the next chapter). It was thought that education amongst the masses could initiate a rise of Polish patriotism, leading to independence, although Manya's participation risked imprisonment or deportation.[16] The Zorawskis gave her permission to teach the locals to read and write in Polish, earning the Zorawskis her respect. She did continue her studies, however, and later wrote: "I was as much interested in literature and sociology as in science… However, during these years of isolated work, trying little by little to find my real preferences, I finally turned towards mathematics and physics."[17]

Manya's life soon unfolded like a Jane Austen romance because the Zorawskis' handsome nineteen-year-old son Casimir returned home from Warsaw University and swept her off her feet to become her first love. But like Elizabeth Bennett being shunned by the family of Mr Darcy in Jane Austen's *Pride and Prejudice,* so too was Manya shunned by the Zorawski family when it seemed their love could result in marriage. Manya did not have sufficient social position and money (despite similar backgrounds).[18] Even after being shunned, she needed money to keep her pact with Bronya and later wrote to her brother: "If they don't want to marry poor young girls, let them go to the devil!"[19]

In 1888, she returned to Warsaw to take up another job with the Fuchs family (industrialists), and after spending a year working as a governess, she finally received a letter from Bronya to come to Paris but by then she was ambivalent and replied: "I dreamed of Paris… [but]… now that the possibility is offered me, I do not know what to do."[20] After persuasion by her father and Bronya, she reluctantly agreed providing she be allowed to spend another year in Warsaw before going to Paris. It may have been the now 'old flame' she kept burning for her former love, Casimir, or obligation to stay with her father, but in any event she was thrilled to get the use of a laboratory to conduct experiments through her cousin Joseph Boguski, who ran the Museum of Industry and Agriculture (otherwise a cover to train Polish scientists).[21] Her final word to Casimir was summed up by her: "If you can't see a way to clear up our situation… it is not for me to teach it to you."[22]

As she arrived in Paris in November 1891, she was now twenty-four and it was eight years since her gold medal at high school.[23] After a few months with her sister Bronya, she decided that in addition to the long commute time to the university, she needed to be fluent in French, which would not be easy as she spoke Polish in her sister's apartment.[24] So she moved to a little apartment in the Latin Quarter frequented by students and artists and close to the Sorbonne – the University of Paris – which she planned to attend. It was very cold, and food, beyond bread and hot chocolate, was infrequent. She would comment on her bleak conditions: "My situation is not exceptional… it was the familiar experience of the Polish students whom I knew."[25] However, she was thrilled to be in Paris because she was free to speak or read what she wished. When she had enrolled in the Sorbonne in 1891, she had already decided to change her name to the French name, 'Marie', so 'Manya' was now in the past. She was now Marie Sklodowska.

She was very fortunate to be taught physics by Professor Gabriel Lippman (later a Nobel Prize winner) and mathematics by Henri Poincaré – Poincaré was one of the giants of mathematics and one of the founders of topology. (One of the great unsolved maths problems published by the Clay Mathematics Institute was the Poincaré conjecture in topology which was finally solved almost a century later by a Russian, Grigory Perelman.) Her rise to success began with a first in her physics Masters degree in 1893, followed by a second place in mathematics in 1894.[26]

But her ability was recognised because before finishing her Master's degree in mathematics, she received a scholarship from an industry organisation to undertake a study of the differing magnetic properties amongst various steels.[27] The advantage for Marie was that it required a laboratory, which she was keen to work in, but finding one was more difficult. She was soon introduced by an acquaintance – a Professor Kowalski – to a Pierre Curie, who was laboratory chief at the Municipal School of Industrial Physics and Chemistry in Paris.[28] As she approached the apartment of Professor Kowalski, she saw Pierre through the French window and would later write: "[He was a] tall young man with auburn hair and large, limpid eyes...[29] [and when meeting him]... I noticed the grave and gentle expression of his face, as well as a certain abandon in his attitude, suggesting the dreamer absorbed in his reflection." That day, she had not only met her future husband but together they would achieve one of the great milestones in science. She commented on their immediate rapport: "There was, between his conceptions and mine, despite the difference between our native countries, a surprising kinship, no doubt attributable to a certain likeness in the moral atmosphere in which we were both raised by our families."[30] She described some interesting insights she gained during their courtship: "Soon he caught the habit of speaking to me of his dream of an existence consecrated entirely to scientific research, and he asked me to share that life. It was not, however, easy for me to make such a decision, for it meant separation from my country and my family, and the... [renunciation] ... of certain social projects that were dear to me. Having grown up in an atmosphere of patriotism kept alive by the oppression of Poland, I wished, like many other young people of my country, to contribute my effort toward the conservation of our national spirit..."[31]

Pierre also expressed his life philosophy in noble terms: "It would nevertheless, be a beautiful thing in which I hardly dare believe, to pass through life together hypnotised in our dreams: your dream for your country; our dream for humanity; our dream for science. Of all these dreams, I believe the last alone, is legitimate. I mean to say by this that we are powerless to change the social order. Even if this were not true we should not know what to do... From the point of view of science, on the contrary, we can pretend to accomplish something. The territory here is more solid and obvious, and however small it is, it is truly our possession."[32] Both Marie and Pierre were free thinkers – liberals, as they were known in that age – who were patriotic, from intellectual families, non-religious, were nature lovers and had studied both literature and poetry in between their science and mathematics and so were well suited.[33] In 1895, they married in a small civil ceremony in the small village of Sceaux.[34] Their honeymoon consisted of a bicycle tour of northern France – this was the 'bicycle age' of France.[35] Pierre and Marie moved into 24 Rue de la Glacière, Paris. Pierre was a professor at the nearby School of Physics and Chemistry.[36] Pierre and his brother Jacques had discovered piezoelectricity and had invented a device called an electrometer to measure small amounts of electric current (which would prove very useful to Marie's discovery of radium and polonium).[37] In 1897, their daughter Irène Curie was born and was delivered by Pierre's father, Dr Eugéne Curie. Marie would write: "It was under this mode of quiet living, organised according to our desires, that we achieved the great work of our lives, work begun about the end of 1897 and lasting for many years."[38]

It is at this point that one of the greatest stories in the history of science unfolded, and a detailed account of this story is presented in Chapter 16, 'Marie's Science'. In short, after four long years she had:

- Coined the term 'radioactive'
- Developed new methods for measuring radioactivity
- Established the technique of using radioactivity to discern different elements
- Found radioactive rays emanating from Thorium
- Found and named a new element Polonium after her native Poland
- Found and named Radium as a cornerstone radioactive element which would found the age of nuclear medicine
- Changed the face of scientific understanding of the atom, which will be explained in Chapter 17, 'Marie's Reality'
- Laid the foundations for scientists such as Rutherford to discover the inner workings of the atom.

After she wrote to her father about her success in discovering radium and polonium, he remarked it was a shame that her work seemed to have no practical use (!). Sadly, her father died not long afterwards (14 May 1902). Indeed, he would later be proved wrong because a principal use of radium has been shown to be cancer therapy around the world. Marie would later write: "It may be easily understood how deeply I appreciated the privilege of realising that our discovery had become a benefit to mankind, not only through its great scientific importance, but also by its power of efficient action against human suffering and terrible disease. This was indeed a splendid reward for our years of hard toil."[39]

Her final requirement for completion of her thesis to obtain her PhD was to present the result in front of a committee of professors. On 25 June 1903, she entered a packed room, including members of the public –for her fame was growing now – and delivered her dissertation entitled 'Researches on Radioactive Substances'. After answering the professor's questions, they took a few minutes and announced: "The University of Paris [the Sorbonne] accords you the title of doctor of physical science with the mention of *très honourable* [high honours]." They also stated that she had delivered the 'greatest scientific contribution ever made in a doctoral thesis'. On hearing this, the crowd burst into resounding applause.[40]

Then later that year, in early November 1903, the Royal Society of London awarded Marie and Pierre the Humphry Davy Medal for the most important discovery in chemistry. On 14 November 1903, Marie and Pierre and Henri Becquerel were awarded the Nobel Prize in Physics. The prize money of 70,000 francs (about US $20,000) was a welcome boost to their research effort, but the fame that came with a Nobel Prize would prove quite challenging in unexpected ways.

Marie almost did not win the Nobel Prize. Four members of the French Academy of Science, including Marie's own professors, Gabriel Lippman and Henri Poincaré, sent a letter to the selection committee attributing sole credit to Pierre for the work. Fortunately, this sexual discrimination (or possible 'foreigner discrimination') was prevented by the intervention of a Swedish mathematician, Magnus Goesta Mittag-Leffler, who supported women scientists. The Swede had written to Pierre advising of the problem, and Pierre had written back confirming Marie's central role in the work on radioactivity and that it would be a travesty of justice for her not to be awarded the Nobel Prize.[41] Marie would later describe the aftermath of the Nobel Prize: "This event greatly increased the publicity of our work. For some time there was no more peace. Visitors and demands for lectures and articles interrupted every day..."[42]

On 19 April 1906, tragedy struck. Pierre had been to a luncheon of professors and was on his way in heavy rain to a publisher and to visit a library. As he was braving the heavy rain with his umbrella, he arrived at his publisher's but the doors were locked. He turned to cross the street but was run over by a horse-drawn wagon with a load of military supplies weighing 6 tonnes and was killed instantly. The lab assistant, Pierre Clerc, who identified the body, said: "He wasn't careful enough when he was walking in the street, or when he rode his bicycle. He was thinking of other things." In a similar vein, his father awkwardly said: "What was he dreaming of this time?"[43]

So much success was now met with tragedy, which floored Marie. Her dear husband, who had come through the painstaking hard years to ride success with her, was gone. Marie would later write: "Crushed by this blow, I did not feel able to face the future. I could not forget, however, what my husband used to say, that even deprived of him, I ought to continue my work." Her sisters came from Poland to console her and it was only thoughts of her daughters that spurred her to soldier on. In her own words: "I'm young enough to earn my living and that of my children."[44] A first kind gesture came on 14 May 1906 when the Sorbonne offered her a professorship where she would replace the position held by Pierre. She held mixed emotions at accepting this offer as she explained: "The honour that now came to me was deeply painful under the cruel circumstances of its coming. Besides, I wondered whether I would be able to face such a grave responsibility."[45]

Later that year, Lord Kelvin, then a leading scientist in England (who, amongst other things, invented the Kelvin scale for temperature), laid down another challenge for Marie. He wrote a letter to the *London Times* claiming that the Curies had been mistaken in claiming radium as a new element but that instead it was a compound of helium and lead. Lord Kelvin had 'form' in this regard because he had also claimed Röntgen's X-rays were a hoax. However, some considered him England's greatest scientist and so Marie had to refute the challenge. This would require recovering metallic radium which she was now determined to do (but it would take years).[46] More support was received the following year when American steel magnate Andrew Carnegie pledged an annual donation of $50,000 for scholarships which she applied in expanding her laboratory staff with student researchers, including many young women.[47]

Further challenges loomed. In November 1910, Marie offered herself as a candidate to the French Academy of Sciences to fill the seat that had become available. However, a rival candidate, Edoard Branly, who was supported by French Catholics and who had also received an honour from the Pope, sparked a media campaign where newspapers smeared her as a foreigner and a woman and even presented pictures of her that appeared as sinister 'mug shot' images. The campaign did its damage and she lost her bid for the seat.[48]

By the end of 1911, it seemed that Marie had been having an affair with Paul Langevin, who had been a pupil of Pierre's. Apparently, his present wife was not able to satisfy his intellectual needs, but it was just a matter of time before she filed for divorce, with consequences.[49] Marie had been attending the famous Solvay conference in November 2011 where such people as Ernest Rutherford and her friend Albert Einstein were in attendance. As she left the lobby of her hotel for the morning session of the conference, she was confronted with reporters waving a copy of *Le Journal* in her face, which accused her of having an affair with Paul Langevin – a married man with four children. She and her daughters were later forced to take shelter with friends.[50] The *Le Journal* story included this line: "The fires of radium which beam so mysteriously… have just lit a fire in the heart of the scientists who studies (sic) their action so devotedly; and the wife and the children of this scientist are in tears…" *Le Journal*, 4 November 1911.[51]

At the height of this scandal, she was awarded a second Nobel Prize for Chemistry and, despite this making her the first and only woman to receive two Nobel Prizes, the press shunned her by paying hardly any attention to her.[52] She became aware that members of the Swedish Academy of Sciences, who administer the Nobel Prizes, were urging her not to attend the ceremony amidst the controversies plaguing her. She wrote in reply: "In fact, the prize was given to me for the discovery of radium and polonium. I think there is absolutely no connection between my scientific work and the facts of my private life, which uninformed and disreputable people use against me".[53]

Despite being ill from kidney problems and depression following the stress of the scandal, she travelled the forty-eight-hour journey to the Nobel ceremony with her sister Bronya and daughter Irène, like Daniel into the lion's den. Yet, the expected roar of protest she feared did not eventuate and after her tribute to Pierre there was a tumultuous applause to her

Nobel speech. The scandal seemed to be subsiding, for even during her stay, the Stockholm newspaper story of the Langevin divorce did not mention her name. Upon her return from Sweden, she had an operation to repair her kidneys, which nearly killed her.[54]

By 1914, the Great War, now known as World War I, had broken out, which prompted the French government to move to Bordeaux in case the Germans overtook Paris. As a consequence, Marie was asked to take her store of radium (the country's only supply of radium, worth 1 million francs) also to Bordeaux to a bank vault for safekeeping, which she did. But Marie wished to do more and noticing that French hospitals lacked X-ray equipment, she devised a plan for both new radiology facilities in hospitals (including university equipment) and mobile radiology stations to be manufactured.[55] Marie described this initiative: "These stations rendered great service during the Battle of the Marne. But as they could not satisfy the needs of all the hospitals of the Paris region, I fitted up, with the help of the Red Cross, a radiologic car."[56]

She would later write: "The use of the X-rays during the war saved the lives of many wounded men; it also saved many from long suffering and lasting infirmity."[57] Marie continued to undertake war-related work until about 1919, wrote a book *Radiology in War* and offered courses on radiology to American soldiers who remained in France.[58]

In 1920, Marie was interviewed by an American journalist, Marie 'Missy' Meloney, who was an editor of *Delineator*, a prominent women's magazine of the day.[59] Marie had formed the Radium Institute and was in search of funding and so stressed both the needs and benefits of the Radium Institute. She emphasised that despite her team leading the world in radium research, America had fifty times as much radium as the one gram they held. Marie Meloney supported the cause and organised a 'Marie Curie Radium Campaign', culminating in a visit by Marie Curie to meet President Warren G. Harding in 1921 and to return to Paris with her much-needed funds for pitchblende, more equipment and research.[60] The Radium Institute would be her passion for the rest of her life. After worsening health, she finally died in a sanatorium in the French Alps on 4 July 1934, with her daughter Ève by her side.[61] Her daughter Ève gave testament to her Nobel Prize-winning parents: "By the most desperate and arid effort they discovered a magic element, radium. This discovery not only gave birth to a new science and a new philosophy: it provided mankind with the means of treating a dreadful disease."[62] Marie Curie's family won five Nobel Prizes in all, with Irène Curie and her husband sharing the 1935 Nobel Prize in Chemistry. Ève did joke that she was the only one not to win a Nobel.[63]

In 1995, it was decided that both her and Pierre's remains should be moved from Sceaux, the village they loved and had cycled through in their times of leisure, to the more stately national mausoleum, the Panthéon, in Paris. France's then head of state, President Mitterand, explained: "By transferring these ashes of Pierre and Marie Curie into the sanctuary of our collective memory, France not only performs an act of recognition, it also affirms a faith in science, in research, and its respect for those who dedicate themselves to science, just as Pierre and Marie Curie dedicated their energies and their lives to science."[64] President Mitterand also praised Marie Curie's 'remarkable battle'.[65] Another historical milestone saw the University of Paris VI renamed in 1974 as the Université Pierre et Marie Curie (which merged with Sorbonne University in 2018).

The Radium Institute has grown into a substantial organisation with staff and treatment facilities comparable to a modern hospital. During its initial growth in the 1920s, it was realised that radium was both a treatment and a danger depending on dosage, and sadly many who handled radium (including the Curies) suffered from the adverse health effects – cancer being the most significant due to the bone marrow being affected, causing diseases of the blood like leukaemia or anaemia.[66] Fortunately, through standards introduced by Marie herself, treatment using radium is now highly successful.

Einstein once said of her: "Marie Curie is, of all celebrated beings, the only one whom fame has not corrupted."[67]

Chapter Fifteen Plate

Marie Curie and other greats of science at the famous 1927 Solvay Conference on quantum mechanics. Albert Einstein, Max Planck, Paul Dirac, Werner Heisenberg, Niels Bohr, Max Born and many other Nobel prize winning scientists are featured

Source: Author: Benjamin Couprie, Institut de Physique de Solvay

FIFTEEN
CURIE'S WORLD

Maria (or Manya as she was known – later Marie) was born into a Poland that had a complicated history. In late medieval times (the 16th century), Poland was at its zenith as a great state spanned by Prussia, the Baltic Sea, Russia and the Black Sea.[68] In the 17th century, Poland's quest to dominate north-eastern Europe saw a loss to Sweden. During the Swedish-Polish War, over one third of Poland's population fled or died.[69] The Polish government, weakened from the strain of war and surrounded by hostile powers – Russia, Turkey, Sweden, Prussia and Austria – succumbed to conquest and partition by Russia, Austria and Prussia between 1772 and 1795. In 1807, Napoleon established the Grand Duchy of Warsaw, but the Battle of Leipzig in 1813[70] caused major changes again. (Poland was effectively a bystander in the clash between the massive forces of Napoleon –185,000 troops – against the combined forces of Austria, Prussia, Russia and Sweden – 350,000 troops).[71] The result – apart from Napoleon losing the battle[72] – was a smaller Kingdom of Poland, established in 1815 but controlled by Russia where Poland effectively became a protectorate of Russia.[73] This was broadly the state of Poland that Marie grew up in.

Napoleon's decisive loss at the Battle of Waterloo that same year (1815) ended his rule of France.[74] Napoleon had taken power in a *coup d'état* in 1799.[75] Although the French Revolution saw the monarchy lose power (despite periods of comebacks) which eventually resulted in France becoming a democratic republic, the ideas that caused the French Revolution were part of a mindset of the age dominated by a philosophy called 'positivism' that was also prevalent in the defiance of native Poles like Manya as she then was. In the last chapter, mention was made of Marie and Pierre's small civil ceremony in the small village of Sceaux. Her daughter Ève commented: "There would be no religious ceremony: Pierre was a free thinker and Marie, for a long time past, has ceased the practices of religion".[76] Ève also writes of the oppression in Poland from Russia and how the 'mystic dreaming' of the Catholic religion became a 'resource' or 'force of resistance', but "[t]he mystic dream no longer dwelt in Manya [Marie]. By tradition and convention she remained a practicing Christian, but her faith had been shaken by [her mother's] death; little by little it had now evaporated."[77] And later Ève writes: "...[Marie] allied herself with some ardent positivists".[78] For instance, when Manya/Marie presented her sister's friend Marya Rakovska with a photograph of herself and her sister Bronya, she wrote across it: "To an ideal positivist – from two positive idealists."[79]

Apart from her grieving over her mother's death, did positivism cause Marie, from a devout Catholic country, to shun her inherited faith? In Chapter 23, I quote from Stephen Hawking's wife, Jane, who lamented 'the positivist approach' of modern scientists (such as her then husband) who saw God as an obstacle to scientific pursuits despite the mathematical beauty of the universe.[80] What was positivism and its origins?

Positivism emerged after the French Revolution but was more part of the development of ideas that shaped European history in that age. Isaiah Berlin doubts that a neat list of causes can be ascribed to the French Revolution, whilst acknowledging that the strongest ideas that influenced the event came from Rousseau and Voltaire[81], who were both thinkers of the Enlightenment. The Enlightenment included the Scientific Revolution shaped by Newton, Leibniz, Galileo, Kepler and others, but this new type of thinking extended outside science. Locke, who is seen as the father of British empiricism, also wrote on political philosophy and thought that we only get our knowledge from the physical world through what comes through our senses. Leibniz (and Descartes) were rationalists and held that we can get knowledge from the metaphysical realm prior to getting it through our senses. These concepts were combined by Kant, regarded as one of the greatest philosophers, who expounded 'idealism', which involved a blend of both rationalism and empiricism. Kant saw that our minds were structured to perceive things – in time and space. (We could have just as easily detected reality as a fog of waves and frequencies foreign to our present crisp image of reality.) In the same way, our minds know what a substance. For example, as we saw in Chapter 10, in discussing Plato's theory of forms or Aristotle's essence of an object or its mechanical purpose, we can perceive that an ashtray can take many forms. That is, our mind still knows what it is. This led to a more difficult question of just what 'things in themselves' were, ie. the nature of objects when we aren't looking. This question has taken on greater significance with quantum mechanics due to the observer problem with wave collapse. Therefore, it seems that if a particle were the 'thing in itself', it could have no solid existence without our observation as its position and momentum are only known on wave collapse when we 'intervene' to observe it.

Philosopher Roger Scruton sees both Kant and Hume as two of the greatest philosophers. We discussed Hume's fork in Chapter 10. Scruton comments on the period of the Enlightenment covering the philosophers mentioned above: "From the point of view of the historian it is perhaps the richest and most exciting of all intellectual eras, not because of the content, but because of the influence of the ideas that were current in it."[82]

Locke, Hobbes, Voltaire and Rousseau were amongst the philosophers who laid out new political philosophies that influenced both the American Revolution and the French Revolution. How then did positivism (also known as scientific empiricism) become wedged in countries such as Poland as an ideology mixed up in politics and revolutions?

Comte's view that society was changing from the partnership of military and theological values to the partnership of scientific and industrial ones was highly relevant to the conflict between Russia and Poland.[83] Russia's conquests of other countries involved militaristic expansion strongly influenced by religion – for example, it saw itself as a future safe haven for Christendom in its unsuccessful aims to conquer Constantinople. However, of more relevance to Poland was that Russia, Austria and Prussia had partitioned Poland in 1772 (during the reign of Catherine the Great of Russia – a Prussian princess).[84] Positivism, being a revolt against German philosophy and Russia's monarch being Prussian (German), the general philosophy of positivism as a movement against militarism and religion as opposed to science and reason became a natural ideology for rebellion during Russia's joint occupation of Poland. Thus, science and reason were immediate goals for young Poles to challenge the divine monarchs of Russia, Prussia and Austria. Voltaire's rebellion against divine monarchs and religion exemplified this type of thinking.

We saw Voltaire's support for Newton. Voltaire saw that science could greatly advance human affairs, and his hopes extended to great progress in science and education in human affairs – even a moral science.

Rousseau, aware of Locke's empiricism, took an unexpected intellectual leap. Bertrand Russell describes it well: "His first literary success came to him rather late in life. The Academy of Dijon offered a prize for the best essay on the question: Have the arts and sciences conferred benefits on mankind? He contended that science, letters, and the arts are the worst enemies of morals, and by creating wants, are the sources of slavery; for how can chains be imposed on those who go naked… he admired the noble savage… Science and virtue, he held, are incompatible, and all the sciences have an ignoble origin. Astronomy comes from the superstition of astrology; eloquence from ambition; geometry from avarice: physics from vain curiosity; and even ethics has its source in human pride. Education and the art of printing are to be deplored; everything that distinguishes civilized man from the untutored barbarian is evil."[85] Four years later, he did not win a prize with a second essay which included his insight: "man is naturally good, and only by institutions is he made bad".[86] Bertrand Russell saw this as "the antithesis of the doctrine of original sin and salvation through the Church."[87]

Although Rousseau and Voltaire at first corresponded amicably, they eventually fell out with much acrimony. Voltaire, who had ridiculed Leibniz's theodicy, was especially incensed by the Lisbon earthquake, even regretting his more optimistic worldview held by him earlier in his life. John Wesley gave a sermon, 'The Cause and Cure of Earthquakes'; "sin is the moral cause of earthquakes, whatever their natural cause may be… they are the effect of that curse, which was brought upon the earth by the original transgression of Adam and Eve."[88] Historians Will and Ariel Durant describe Voltaire's reaction to this and other explanations: "Voltaire fumed at these explanations, but he himself, could find none to reconcile the event with his faith in a just God. Where now was Leibniz's 'best of all possible worlds'? ...In any angry reaction against his own early optimism, Voltaire composed (1756) his greatest poem – *On the Lisbon Disaster, or An explanation of the Axiom 'All is Well'*. The last few lines of the poem (English translation) read:

> "*...Buried beneath their roofs, end without help*
> *Their lamentable days in torment vile!*
> *To their expiring and half-formed cries,*
> *The smoking cinders of this ghoulish scene,*
> *Say you, 'This follows from eternal laws*
> *Binding the choice of God both free and good'?*
> *Will you, before this mass of victims say,*
> *'God is revenged, their death repays their crimes'?*"[89]

The authors provide further thoughts of Voltaire: "But what crime, what fault those infants committed who lay crushed and bloody on their mother's breasts? Had London or Paris less vice than Lisbon? Yet Lisbon is shattered, and Paris dances. Could not an omniscient God have made a world without such meaningless suffering? 'I respect my God, but I love mankind'. The poet looks upon the world of life, and sees everywhere, in a thousand forms, a struggle for existence, in which every organism, sooner or later is slain." Referring to 'This bitter summary of biology', they quote further from Voltaire including this extract: "Thus all the world in all its members groans, all born for suffering and for mutual death. And in this fatal chaos you will compose, from the misery of each part, happiness of the whole! What happiness?... You cry out in mournful tones that 'all is well'; the universe gives you the lie, and your own heart refutes a hundred times the error of your mind. The elements, and animals and men – all are in war. Let us confess it: evil strides the earth."[90]

The emphasis is on 'natural evil' or the chaos arising from the eternal laws of the universe that God chose when he chose the best of all possible worlds according to Leibniz. It is easy to see now why Voltaire went on then to write *Candide* as a satire on Leibniz's 'best of all worlds' theodicy. Voltaire exemplifies the utter despair we all suffer in viewing the chaos of the world. Looking through the lens of the knowledge we have today, it can be seen that this despair arises from viewing the world as a Clockwork Universe and of course like a clock it can be made perfectly by God – at first sight. Leibniz saw that even the clock has impossibilities by the very foundations of logic – it cannot have 'Circuit A and Circuit Not A' or 'Circuit A and Circuit B' if impossibilities result. Voltaire believed in 'mechanism' and all the new science as to how the world worked, yet when dealing with how God created the world, it seems the despair arose from the simple assumption that 'God can bypass any imperfection in that mechanism'. But Voltaire himself said, "Perfect is the enemy of the Good" – not to address that problem but perhaps a corollary of that statement may have lessened his despair. In summary, the insight that mechanism explained the world (as shown by science) may indicate that God must use a mechanism to create reality. In Chapter 13, 'Is a Perfect World Possible?', the implications of 'mechanism' were examined in the light of what we now know about the mathematics of chaos and what we have called the Hippasus Dichotomy. Voltaire's despair underscores the importance of these very human issues against the spectre of the natural evil inherent in nature and the bitter reality of biology that is apparently required to construct human existence. Let us hope that this is the best of all worlds. Voltaire's final word in his poem was "What must we do, O mortals? Mortals, we must suffer, submit in silence, adore, hope and die."[91]

Rousseau responded to Voltaire's poem in these terms:

> ***"Voltaire, in seeming always to believe in God, never really believed in anybody but the devil, since his pretended God is a maleficent Being who according to him finds all his pleasure in working mischief. The absurdity of this doctrine is especially revolting in a man crowned with good things of every sort, and who from the midst of his own happiness tries to fill his fellow creatures with despair, by the cruel and terrible image of the serious calamities from which he is himself free."***[92]

Bertrand Russell summarised their subsequent relationship: "[This caused]… a bitter enmity between Voltaire and Rousseau… Voltaire treated Rousseau as a mischievous madman; Rousseau spoke of Voltaire as 'that trumpet of impiety, that fine genius, and that low soul.'[93] On the importance of Rousseau, he said: "There are only two parts of his thinking that… [concern the history of philosophical thought]… these are, first, his theology, and second, his political theory. In theology, he made an innovation which has not been accepted by the majority of protestant theologians… modern Protestants who urge us to believe in God, for the most part, despise the 'old proofs'; and base their faith upon some aspect of human nature – emotions of awe or mystery, the sense of right and wrong, the feeling of aspiration, and so on. This way of defending religious belief was invented by Rousseau."[94]

Rousseau's other great contribution was his work *The Social Contract* (1762), which was reprinted every four months in the ten years after the Reign of Terror of 1789.[95] Its influence upon the French Revolution is clear when considering this extract from it:

> ***The 'Social Contract' in Rousseau's words is "the total alienation of each associate, together with all his rights, to the whole community; for, in the first place, as each gives himself absolutely, the conditions are the same for all; and this being so, no one has any interest in making them burdensome to others."[96] According to Russell, Rousseau's social contract is analogous to Locke's model and more akin to Hobbes who also spoke of man's natural state in Leviathan.[97] Clearly, we can see now the origin of the principle of society defending the individual rights and freedoms in democracies today.***

Both Rousseau's and Voltaire's philosophies called for social reforms and the French Revolution, but an ideology was forming, namely, that of a new science – a social science. Sieyés (who Napoleon joined in the *coup d'état* in 1799) and Condorcet (a champion of France's then new system of weights and measures, helping create the new order of things) both coined the term 'social science'.[98] France saw the first 'social scientist' in the form of Claude Henri de Saint-Simon, who used the Enlightenment approach of applying the new scientific methods to human affairs – he asked whether there were undiscovered laws for social conduct.[99] He used the term positivism to describe society moving on from metaphysical explanations or in other words largely moving on from religion or politics to guide how society should be ordered.[100]

It was then left to another Frenchman, Auguste Comte (1798–1857), to found the positivism movement, following closely on the heels of both the French Revolution and the Industrial Revolution. Historian Peter Watson describes his philosophy: "Comte understood that society around him was changing in a fundamental sense: what he called 'theological' and 'military' values were giving way to 'scientific' and 'industrial' ones. In such a world, he said, industrialists replaced warriors, and scientists replaced priests".[101] Comte wrote a book *Cours de Philosophie Positive (Course of Positive Philosophy)*, where he expounded a theory where humanity and science pass through three stages:

1. A theological stage where human beings attribute natural phenomena to a deity – this certainly was evidenced anthropologically when ancient civilisations worshipped 'nature gods'.

2 A metaphysical stage where human beings attribute phenomenal causes to unseen forces or forms.

3. A final stage where human beings dispense with searching for 'final causes' and merely systemise observations of phenomena according to methods akin to the scientific method.[102]

Then Comte's philosophy took a bizarre turn. He admired the benefits of organised religion and attempted to form his own (non-faith) religion to enshrine ritual in a new social order with objectives to 'live in love on the basis of positive knowledge'. Peter Watson commented that the subsequent decline in Comte's positivism accompanied his failure to allow for psychology or the role of conflict in society.[103] A later variant of positivism 'logical positivism' – relying again on the purely empirical knowledge and dismissing metaphysical knowledge – also declined because the division between physical and metaphysical knowledge is blurred, which really restores both realms of knowledge to importance.

Thus, positivism really gave hope to people that they were moving forward, away from the divine order where monarchies kept people oppressed towards the greater number of rights and freedoms based on new orders of society grounded in the social sciences. People like Marie and her fellow patriotic Poles could cling to this philosophy as it was a movement for change. However, in the process, things like psychology and faith were ignored, which remain vitally important for individuals in their lives.

It is no surprise therefore that opposing views to Comte's philosophy were voiced, with one vivid example from Dostoevsky in *The Brothers Karamazov*:

> ***"...that the science of this world, having united itself into a great force, has ...examined everything heavenly that has been bequeathed to us in sacred books, and, after hard analysis, the learned ones of this world have absolutely nothing left of what was once holy. But they have examined the parts and missed the whole, and their blindness is even worthy of wonder... Even in the movements of the souls of those same all – destroying atheists... [the whole]... lives, as before immovably!... for until now neither their wisdom nor the ardour of their hearts has been able to create another higher image of man and his dignity than the image shown of old by Christ. And whatever, their attempts, the results have been only monstrosities."***[104]

Philosopher Julián Marías comments that "Almost all nineteenth-century philosophy is essentially dominated by positivism and shows its influence one way or another."[105] He summarises it as follows: "What then is the philosophy of positivism? Apparently, it is a reflection upon science. After the sciences are exhausted, there remains no independent object for philosophy but the sciences themselves; philosophy becomes a theory of science. Thus positive science acquires unity and consciousness of itself. But it is clear that philosophy disappears; and this is what happens in the positivist movement of the 19th century, which has very little to do with philosophy."[106] Clearly, the positivist movement approached scientism. which merely elevates science to faith despite science merely being the scientific method and the body of knowledge it acquires. However, the insight here is that something like positivism or scientism became mixed with humanist goals (whether faith-based or secular), resulting in a cultural movement independent of proper foundations, which is really the point that Marías makes. A philosophy serving human goals must originate and be set on proper foundations and principles to endure.

Marías refers to this positivism movement as the 'mathematisation of thought' where 'metaphysics is impossible... [or]... meaningless'.[107] Wittgenstein, one of the greatest philosophers of the 20th century[108], held the view that it was "a vulgar illusion that science could generate a description of all those things with which our humanity... is mingled."[109] This view was really exemplified with the Romanticism movement, championed by Rousseau[110] and, according to Isaiah Berlin, expanded by such philosophers as Herder: "I am not here to think, but to be, feel, live!"[111] Hamman: "The tree of knowledge has robbed us of the fruit of life,"[112] with Fichte as the true founder: "The essence of man is freedom, and although there is talk of reason, harmony, the reconciliation of one man's purpose to that of another in a rationally ordered society, yet freedom is a sublime but dangerous gift: 'Not nature but freedom itself produces the greatest and most terrible disorders of our race; man is the cruellest enemy of man'."[113]

It seems inexplicable that after the brilliant advances of the Scientific Revolution, Romanticism should emerge as a major force. Even some of the greats such as Immanuel Kant, a scientist as well as a philosopher, contributed to Romanticism – "[reason] 'as a light that illuminates nature's secrets' is inadequate... as an explanation... the process of birth is a better metaphor for it implies human reason creates knowledge."[114] The emphasis upon creating moved art and literature alongside science in prominence. Historian Peter Watson concludes: "Romanticism was a massive revolution in ideas... [and]... French Romanticism was essentially a reaction to the French Revolution... We are still living with the consequences of this revolution... [of ideas]." The rival ways of looking at the world – the cool, detached light of scientific reason, and the red-blooded, passionate creations of the artist – constitute the modern incoherence."[115] Fichte also said: "There are two worlds and man belongs to both... There is the material world, 'out there', governed by cause and effect, and there is the spiritual world, 'Where I am wholly my own creation'." Peter Watson comments that this underscores how Fichte elevated the will to be a major role in psychology. Thus, the hope that science would dictate the perfect life for humans gave way to the individual's freedom of will which creates human reality.

In Chapter 8, Thomas Kuhn's landmark book *The Structure of Scientific Revolutions* was mentioned to show that science is not immune from the 'psychological' will of scientists. Indeed, Kuhn's books evoked emotional responses in the form of criticism that Kuhn wrongly held science was not based on reason but on 'faith and peer pressure' or was 'a matter for mob psychology'.[116] But this is but one instance of the greater principle at play here. Kant, who was quite aware of the scientific method exemplified with Newton (and indeed Kant himself contributed to science – recall his comment about the right- or left-handedness in possible universes referred to in Chapter 10) held that one can discover significant truths about reality by using pure reason without empirical observation or experiment (Chapter 12) and, as observed above, reason can be used for human creativity. It is this fusion of human factors and science that can bridge the modern incoherence between science, art and literature.

Living at the same time as Marie was Nobel Prize winner and French philosopher Henri Bergson, influential during her life, and writer of *Creative Evolution.* Rather than nature being designed by God, it is a 'work in progress' or as Bergson says: "It is a creation that goes on forever in virtue of an initial movement. This movement constitutes the unity of the organized world – a prolific unity, of an infinite richness, superior to any that the intellect could dream of, for the intellect is only one of its aspects or products".[117] The problem of the origin of creativity was also tackled by author Arthur Koestler in his book *The Act of Creation,* which contains an initial overview on page iii of the book: "The problem of creativity is fundamental to the assessment of man's condition. I believe," Koestler writes, "that view to be depressingly true – but only up to a point... There are two ways of escaping our more or less automated routines of thinking and behaving. The first is the plunge into dreaming or dream-like states, where the rules of rational thinking are suspended. The other way is also an escape – from boredom, stagnation, intellectual predicaments and emotional frustrations – but an escape in the opposite direction; it is signalled by the spontaneous flash of insight which shows a familiar situation or event in a new light".[118] (Indeed, as the 'Infinite Background Problem' outlined in Chapter 13 demonstrates, such 'familiar situations' may be *infinite,* making creativity a necessary tool even in Leibniz's *Best of All Worlds*).

It is no wonder that Marie, faced with a choice between her mother's Catholic faith and the positivist- or scientism-based ideology, lost her faith, as many scientists must have done (I do not believe it is as simple as stating 'Marie lost her faith because she lost her mother at the age of eight', as the complexities of Leibniz's theodicy has shown). Some further insights can be gained from a letter to her cousin Henrietta Michalovska as related by her daughter Ève: "Manya to Henrietta [who had just given birth to a dead child], April 4, 1887: '...What suffering it must be for a mother to go through so many trials for nothing! If one could only say, with Christian resignation, 'God willed it and his will be done!' Half of the bitterness would be gone. Alas, that consolation is not for everybody. I see how happy are the people who admit such explanations. But strangely enough, the more I recognise how lucky they are, the less I understand their faith, and the less capable of sharing their happiness... Forgive me for these philosophical reflections: they are caused by the backward and conservative spirit of the town where you live. Do not judge it too harshly, for social and political conservation usually comes from religious conservation and the latter is a happiness – even though for us it has become incomprehensible. So far as I am concerned, I should never voluntarily contribute towards anybody's loss of faith. Let everybody keep his own faith, so long as it is sincere. Only hypocrisy irritates me – and it is as widespread as true faith is rare... I hate hypocrisy. But I respect sincere religious feelings when I meet them, even if they go with a limited state of mind...'''[119]

Marie laments the vacuum she feels in her spiritual worldview. Recall Marias's observation about positivism: 'metaphysics is impossible... [or]... meaningless'.[120] The problem is that whilst science had developed a grand scheme of thought, satisfying rational minds like Marie Curie, religion developed no comparable grand scheme of thought which could pass the various tests of the analytical mind. (Theology is based on revelation and the authority of scripture rather than reason, analysis and experiment – however, there seems no obstacle – except cultural paradigms – to the construction of such a new grand scheme of theology or philosophy of God.) Each paradigm of science has these tests but, as Kuhn observed, paradigms come and go. Moreover, her fellow countryman Alfred Korzybski pronounced one of the great insights of science – 'the map is not the territory'. The reason why this truism repeatedly applies to themes in this book is that it strikes at the heart of any scheme of reality and in fact is famous because it prescribes the difference between belief in an abstract system versus the actual reality that exists. Thus, to avoid the conundrum of paradigms coming and going, perhaps religion, despite the absence of its own grand scientifically coherent scheme, can become an ageless scheme which, given the right scheme of thought, can pass the analytical tests of a rational woman.

Yet her depth of knowledge beyond maths and science was astonishing.

Ève says, "Her notebook reflects the inner life of an over eager young being, bewildered by the diversity of her gifts,"[121] and relates how she read the greats of literature; Dostoevsky, Goncharov, German and French poetry and much more. Even three pages from Renan's *Life of Jesus* appeared with Marie's note: "Nobody ever made the interest of Humanity predominate in his life over worldly vanity as He did..." as well as Russian philosophical essays, and so on. Art in the form of literature is great at posing questions but not so good at answering them, as might be needed in any grand scheme of thought as discussed above. And so she focused on science despite its inability to provide the emotional and spiritual support that she acknowledged was so powerful an aid in tragic circumstances – support that she greatly needed in her inner life. How you shape your inner life in terms of beliefs, convictions and worldviews is vital to your persona and mindset.

In summary, Isaiah Berlin describes the opposing ideologies in the age of Marie Curie: "Men's beliefs in the sphere of conduct are part of their conception of themselves... and this conception in its turn, whether conscious or not, is intrinsic to their picture of the world... A man who, like Aristotle or Thomas Aquinas, believes that all things are definable in terms of their purpose... is committed to the view that the end of human life consists in self-fulfilment, the character of which must depend on the kind of nature that a man has, and the place he occupies in the harmonious activity of the entire universal, self-realising enterprise... [and it follows that the political philosophy and such purposes will be]... radically different from... someone who has learned from Hobbes or Spinoza or any modern positivist that there are no purposes in nature, that there are only causal (or functional or statistical) laws..."[122]

Unlike religion or philosophy, science struggles to provide meaning to life,[123] and yet Marie gravitated towards the limited grand scheme of thought that science provided. Perhaps the theologian or philosopher might concede that meaning is created by fulfilling (metaphysical) goals that seem irrational to the scientist (but which might not be irrational!). If so, in this sense, there can be no greater exemplar than the meaning of life apparent in Marie's great life in overcoming suffering and tragedy to heroically achieve her humanitarian goals.

SIXTEEN

CURIE'S SCIENCE

The scientists of Marie's age saw the assembly of the periodic table of elements in progression. It did not matter that the atom was an impenetrable mystery because the periodic table had covered the field, and only the gaps in the periodic table needed to be filled. This activity, to fill the gaps, became the ideal scientific quest in her age. Atomic pioneer and chemist John Dalton (1766–1844) had experimented with the composition of gases to determine why, say, x amount of oxygen and y amount of carbon combined in certain proportions to form 'compound atoms'. Dalton, despite some flaws in his theory, founded the atomic chemistry that forms part of today's science.[124] In 1803, he presented four main principles of atomic theory[125]:

1. Elements are made of indivisible small particles or atoms.

Chapter Sixteen Plate

Marie Curie and her husband Pierre in their laboratory in Paris

Source: Author: Vitold Muratov

2. All atoms of the same element are identical and different elements have different atoms.

3. Atoms are never created or destroyed.

4. Different atoms of different elements join in simple ratios to form 'compound atoms'.[126]

Dalton was using the atom as a counting unit when applying statistics to gas and steam pressures.[127] He allowed chemists to know the ratios of elements amongst numerous compounds based on the atom as the most fundamental unit of matter – despite not knowing what it was.[128]

Russian chemist Dmitri Mendeleev published his famous periodic table of the elements in 1869.[129] By this time, scientists had identified over sixty elements from the over 110 elements we now know today that exist.[130] He arranged his chart in rows and columns but noted that certain elements had similar properties. Thus, he had lined up these elements in columns. The second column, which was completed by Marie and Pierre, contains the alkaline earth metals (being oxides of the elements in that column). For example, the first element in that column is beryllium, the second is magnesium, whilst the last one is radium. When Mendeleev ordered his table, blank squares appeared, allowing scientists to predict the next unknown substance that might be discovered.[131] In those days, it was the greatest achievement of scientists to discover and place a new element in the periodic table. For example, gallium was discovered by a French scientist in 1875, and he named it in view of Gallia, meaning France in Latin. Scandium was discovered in 1879 by a Scandinavian scientist, and germanium was discovered in 1886 by a German scientist.[132] Marie had named polonium as a matter of national pride, especially in view of the Russian domination of Poland at that time.

Discovery of each new element meant identifying its unique chemical properties (such as its atomic weight, so well used by Dalton in constructing his chemical system). However, radioactive metals (to use the term Marie would later coin) were awkward as they could only be extracted in minute quantities, making analysis difficult.[133] Thus, using the type or strength of radioactivity was one of Marie's innovations despite some scientists' reluctance to accept such new elements (recall Lord Kelvin's challenge to Marie Curie in Chapter 14). Now the arrangement of elements in Mendeleev's periodic table was arranged according to atomic weight, but something was not quite right about the ordering – it seemed approximately right – but it was not until after Marie's work that the key to unlocking this and other mysteries of the atom could be solved.

In 1895, a German physicist, Röntgen, discovered a new form of radiation which we now know as X-rays. Röntgen won the first-ever Nobel Prize in Physics for this discovery.[134] How he discovered it was remarkable, described by *Physics Today* and quoted in *Time* magazine:

> *"One day in late 1895, the German physicist was preparing to begin an experiment with cathode rays... In his darkened lab, he covered the tube with black cardboard to hide its glow, but noticed a glimmer of light on a fluorescent screen across the room. Curious, Roentgen placed a sheet of black cardboard between the screen and the tube, then another, then a book of 1000 pages, then a wooden shelf board more than two and a half centimeters thick, according to a story in the journal* Physics Today. *'The glimmer remained'. At some point, he held up a small lead disk, and cast a terrifying shadow on the screen: the dark shape of the disk itself, along with the skeletal outline of the bones in his hand. According to* Physics Today, *Roentgen was very late to dinner with his family that night. When he did show up, 'he did not speak, ate little, and then left abruptly' to return to his lab. Afraid that he might have imagined the whole thing, he cautiously told a friend, as quoted by the journal* Resonance, *'I have discovered something interesting, but I do not know whether or not my observations are correct.' Eventually he summoned the courage to tell his wife what he'd seen, and enlisted her help in a follow-up experiment. Just before Christmas that year, he replaced the fluorescent screen with photographic paper and took the world's first X-ray, a clear image of the bones and wedding ring on his wife's left hand. She found the experience as unnerving as he had, exclaiming, "I have seen my death."*"[135]

Many scientists began to study X-rays, which are high energy electromagnetic waves which fortunately do not pass through dense materials but do pass through not-so-dense materials (later, the discovery that X-rays had tiny wavelengths compared to light heralded X-ray spectroscopy, which was instrumental for Watson and Crick to discover DNA). One such scientist, Henri Becquerel, also discovered another type of radiation emanating from uranium compounds or salts. He was testing responses to sunlight when during cloudy weather he noticed the uranium salts still gave off radiation which could pass through metal foil and make impressions on photographic plate bordered by black paper.[136] The Becquerel rays forming this radiation had ionised the air, allowing it to conduct electricity.[137]

Marie wisely chose the Becquerel rays to study for her PhD thesis. For this work, she was permitted to use a cramped storeroom at the Paris Municipal School of Industrial Physics and Chemistry where Pierre worked as professor of physics.[138] When she began her experiments towards her thesis, Marie noted that the Becquerel rays ionised the air and saw that measuring the electricity in the air would be a different way of measuring the rays than merely observing them fogging up a photographic plate. Pierre's invention of the electrometer was ideal for this task as it could measure minute electrical current. Marie later explained: "Instead of making these bodies act upon photographic plates, I preferred to determine the intensity of their radiation by measuring the conductivity of the air exposed to the action of the rays."[139] Her experiments with different uranium compounds confirmed some of Becquerel's work but she started to develop a groundbreaking hypothesis, namely, that the rays emanated from the atoms of uranium rather than being some other cause, such as stored sunlight. Marie's insight that this was a property of the atom of the radioactive element itself and that radioactivity was the emission of particles led to the insight that there were subatomic particles which would break the atomic paradigm of the day – in other words, the atom was not the smallest unit of matter.[140] She later discovered that a second element, thorium, emitted Becquerel rays and this prompted her to coin the word 'radioactivity'.[141] She did not realise she was becoming one of the founders of nuclear physics. (Note today that X-rays are considered electromagnetic radiation that emanate from the electron cloud of the atom and are not termed radioactive. Radioactivity is reserved for particle radiation that originates from nuclear processes. For example, gamma radiation is considered to be part of radioactivity because it originates from the nucleus.)

Pierre soon joined her project as he was fascinated with her progress (suspending work on his piezoelectric crystals). She had discovered something curious, namely, that two uranium ores, pitchblende and chalcolite, displayed more radioactivity than elemental uranium. Marie believed there was something else in the uranium ores that was doing this.[142] Marie described her next step: "I undertook next to discover if there were other elements possessing the same property, and with this aim I examined all the elements then known, either in their pure state or in compounds. I found that among these bodies, thorium has an intensity of the same order as uranium, and is, as in the case of uranium, an atomic property of the element… Certain ores proved radioactive; these were containing uranium and thorium; but their radioactivity seemed abnormal, for it was much greater than the amount I had found in uranium and thorium… This abnormality greatly surprised us… I then made the hypothesis that the ores uranium and thorium contain in small quantity a substance much more strongly radioactive than either uranium or thorium. This substance could not be one of the known elements, because these had already been examined; it must, therefore, be a new chemical element… Pierre… [joined me]… in the search for this unknown substance. We chose, for our work, the ore pitchblende, a uranium ore… Since the composition of this ore was known through careful chemical analysis, we could expect to find, at a maximum, 1 per cent of the new substance. The result of our experiment proved that there were in reality new radioactive elements in pitchblende, but that their proportion did not even reach a millionth per cent!"[143] In other words, it was a substance so tiny in quantity yet so mighty in power.

Marie found that the pitchblende gave off radiation four times greater than uranium, which caused her to double-check everything to eliminate the possibility of errors. But after checking, she wrote in excitement to her sister Bronya: "The radiation that I couldn't explain comes from a new chemical element. The element is there and I've got to find it. We are sure!"[144] They also added a chemist to their project, Gustave Bémont, a laboratory chief from where Pierre worked. Bémont advised that they should use a process whereby the pitchblende should first be ground to powder and then dissolved in acid repeatedly to separate the up to thirty elements in the pitchblende.[145]

Pitchblende was made up of about thirty different elements and so to separate the mysterious substance required grinding it to a powder, then applying different chemical separation techniques such as acid, heating and gassing, each time taking the residue of that stage and moving to the next stage. The final residues resulted in two fractions – one of bismuth and one of mainly barium, which were both highly radioactive. This reflected their proximity in the periodic table. The bismuth fraction contained a new substance which they named polonium after Marie's native Poland, which they wrote up and published. Polonium is close to bismuth on the modern periodic table and shares the same row, indicating the same number of electron shells. That was July 1898 but by December 1898 they had released a second publication with the surprising conclusion that a second substance was present in the barium fraction, which they named radium.[146] Radium is close to barium on the modern periodic table and is in the same column, indicating the same number of outer electrons giving similar bonding properties.

However, the discoveries had to be proven. Polonium did not prove to be a good candidate for separation (polonium's shorter 'half life' – the period of radioactive decay – may have been the reason) and indeed was not the most powerful radioactive substance. Whereas polonium was found to be 400 times more radioactive than uranium, radium was found to be about 2.7 million times more radioactive. They had to now move to a project of greater scale – they would need to process tonnes of pitchblende to get enough of the tiny amounts of radium that they had detected. The Austrian government owned the mine in Bohemia (now the Czech Republic) that was supplying the pitchblende ore to Marie and Pierre and was keen to see if the waste pitchblende could yield more value. Researchers such as Marie and Pierre could deliver this type of research as to further possible commercial byproducts.[147] Thus, their supply of ore for the operation was secured.

Marie's 'storeroom' lab was too small for this larger processing operation and so an abandoned shed (previously used for medical school dissections) was located across the courtyard. It was not weatherproof or well ventilated. Some of the immense processing burden was then offset by Pierre arranging for the Central Chemicals Products Company to do some processing – this company made Pierre's scientific instruments.[148] In return, the company was to receive a share of radium sales (and indeed eventually profited handsomely from this revenue).[149] Although they were entering what many regard as the romantic scientist's dream period, Marie would later call this the 'miserable little shed'.

Amidst the vats, pots, scientific apparatus and an iron stove to minimise the bitter cold, Marie and Pierre would then labour for over three years to extract the radium, creating one of the great stories of endurance in science. Marie describes the experience:

> ***"Yet it was in this miserable old shed that we passed the best and happiest years of our life, devoting our entire days to our work. Often I had to prepare our lunch in the shed, so as not to interrupt some particularly important operation. Sometimes I had to spend a whole day mixing a boiling mass with a heavy iron rod nearly as large as myself. I would be broken with fatigue at day's end. Other days, on the contrary, the work would be a most minute and delicate fractional crystallisation, in the effort to concentrate the radium. I was then annoyed by the floating dust of iron and coal from which I could not protect my precious products. But I shall never be able to express the joy of the untroubled quietness of this atmosphere of research and the excitement of actual progress with the confident hope of still better results. The feeling of discouragement that sometimes came after some unsuccessful toil did not last long and gave way to renewed activity… One of our joys was to go into our workroom at night; we then perceived on all sides the feebly luminous silhouettes of the bottles or capsules containing our products. It was a really lovely sight and one always new to us. The glowing tubes looked like faint fairy lights."***[150]

The operation in the shed started in early 1899 and three years later they were still going. In 1902, a physicist, George Sagnac, joined the team in the shed and the French Academy of Sciences awarded them a grant of 20,000 francs (approximately US$4,000 at the time). They were able to pay the costs of gaining more pitchblende as their quest towards a sufficient quantity of radium came closer. Marie wrote: "At last the time came when the isolated substances showed all the characteristics of a pure chemical body." She needed to test the sample of radium and this task fell to Eugène Demarçay, who used spectroscopy to determine its atomic weight. Despite initial difficulties with its radioactivity interfering with his instruments, he advised her that the sample – only one tenth of a gram – had an atomic weight of 225.93. Marie wrote: "It had taken me almost four years to produce the kind of evidence which chemical science demands, that radium is truly a new element… The demonstration that cost so much effort was the basis of the new science of radioactivity." [151] During this period, Rutherford had decided to investigate Marie's discoveries, including radium and thorium, to determine the nature of the atom. He concluded that radioactivity was a process where one radioactive atom was transforming to another by the decay of the radioactive particles being emitted.[152] After discerning radioactivity to be alpha and beta particles, between 1900 and 1903, Rutherford worked out that alpha particles and beta particles carried a charge and were probably a stream of particles such as electrons – in fact, beta particles turned out to be electrons or positrons and alpha particles were helium-4 nuclei.[153] Rutherford, whose work eventually led him to pioneer the correct model of the atom with nucleus and electron, would later get the Nobel Prize for discovering the mechanism of radioactive decay.[154]

Marie and Pierre's perseverance in the discovery of radium heralded the birth of nuclear physics and atomic energy.[155]

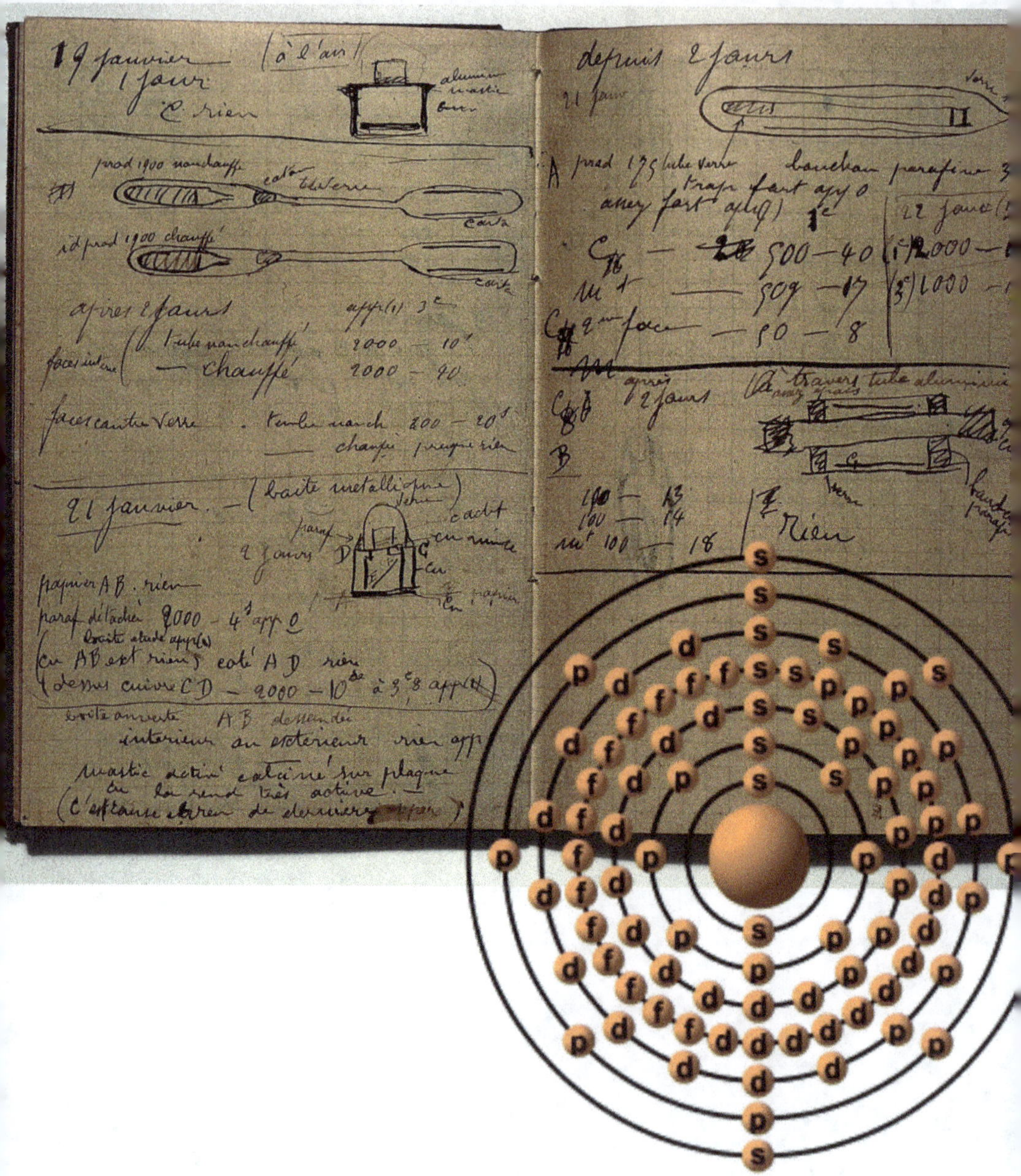
19 janvier
1 jour
21 janvier
depuis 2 jours
rien

SEVENTEEN

CURIE'S REALITY

Marie not only founded the science of radioactivity and the birth of atomic energy but pioneered something deeper – she had found the key to unlocking the mysteries of the composition of the universe.[156] It was the unlocking of the atom that was one of the greatest milestones in the history of science in which Marie played a major role.

Atoms represent an excellent example of metaphysics remaining important to physics. Indeed, we saw in Chapter 10 how the discovery of quarks

Chapter Seventeen Plate

Pages from Marie Curie's notebook, overlaid with diagram for electron configuration for the radium atom which was only known many years after Marie Curie's discovery of radium after the work of Rutherford, Bohr, Heisenberg and other pioneers of atomic theory and quantum mechanics.

Source: (Notepages) https://wellcomeimages.org/indexplus/image/L0021265.html
Source: Author: Peo at da.wikipedia (radium atom overlay) &
*https://creativecommons.org/licenses/by-sa/3.0/deed.en**
*https://commons.wikimedia.org/wiki/Commons:GNU_Free_Documentation_License**
** or such deed or licence instrument as may be applicable to the licence*

occurred after the theoretical ('metaphysical') work of Murray Gellman and George Sweig applied to the expected mathematical symmetries of subatomic particles predicted the existence of subatomic quarks before they were discovered experimentally.

Marie, being instrumental in unlocking the atom, placed cracks in the 2,000-year-old atomism paradigm. What was this paradigm? Although the founders of atomism, Democritus of Abdera (fl. 410 B.C.) and Leucippus of Miletus (fl. 440 B.C.), had no access to any experimental scientific methods to formulate their theory, it was metaphysics, or 'natural philosophy', that allowed them to arrive at such a formidable and enduring theory. To them, atoms were the smallest units of matter that made up the universe. Historian Peter Watson describes their theory: "...the world consisted of 'an infinity' of tiny atoms moving randomly in an 'infinite void'. These atoms, solid corpuscles too small to be seen, exist in all manner of shapes and it is their 'motions, collisions and transient configurations' that account for the variety of substances and the different phenomena that we experience. In other words, reality is a lifeless piece of machinery, in which everything that occurs is the outcome of inert, material atoms moving according to their nature. No mind and no divinity intrude into this world... There is no room for purpose or freedom."[157] It may be recalled that in discussing Leibniz's metaphysics in Chapter 8, another Greek term, 'monads', described a deeper layer of reality. (Recall they were units of reality but, unlike atoms, monads had no extension, so atoms would be rooted in monads.) In fact, Leibniz was a key critic of Greek atomism for a number of reasons:

1. Why should an atom occupy this position rather than any other? He relied on his principle of sufficient reason and appears also to rely on one of his other key logical principles outlined in Chapter 10 – *The Identity of Indiscernibles*: 'If a has all its properties in common with b, then a and b are one and the same. Hence, if a and b are not identical, then there must be some difference between them'. Unlike Newton, Leibniz did not believe in the concept of absolute space and so the only way to discern one atom from another would be in its position or velocity relative to other atoms.[158]

2. Leibniz held "that they involve discontinuities in nature (density changed

discontinuously at their boundaries); that their cohesion would require a perpetual miracle; and that no theory of their inelastic collisions is tenable."[159] Again, Leibniz, in his work as a physicist (who was well aware of kinetics as he founded kinetic physics), realised a significant problem if there was an inelastic collision and indeed he has been proven to be right because atoms smashing into atoms is not inelastic in the sense that great changes can occur (as seen in particle accelerators). Indeed, Leibniz, with Boscovich and Kant, is seen as influencing the founding of field theory famously developed by Faraday and other great scientists.[160]

3. Leibniz further held that whatever had extension was divisible (unlike monads, which were not divisible).[161] Again, Leibniz was later shown to be right with respect to atoms for they are divisible. In fairness to the Greeks, what we call atoms today may not be 'atoms' to the Greeks, who may have favoured subatomic particles (or even monads if the strangeness of quantum mechanics was known to them).

It is also interesting to see the clash between the Greek concept of no divine intervention with atoms compared to Leibniz's monads, which were pre-programmed by God as a type of intervention, yet there is important common ground. Like the Greeks, Leibniz saw the world as having no divine intervention; otherwise, the laws of the universe would be perpetual miracles. Thus, creation of the universe by God ended God's intervention. But like the Greeks, the urge to retain determinism – God maintaining control – seems present in Leibniz for, as we saw, Leibniz put forward the public position that all monads are pre-programmed perhaps to appease religious authorities (despite the doubts I expressed in Chapter 10 that Leibniz may have been 'tongue in cheek'). All this produces a further clash of ideas, both with the Greeks and Leibniz.

God remaining in control is inconsistent with Leibniz's concept of 'detached' creation. Furthermore, the above quoted Greeks' concept of a universe of atoms excludes purpose or freedom, raising the issue of whether humans can have free will in their materialist universe – especially one that is now known to be chaotic and uncertain after the discovery of Heisenberg's Uncertainty Principle. Yet, as commented on in Chapter 10, it is quite plausible for the theologian to resolve this clash of ideas by proposing that Heisenberg's Uncertainty Principle was bundled into the 'detached' or material universe by God on purpose. Why? The theologian would hold that God created this uncertainty mechanism as a feature of the universe to prevent a world of pre-programmed human puppets and thus create true free will by banishing determinism (contrary to the then religious doctrines).

Newton also clashed with Leibniz on this concept of the 'detached' universe during their dispute on calculus and other matters. (Recall their argument on differing viewpoints on space and time in Chapter 10.) Although Newton saw theology as not part of physics, he certainly saw it as part of metaphysics, and using Professor Illiffe's words: "…God was a real ubiquitous and infinite spirit that had an intimate relationship with the vacuous parts of the cosmos… [which Newton saw as the core of his metaphysics]."[162] When rewriting his *The Principia* with the help of Professor of Astronomy Roger Cotes, Newton noted that discoursing about God "does certainly belong to experimental philosophy."[163] Leibniz had attacked Newton in November 1715, accusing Newtonians of believing "space was God's body by which He perceived what was going on in the Cosmos."[164] And in reply to Leibniz, Cotes again assisted Newton in writing a preface (in colourful language) in which "he termed a 'miserable reptile' anyone who thought one could derive the system of the world by thought alone, or who believed God created a cosmos whose perfect working effectively denied a role for free will or supernatural intervention".[165] This clearly targeted Leibniz's 'detached' universe and pre-programmed monads that God had created like a giant clock no longer requiring intervention.

What was Newton's view in relation to atoms? In his early studies, he wrote essays in a collection of papers called *Philosophical Questions*.[166] Newton, after studying the Greek atomism and other writers of his day, made some incisive insights on these 'units of reality'. First, "Prime Matter could not have been a homogenous differentiated mass at the outset".[167] Second, the fact that matter could be condensed but separated by space indicated that it was made up of these tiny parts – atoms.[168] Third, they were tiny parts and had extension "but were so little that 'theire can [not] be a place too little for y^m to creepe into" and that these were atoms.[169] Fourth, a "finite portion of matter – indeed, all existing matter – could not by itself be composed of an infinite number of parts, for then it would itself be infinite, which was a contradiction. He added that whilst matter was not infinitely extended, the vacuum was."[170] Indeed, we saw in Chapter 10 how Zeno's paradox of an infinite number of dimensionless points can make up a finite distance remains a difficult issue. Grünbaum's resolution of it involved dispensing with distance in favour of a 'system of relations' bearing more resemblance to topology ran contrary to theorists like Kant, who observed that space is at least right-handed or left- handed. Yet atoms are discrete and, as Newton observed, solve the problem. The question remains whether the subatomic particles which appear more as 'clouds of probability' are discrete enough to escape Zeno's paradox and dispense with the metrics (ie. distance).

According to the author of the book *Atom*, Piers Bizony, "...mankind's discovery of atoms ranks as being the most important in the history of science... it is hard to imagine what little knowledge Marie and Pierre Curie started with when they embarked on their quest to explain radioactivity or what a great leap of intuition Rutherford needed to make to figure out what the inside of an atom looks like. What is even more remarkable is that in cracking the code of the atom, physicists needed to invent a new way of thinking, a mathematical construct far stranger than anything they could have dreamt up – ideas so powerful that they form today the most powerful scientific theory ever produced: quantum mechanics."[171]

What then were Marie's major achievements in unlocking the nature of reality?

1. She developed a new technique for measuring and extracting radioactive substances (indeed, she coined the term 'radioactive').

2. She developed the technique of discerning different elements based on their differing radioactivity – later scientists like Mosely would use the differing X-ray emissions from elements to distinguish between those elements.

3. Her work showed a paradigm shift in the understanding of the atom because once it was realised that radioactivity involved the emission of alpha, beta and gamma particles, scientists knew the atom could sometimes be unstable and that subatomic particles *existed*. This would

become vital in the journey to discover the fundamental 'building blocks' of matter – elementary particles, which would gradually become the *Standard Model of Particle Physics* (discussed in Chapter 21 Einstein's Reality.)

4. The heat from radium was initially thought to be radiant heat that could collect in any element, but Marie's great insight was that it was an atomic property. This fact, coupled with the fact that radioactive emissions were particles, indicated to her that atoms were not immutable and unchanging.

5. She had the insight that if atoms were emitting particles they were changing into something else – be they isotopes or new elements. We now know that radioactive decay of radioactive elements eventually sees those elements transformed into lead. This was astonishing to scientists who had long dismissed the alchemists' dream of the transmutation of metals (preferably into gold rather than lead!).

6. Her discoveries led to a whole new realm of nuclear medicine to treat cancer.

7. The discovery of radium even impacted other realms of science. For example, in calculating the age of the earth, a problem had been that the mass of the sun would take 8,000 years to burn away on some older accounts or only 20 million years using Lord Kelvin's gravitational contraction method, but this was only one tenth of the 200 million years needed for Darwinian evolution. The discovery of nuclear processes in the sun extended the earth's age by billions of years, adding to both geological and biological ages transforming these sciences.[172]

Marie had also achieved a huge milestone for women in winning two Nobel Prizes. Marie had not only unlocked some secrets of the universe and provided a gateway to others but had sparked the ultimate paradigm shift in science – a scientific revolution in atomic theory.

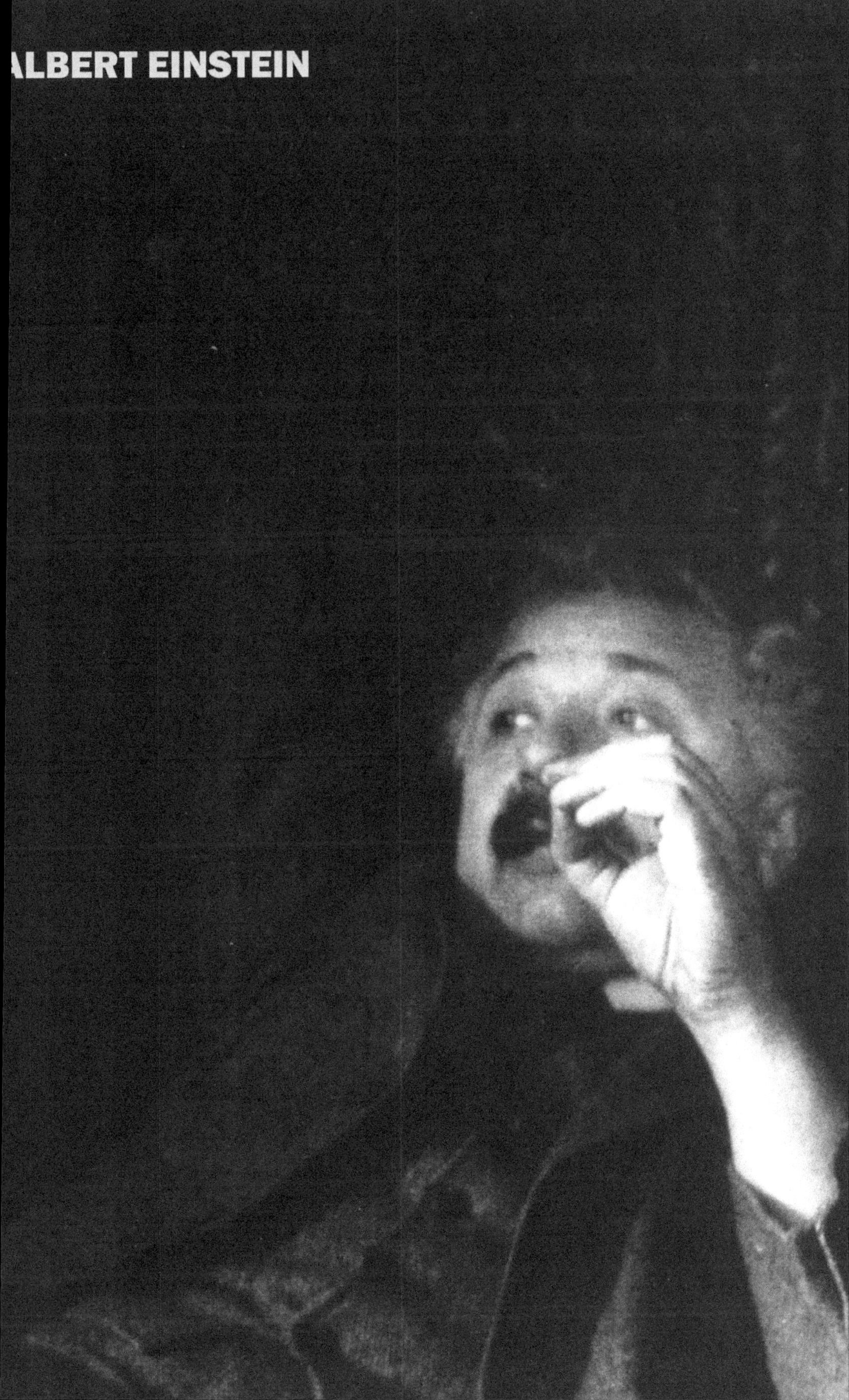
ALBERT EINSTEIN

EINSTEIN

Chapter Eighteen Plate

Photo of Niels Bohr (left) and Albert Einstein (right) in 1925 at the home of physicist Paul Ehrenfest (who took this photo)

EIGHTEEN

ALBERT EINSTEIN

Albert Einstein was born on 14 March 1879 and grew up in Munich.[173] He was educated under the German system of *Bildung*, meaning 'formation', under which physics, mathematics and philosophy were blended to allow 'the individual to mature into independence and responsibility'[174] (see next chapter, 'Einstein's World', for more on *Bildung*). His father and uncle were electrical engineers.[175] Perhaps the brain of a genius takes longer to develop as he did not speak until he was aged three.[176] His greatest pleasures were the violin and mathematics yet he hated school.[177] Einstein was Jewish and his family naturally felt Judaism was a heritage to be proud of even though he came to believe that many biblical customs were based on superstition.[178] After his family's business lost electrical contracts due to rising anti-semitism, they moved to Pavia near Milan, Italy in 1894 in an endeavour to gain more business, but Einstein stayed behind with another family to pursue his studies.[179] "The teachers... seemed to me to be like drill sergeants," Einstein would comment many years later.[180] He was not an ideal student despite his genius and ironically one teacher said to him, "Einstein, you'll never amount to anything!"[181] Although he had excellent grades in maths and physics, he dropped out of school at age sixteen and rejoined his family in Italy.[182] After an initial attempt to enter the Swiss Polytechnic which was not successful (even though no high school diploma was required), he returned to Switzerland to stay with yet another family – the Winteler family – to brush up his studies to try again.[183] Einstein developed an attachment for the daughter of the family, revealed in a love letter intercepted by her mother where Einstein wrote: "Beloved sweetheart… you mean more to my soul than the whole world did before…"[184] Einstein also appears to have engendered a good relationship with her mother, writing to her on one occasion: "Strenuous intellectual work, and looking at God's nature are the… angels that shall lead me through all of life's troubles… What a peculiar way this is… One creates a small little world for oneself, and as lamentably insignificant as it may be in comparison to the perpetually changing size of real existence, one feels miraculously great and important."[185] Einstein's broad education in *Bildung* seems to have produced a rich philosophical repertoire in his approach to both science and life, and it is arguably this overview or ability to view reality from a higher level – lateral thinking – that produced such spectacular results. On his second attempt, he was admitted to the Swiss Polytechnic to study mathematics and physics before working as a temporary maths teacher after he graduated (as the Polytechnic trained him to be).[186] Einstein was critical of his time at the Polytechnic, commenting that their teaching methods "strangled the holy curiosity of inquiry."[187] Indeed, one of his teachers, Professor Weber, was very pedantic and made him rewrite an entire research report again because it was not written on paper of the requisite size.[188] He then sought to annoy Professor Weber by calling him 'Herr Weber', behaviour that later worked against him in gaining employment.[189] Einstein made some new friends at the Polytechnic such as Marcel Grossman, who shared his lecture notes with him.[190] It was at this time that he met his future wife, Mileva Marić, who was also studying at the Polytechnic. During his studies, he found time to frequent the cafés and bars of picturesque Zurich to discuss the big questions in life with Mileva, Marcel and other friends.[191] His romance with Mileva blossomed and in the colourful correspondence between them, he wrote to her: "Without you, I lack self confidence, pleasure in work, pleasure in living… we shall be the happiest people on earth together, that's for sure."[192] Einstein was fascinated with magnetic or electric fields from an early age and studied Faraday, Maxwell and JJ Thomsen. His father had given him a compass at the age of four or five and becoming excited with its mystery, he trembled and went cold.[193] It was a tremendous period of scientific discovery with

Röntgen's discovery of X-rays, Marie Curie's discovery of radium and polonium, Rutherford's unmasking of the atom and other breakthroughs, which we saw in the chapters on Marie Curie.

However, Einstein's lack of tact, skipping of classes and defiance led to a refusal by Professor Weber to write him a letter of recommendation, weakening his ability to get a position and in 1901 his many attempts to do so failed.[194] This prompted him to return to his family in Italy when his father, seeing his despondency, wrote to a famous scientist friend but to no avail.[195] His romance continued by correspondence with him writing to her in these terms: "My dear dolly! This evening I sat two hours at the window and thought about how the law of interaction of molecular forces could be determined. I've got a very good idea. I'll tell you about it on Sunday… Ah, writing is stupid. Sunday I am going to kiss you in person. To a happy reunion! Greetings and hugs from your… Albert PS Love".[196] They later met at Lake Como and Mileva wrote to a friend regarding the encounter: "We rented a very small… [horse-drawn]… sledge, the kinds they are using there, which has just enough room for two people in love with each other, and the coachman stands on a little plank in the rear… and calls you 'signora' – could you think of anything more beautiful?… I held my sweetheart firmly in my arms under the coats."[197] By May 1901, she was pregnant, and after Mileva had a little girl he wrote to her in early 1902: "It has really turned out to be a little girl, as you wished! Is she healthy and does she cry properly? What kind of little eyes does she have? Is she hungry? I love her so much and don't even know her yet!"[198] Sadly, it appears, she was given up for adoption due to the stigma of illegitimacy.[199]

Fortunately, efforts by the father of his friend Grossman to secure a position for him succeeded and he became a 'technical expert – third class' at the Swiss patent office in the smaller Swiss city of Bern (despite his wish to obtain 'technical expert – second class').[200] In 1903, he married Mileva – he was twenty-four and she was twenty-eight – and they moved into an apartment in Bern overlooking the Alps.[201] In 1904, they had their first child – a boy named Hans Albert.[202] Mileva, despite her qualification from the Polytechnic, did not secure employment and became disenchanted with being at home. She knew that husband and wife teams could exist as Marie and Pierre Curie had done, but this did not seem to be happening with Albert and herself. Albert was too immersed in deciphering the puzzle between mass and energy. After Newton, mass was seen as an object having resistance to acceleration or something that housed kinetic energy as Leibniz had surmised. Electricity was totally different and linked somehow to fields. Magnetic fields could move mass so there was a connection, but what was it? These and many other things puzzled Einstein. As we saw in Part 1, Newton's approach was to view the world as a Clockwork Universe wound up by God but, with schools of thought such as positivism pervading society, God's role was gradually written out of scientific thought. Yet 'God' remained in the vocabulary of Einstein (and, as we shall see later in this book, firmly in Hawking's vocabulary despite his apparent 'agnosticism' displayed in the Preface to this book). On this issue, Einstein, whilst cynical of religion, held a reverence for the Cosmic God.

In the Foreword to Einstein's 1938 book *The Evolution of Physics* (co-authored with Leopold Infeld),[203] Walter Isaacson referred to this 'reverence' for the universe as his religious faith, as seen in Einstein's 1936 essay quoted by Isaacson: "The very fact that the totality of our senses experiences is such that, by means of thinking, it can be put in order, this fact is one that leaves us in awe. The eternal mystery of the world is its comprehensibility... The fact that it is comprehensible is a miracle." Einstein's use of the word 'God' in such phrases as "God does not play dice with the universe" or "God does not care about our mathematical difficulties He integrates empirically", and so on, seems to suggest that God was more than a metaphor to Einstein. Yet in his famous 'God letter', written after the Second World War, he dismissed religions as superstitious,[204] saying: "The word 'God' is, for me, nothing but the expression and product of human weaknesses". This seems to rule out belief in a personal God as opposed to an intelligent cosmic force.

Recent books on Einstein describe his faith as Spinoza's God[205] (more pantheism[206]). Einstein further said: "...the problem involved is too vast for our limited minds."[207] He certainly was not an atheist as shown by this further quote from Isaacson where Einstein was discussing our ability to analyse the universe with his friend Maurice Solovine (who had studied Hume and Mach), where Einstein said: "Well, *a priori,* one should expect a chaotic world which cannot be grasped by the mind in any way. There lies the weakness of positivists and professional atheists."[208] According to Isaacson, despite Solovine having an aversion to such language, Einstein continued: "I have no better expression than 'religious' for this confidence in the rational nature of reality and in its being accessible, to some degree, to human reason. When this feeling is missing, science degenerates into mindless empiricism."[209]

In a collection of papers made into the book *Science and Religion*, a paper entitled 'Physics' by Richard Olsen provides stunning insights into Einstein's beliefs as he explains: "After the professionalism of natural science during the nineteenth century... few reported turning to scientific study primarily out of religious motivations... One remarkable exception was Albert Einstein (1879–1955) for whom self-reported religious reasons played a major role in motivating his own scientific work and in his interpretations of the scientific work of others". He then quotes some of Einstein's ideas about God: "I cannot conceive of a God who rewards and punishes his creatures. Neither can I, nor would I, want to conceive of an individual that survives his physical death: let feeble souls, from fear or absurd egoism, cherish such thoughts."[210]

As Einstein would overturn Newton with relativity, so too he would overturn the main dogma of the world's religions! Perhaps he seeks to overturn the concept of the soul – religious or non-religious. Peter Watson, in his epic book of the history of ideas – *Ideas – A History from Fire to Freud* – concludes that the three most important ideas in history were the experiment, championed after the Scientific Revolution and responsible for incredible advances in society, the idea of Europe, drawing on its momentous history with its global influences and the soul (with or without faith), which he explains:

> ***"The transition from the world of the soul (including the afterlife) to the world of the experiment (here and now), which occurred first and most thoroughly in Europe, describes the fundamental difference between the ancient world and the modern world, and still represents the most important change in intellectual authority in history."***

He then argues that the soul "has evolved beyond God, beyond religion, in that even people without faith – perhaps *especially* people without faith – are concerned with inner life."[211]

Just as Einstein was surprised and could not accept the uncertainty element in quantum mechanics – displayed in the classic debates with Niels Bohr and Werner Heisenberg which we will see in Chapter 21 – so too he may have been surprised with a development in medical science that has gone largely unnoticed by the scientific community. Just as Hawking commented on the implications of the Big Bang that: "Many people do not like the idea that time has a beginning, probably because it smacks of divine intervention",[212] so too many may have viewed cardiologist Dr Pim van Lommel's scientific study similarly.

The landmark study by Dr Pim van Lommel MD was reported in his 2010 book *Consciousness Beyond Life – The Science of the Near-Death Experience.*[213] This followed a clinical study by Dr van Lommel published in the prestigious medical journal *The Lancet.*[214] This study involved 344 patients from ten different Dutch hospitals interviewed over a period of thirteen years regarding near-death experiences, including visions of rising up a tunnel, out-of-body experiences, life reviews, meetings with deceased persons, etc. As the title to the book proclaims, the study demonstrates 'consciousness beyond life' or, on one view, that the human soul exists. One can interpret this in many ways. One avenue arises from the concept of *emergence* in science – that some *whole* arises from the *parts* not discernible from those ***parts***. (In fact, author Nancy Ellen Abrahams, using the science of emergence, has written a book on emergence generating a 'God' – see *A God That Could Be Real – Spirituality, Science and the Future of Our Planet.*[215]) Dr van Lommel rules out the influence of drugs and it is interesting that a leading critic, Dr Karl Jansen MD PhD, who in 1997 published a paper 'The Ketamine Model of the Near-Death Experience: A central role for the N-methyl-D-aspartate receptor'[216] (proposing the drug ketamine as an example of drug-induced near- death experiences), shifted position in writing a book in 2000, *Ketamine: Dreams and Realities,* where he made the following statement:

> ***"The better option in our attempt to understand these states of being may be to consider both the scientific and spiritual issues, rather than ignoring one or the other. Niels Bohr observed that: 'The opposite of profound truth may be another profound truth. The divide between what is labelled as scientific understanding and a spiritual understanding has been inconstant and has tended to change over time. We can view science as that which can be measured and spirit as that which must be experienced directly."***[217]

Perhaps more striking is the evidence presented by two authors Kenneth Ring and Sharon Cooper in their book *Mindsight: Near-Death and Out-of-Body Experiences in the Blind.*[218] The authors identified thirty-one legally blind people, including fourteen who were blind from birth, who had experienced an NDE or out-of-body experience and were able to visually describe what transpired.

In concluding this discussion of Einstein's concept of soul and afterlife, it is interesting to see Stephen Hawking's requirements for good science: "A theory is a good theory if it satisfies two requirements. It must accurately describe a large class of observations on the basis of a model that contains only a few arbitrary elements, and it must make definite predictions about the results of future observations."[219] It seems a soul can be predicted.

In 1905, Einstein submitted five papers to the *Annalen Der Physik (Annals of Physics)*, a German physics journal, the first of which concerned an explanation for the photoelectric effect, which had puzzled scientists for a long time. When light strikes certain types of material – metals display the effect better than other materials – an electric current is emitted, ie. electrons are given off. Not only light but other types of radiant energy such as infrared, ultraviolet light, X-rays or gamma rays will also produce the effect.[220] Einstein extended the principle of quantising energy that Max Planck had introduced and called light quanta *photons*.[221] He submitted a second paper concerning the size of molecules for which he received his doctorate by the Swiss Polytechnic. His third paper related to another mystery – the reason for the agitation of pollen, called Brownian Motion. Einstein's breakthrough was to state that the motion was due to the heat, ie. molecules fluctuating and agitating or bombarding the pollen molecules.[222]

The fourth paper was entitled 'On the Electrodynamics of Moving Bodies'. In this paper he introduced his famous principle of Special Relativity, the idea that space and time are relative. After work by Galileo, Roemer and Fizeau, it was realised that light had a speed (~186,000 miles per second or 299,792,458 metres per second) and this gave Einstein a great insight.[223] Our whole reality is dependent on light and indeed in the 19th century Hertz had discovered that electromagnetic waves also travel at the speed of light.[224] Einstein's great insight was the fact that light and other transmissions of energy having a speed meant measurements could be distorted depending on where those measurements were observed. He even thought of himself as a photon (or even riding a bicycle on a light beam[225]) in his thought experiments, perceiving another photon.[226] His fifth paper built on this insight and with some elegant mathematics he was able to deduce $E= mc^2$ (which is more fully explained in Chapter 20)[227]. It followed that this relativity of *where* a measurement was observed meant the object's mass, length and time frame could vary with speed, producing weird effects at close to the speed of light. For instance, if a twin could travel at the speed of light around our galaxy and return, say, twenty years later (earth time), the travelling twin would be younger than the earthbound twin because the travelling twin's clock slows down at close to light speeds. Einstein also gives the example of a stick's length shrinking to no length as it reaches the speed of light. Mass could become infinite at the speed of light.

One aspect of the theory was that it explained why so much radiation was emitted from radium, ie. the lost mass of radium atoms were being emitted in the form of radiation energy governed by $E = mc^2$. Of course, we do not see these weird effects and still 'enjoy' Newton's laws because all our experience of moving objects involve non-light speeds. Einstein would later reminisce: "When I talked about experiments with clocks at different parts of a train, I still only possessed one clock!"[228] Newton's concept of absolute time and space had been swept away. Time and space depend on the position and velocity of the speeding object. German mathematician Minkowski had suggested a system of four dimensions of space time, t, x, y, and z, which was perfect for Einstein to adopt to put some mathematics to his new theory. Three years after Einstein's landmark 1905 paper on Special Relativity, Minkowski addressed the 80th Assembly of German Natural Scientists and Physicians and said:

> ***"Space by itself, and time by itself, are doomed to fade away into mere shadows, and only a kind of union of the two will preserve an independent reality."***[229]

Einstein's theory of special relativity was also consistent with the absence of the ether, demonstrated by the Michelson Morley experiment performed in 1887[230], disproving the hypothesis that the speed of light varied according to the medium of the ether.[231]

The theory has far-reaching effects and even affects theories for the origin of the universe. For example, Stephen Hawking considered a number of different theories for the origin of the universe, with the Big Bang theory being central to his analysis. In modelling the early universe, he said that a limiting factor to any theory is that galaxies that are now far apart but which share common features must have been closer together in the early universe because the 'information' containing such features cannot travel faster than the speed of light.[232] This is confirmatory of the Big Bang just as the discovery of microwave background radiation by Penzias and Wilson in 1965 (their discovery, for which they received the Nobel Prize) was that a background radiation occurred throughout the universe, indicating that this was the after-effect of an event such as the Big Bang. (Stephen Hawking's views, as expressed in his book *A Brief History of Time,* indicate that any origin of the universe theory must account for this microwave radiation.)[233] The *Oxford Dictionary of Physics* describes this relativistic problem as the universe updating itself through each new light wave of reality.[234] The same authors state that no experiment can distinguish between the two frames of reference (as Leibniz had, in effect, thought).[235]

Despite Einstein's 1905 papers revealing some of the greatest theories in the history of science, nothing happened when Einstein's theories were published. It is a sad fact that our modern world has developed silos – vast columns within societies of professionals within certain sectors who are experts within those sectors. Outsiders are normally not 'experts' but some are, and some make brilliant discoveries which these silos deliberately shun in a ruthless quality control culture. These theories include theories that break the existing scientific paradigm in the sense used by Thomas Kuhn in his landmark book *The Structure of Scientific Revolutions* to produce a paradigm shift to the next theory. Even insiders of the silo can be labelled 'outsiders' by silo authorities. An instance of this insider 'silo effect' can also be seen with cases such as Rupert Sheldrake's clash with the editor of *Nature* over his book *A New Science of Life*. Editor of *Nature*, Sir John Maddox, effectively labelled the book heresy when he said: "This infuriating tract… is the best candidate for burning there has been for many years"[236]. Grigori Perelman, who solved the Poincaré conjecture, is perhaps a closer situation to Einstein. Perelman, a mathematician, who had prior roles with universities – powerful silos of the scientific world – lodged his proof of the famous Poincaré conjecture in 2002 on *arXiv*, an online archive for new proofs of mathematical or scientific nature. Unfortunately, nothing happened at first for him and Perelman grew disillusioned. The years passed by and in about 2006, when the mathematics community accepted the proof, it was decided to award the Fields Medal and a Millenium Prize of $1 million to him.[237] However, Grigori Perelman rejected the medal and the prize after so many years. Presumably, the thrill of prizes and prestige thawed over leaving his disenchantment with the mathematics world. It appears on this occasion that the 'outsider' shunned the silo.

So Einstein, the outsider, was not accepted by the scientific silo of the day. However, Max Planck, who was a scientist of great authority within it, had read the Einstein papers with interest and had sent Max von Laue to Bern to see Einstein two years later in 1907. After von Laue made enquiries, he would have been surprised at the address he was given – the patent office in the post office building in Bern. He arrived in the reception and asked for Einstein but expecting a professorial figure ignored the young Einstein when he arrived in reception. Einstein returned upstairs, puzzled. When Einstein was called again to go to reception, it dawned on von Laue that this must be Einstein and they met. The young patent clerk was suddenly elevated from 'outsider status'.[238] They walked the streets of Bern but, despite a good meeting and subsequent discussions, Planck did not offer a position to Einstein, who battled on inside the patent office.[239] Yet this practical ground did him an enormous amount of good because it provided a 'sandbox' of working examples of physics which enriched the repertoire of examples from physics that could populate his mind experiments.

Einstein did apply for a position at the University of Bern and submitted three of his five papers to them to support the application, including the paper proposing $E= mc^2$, but they politely explained that the prerequisites were a dissertation, not mere papers (never mind that they were worthy of Nobel Prizes!)[240]

But 1907 proved more fruitful for ideas because it was in that year that Einstein saw workmen climbing a ladder on a nearby roof and recalled, "...then came to me the happiest thought of my life" because it triggered the thought that if they fell, no one could know where they were without looking at their surroundings.[241] This led to what is possibly the greatest theory in the history of science, inspired by his famous elevator thought experiment. However, it was not until 1915 that he finally submitted a paper on general relativity – the generalisation of his special theory of relativity but with more breakthroughs.[242] In Chapter 20 of his 1916 book *Relativity* he calls it a chest and later calls it an elevator.[243] It led to his equivalence principle – a system under acceleration is equivalent to a system under a gravitational field, giving him a clue that when light crosses such a system it must bend. However, light does not bend so there was only one alternative left – space must bend. Newton had quantified gravity but did not seek to describe what it was. Einstein was going further than Newton because it seemed that he had discovered that gravitational systems curved space.[244] We will expound Einstein's thinking and theory of general relativity in Chapter 20 and his theory of what space is in Chapter 21.

After Einstein had reapplied to the University of Bern with a more conventional dissertation, he was accepted and started lectures in the spring of 1908, with no salary. During his first year of lectures, a professor from the University of Zurich attended one of Einstein's lectures – not out of interest but as a prerequisite for his new position. Finally, in 1909, after seven years in the patent office and unpaid lectures at the University of Bern, Einstein was given a fully paid position as a lecturer at the University of Zurich.[245] Just three years later, he would move again after he was offered a position as Professor of Theoretical Physics with the Zurich Polytechnic (known as ETH). It was 1912.[246] He then spent a brief time at the German University in Prague (now the Czech Republic), where he also became a professor.[247] At the end of 1913, Max Planck persuaded him to join him as a professor at Berlin University, where he only had 'light duties', allowing his research to thrive.

In 1915, he submitted a paper on his great theory of general relativity – a paper that changed the way we perceive the nature of reality. Not only had he swept away Newton's concept of absolute space and time but something radically new was proposed. Space was curved. In fact, space not only was devoid of the ether, it was utterly non-existent and only filled with fields when active, such as gravitational fields. Stephen Hawking described the breakthrough as follows: "Einstein made the revolutionary suggestion that gravity is not a force like other forces, but is a consequence of the fact that space-time is not flat, as had been previously assumed: it is curved or 'warped' by the distribution of mass and energy in it. Bodies like the earth are not made to move on curved orbits by a force called gravity; instead they follow the nearest thing to a straight path in a curved space, which is called a geodesic."[248] Einstein wrote of this breakthrough: "Compared with this problem… [his 1905 Special Theory of Relativity] … is child's play… no one who has gone through the torments, false hopes, could know what it entailed".[249] His friend Arnold Sommerfield said during this time: "Einstein is stuck so deep into gravity that he is deaf to anything else."[250]

By July 1914, Einstein had developed an extra-marital relationship with his cousin Elsa Lowenthal, which took its toll on his marriage to Mileva. Despite Mileva moving to Berlin, Einstein told Mileva that he would give her a threshold of sociability if it was "completely necessary for social reasons". However, not surprisingly, their marriage was broken and Mileva and their two sons left Berlin in a scene portrayed in television dramas where Einstein bid his sad sons and wife farewell.[251] Clearly, the intense period of thought and his new bond with Elsa had placed his life in turbulence. Yet exaltation was his in 1915 on the submission of his paper for the general theory of relativity which he later described: "When a man after long years of searching chances on a thought which discloses something of the beauty of this mysterious universe, he should not be personally celebrated." Yet he also said: "[This]… is the greatest satisfaction of my life."[252]

However, the big test was to validate the theory that space was curved by an experiment. In his book *Relativity,* Einstein explained the background for such an experiment: "According to Newton's theory, a planet moves around the sun in an ellipse, which would permanently maintain its position with respect to the fixed stars… [disregarding their infinitesimal motion at distance]… Thus, if Newton's theory be strictly correct, we ought to obtain for the orbit of the planet an ellipse, which is fixed in reference to the fixed stars. This deduction, which can be tested with great accuracy, has been confirmed for all the planets save one… The sole exception is Mercury, the planet that lies nearest the sun. Since the time of Leverrier, it has been known that the ellipse corresponding to the orbit of Mercury after… [corrections]… is not stationary with respect to the fixed stars, but rotates exceedingly slowly in the plane of the orbit… [and]… the value obtained for this rotary movement of the orbital ellipse is 43 seconds of arc per century [being a deviation differing from that predicted by Newton's equations]. On the basis of the general theory of relativity, it is found that… [in the case of Mercury its rotation]… must amount to 43 seconds of arc per century."[253] This was called Perihelion Precession and in fact before General Relativity explained the effect, one theory was that a mystery planet, Vulcan, was dragging Mercury out of an otherwise Newtonian orbit.[254]

Einstein had struck up a friendship with a young astronomer, Erwin Freundlich, in 1913, and after an initial idea of inspecting astronomical archives failed to produce the data Einstein needed, he explored an idea that Freundlich had about an experiment. This was to construct a daytime experiment at the Mount Wilson Observatory in California. However, the director of that observatory advised that the daytime glare would be too strong to allow a proper comparison between Mercury and the 'fixed' background stars. Finally, Einstein and Freundlich agreed that Freundlich would travel to the Crimea to make an observation during an ellipse (where the background glare would cease during the eclipse). World War I had broken out between Russia and Germany just before Freundlich was travelling to the Crimea in July 1914. Unfortunately, Freundlich, being German and possessing powerful telescopes which could be of great use to spy on the Russian battle fleet at the Crimean port, had great difficulty explaining his innocent expedition. Freundlich and company were imprisoned by the Russians and missed the eclipse (Einstein later obtained their release through a prisoner exchange).[255]

After further ideas and efforts, it seemed Einstein was stuck. A Cambridge astronomer in England, (Sir) Arthur Eddington, became the subject of concern to his employers who feared he could be lost if he was conscripted to war. (The great scientist Henry Mosely had revolutionised the periodic table with his discovery of the atomic number but had died in battle at Gallipoli, Turkey.) When Eddington, a devout Quaker, expressed his religious objections to the war on a form, their concern was intensified because conscientious objectors were sent to prison. It seemed that his mission was to become aligned with the national interests of England by doing what the Germans could not do, ie. prove Einstein's theory right or wrong.[256] (This type of thinking was exemplified in the space race between America and the Soviet Union.) Eddington himself was not anti-German and believed that science transcended war (let us hope it does!).[257] Thus, the unusual idea of a famous astronomer in England co-operating with a famous German scientist (despite Germany and England being at war) emerged. Eddington was thus enlisted, with enthusiasm, on this new scientific mission. Another Quaker, Ruth Fry, commented: "One person who heads an expedition to heal the wounds and desolation of war is stronger than a battalion of men under arms".[258] This time, the mission was not to the Crimea but now to Africa and South America, with two possible locations chosen. Eddington would go to Principe in West Africa, and another party would go to Sobral in the Brazilian jungle. The eclipse was due on 29 May 1919, which was about the time (in mid-1919) that the Versailles peace treaty ending the First World War was signed.

Both parties set off on the good ship *Anselm,* which sailed initially to Morocco where the two parties separated. Eddington arrived in Principe Island, Gulf of Guinea, which had been discovered by the Portugese in the 15th century. Eddington was accompanied by another astronomer, E.T. Cottingham from the Greenwich observatory. After biding time for three weeks, they struggled up the steep slopes and gained a position in a clearing to set up for the observation. On the day of the eclipse, Eddington wrote in his journal: "[In] the morning there was a very heavy thunderstorm from about 10am to 11.30am – a remarkable occurrence at that time of year." At about 2pm – minutes before the eclipse was due – clouds still covered the sun (and the 'still-to-be-seen' Mercury). Yet at 2.13pm, almost miraculously, a gap appeared, sufficient to allow the observation. At 2.13pm, Cottingham, who held the metronome, yelled "Go!" to Eddington, who began taking photographs for five minutes. After it was over, Eddington would later recall: "We had to carry out our programme of photographs in faith." Eddington sent this telegram back to England: 'THROUGH CLOUDS STOP HOPEFUL STOP EDDINGTON'. In Brazil, it seemed that bad luck had given way to good luck. Although the heat had distorted their main telescope, Father A.L. Cortie had insisted on bringing a backup 4-inch telescope, which ultimately was the main telescope used. This produced the best photographs of both expeditions.[259]

It wasn't until November 1919, during which Einstein had become anxious, that the results were announced in dramatic fashion. In a joint session of the Royal Society and the Royal Astronomical Society in Burlington House, Piccadilly, London, the audience waited to see whether Newton or Einstein was right. The great philosopher A.N. Whitehead later recalled:

"The whole atmosphere of tense interest was exactly like that of the Greek drama...There was a dramatic quality in the very staging – the traditional ceremony – and in the background the picture of Newton to remind us that the greatest of scientific generalisations was now, after more than two centuries, to receive its first modification. Nor was the personal interest wanting: a great adventure in thought had at length come safe to shore."[260]

Eddington made the key announcement that the predicted deflection was 1.70" (~1.7 seconds of arc). The Astronomer Royal Sir Frank Dyson said: "After a careful study of the plates I am prepared to say that there can be no doubt that they confirm Einstein's prediction." J.J. Thomson, who had won the Nobel Prize for discovering the electron, announced to the crowd: "This is the most important result obtained in connection with the theory of gravitation since Newton's day. It is... the result of one of the highest achievements in human thought." This moment heralded enormous fame for Einstein after sensational newspaper stories and the throng of journalists teaming up to see him. Yet he was to do so much more, including co-founding quantum mechanics, becoming an activist for world peace and Zionism and even being offered the role of President of Israel, as we will see in the coming chapters.

NINETEEN

EINSTEIN'S WORLD

The Industrial Revolution (1760–1850) had begun after the Scientific Revolution was well and truly under way and began in Great Britain due to its particular economic circumstances in that period.[263] It had spread to Germany and the United States by the time of Einstein's birth and was driven by technological change underpinned by the Scientific Revolution. Newton's genius had changed the scientific culture, encouraging invention for this technological change.[264] Germany was also emerging as a new nation after Chancellor Otto von Bismarck won the Battle of Koniggrätz (1866) and proclaimed Germany as unified (incorporating Prussia and other states).[265] Bismarck was thus emperor of Europe's most powerful nation.[266] This had created an industrial and military colossus with scientific advances becoming 'engines of progress' in Germany. Thus, the education

Chapter Nineteen Plate

Albert Einstein and actor Charlie Chaplin at the opening of City Lights in Los Angeles 1931

Source: https://archive.org/stream/photo40chic#page/n455/mode/2up

*https://creativecommons.org/licenses/by-sa/3.0/deed.en**

**or such deed or licence instrument as may be applicable to the licence*

system needed to foster invention and science. Educators having this vision had introduced *Bildung* (literally 'formation'), where physics and mathematics is blended with philosophy.[267] This principle arose out of the romantic philosophical movement – remember, this movement had little to do with affairs of the heart or the sublime but rather with rejecting the materialism of the Enlightenment and Scientific Revolution. A key feature of the movement was that the individual should develop himself or herself.[268] Its links with literature and poetry were paramount. Author F. Shlegel said: "Concerning Bildung, we speak not of external culture, but the *development of independence*... [and this is why]... Every human being who is cultivated and who cultivates himself contains a novel within himself."[269] This is a great theme as we all admire the great stories and thinking of the individual as a unique reflection of the 'inner self'

This is the type of educational philosophy which empowered Einstein. The romantics had postulated a rift between sentiments and reason and Bildung resolved this challenge.[270] Art and literature were a great outlet to this schism, but philosophy blending with mathematics and science also met this challenge by making reasoned sense of the world. Just as the great philosopher Immanuel Kant had proclaimed the new Enlightenment thinking – 'Dare to Think' – so too he had introduced his principle challenging aesthetics called the Third Critique. This principle saw the sublime – the wonder at absolute grandeur or pure beauty, etc. – as creating tension with reason within the mental faculties. This tension gives rise to alternating feelings of pleasure and displeasure.[271] The influence of this type of thinking can be seen today in the synthesis of this sense of wonder blended with scientific reason, exemplified by the modern scientists, led by Einstein, whose wonder for the universe blended with reason was a hallmark of his thinking. This learning environment was ideal for Einstein's intellect except for one thing – his rebel nature. We have already heard about Einstein's niggling of his teacher, Weber, by calling him 'Herr' rather than 'Professor'. At one stage, Weber said to Einstein: "You're a very clever boy, Einstein. But you have one great fault: you won't let yourself be told anything."[272] Even more remarkable was what his maths professor Minkowski called him – 'lazy dog' – yet later Minkowski must have grown to greatly respect him because he adapted his 4-dimensional mathematics to Einstein's special relativity.[273] Perhaps that is why Einstein was initially reluctant to accept it (but later did).

We also saw in Chapter 15 how Marie Curie, a contemporary of Einstein, grew up in a society whose philosophy was dominated by positivism conceived by the French mathematician and philosopher Auguste Comte.[274] This was a materialistic philosophy that approached scientism but, as observed in the last chapter, Einstein had said to his friend Maurice Solovine, "Well, *a priori*, one should expect a chaotic world which cannot be grasped by the mind in any way. There lies the weakness of positivists and professional atheists."[275] This echoed the romantic philosophy and was similar to Newton's concept of laws of the universe ordered by God. In fact, Einstein and Solovine used to meet another friend, Conrad Habicht, in the cafés of Bern to discuss philosophy and science with such topics as Hume and Poincaré.[276] However, unlike Newton, Einstein had faith in more of a mathematical or cosmic 'God', and as Walter Isaacson comments, with due 'reverence' for the universe akin to a religious faith.[277] These insights portray Einstein as a master of metaphysics and physics without being a materialist or empiricist. He believed in the underlying order of nature and not just a chaotic accidental system.

The German education system increasingly encouraged by these 'engines of progress' allowed theoretical physics professors to become a new feature of education in Germany. Thus, a laboratory for experimentation was not necessary, although it was desirable.[278] Again, Einstein's later thought experiments would have been fostered by such a teaching environment. Some of the greats of science such as J.J. Thomson (who summarised Einstein's achievement so well at the presentation by Eddington[279]) and James Maxwell (whose equations gave science a unifying principle for magnetic and electric fields) had come to see energy as a fundamental principle in the universe. An equation for work done by the energy emitted by a tonne of coal had emerged but short of $E= mc^2$. The principle inherent in these concepts was conservation of energy.[280] Some thought of it in strict scientific terms, whilst some thought only God could create and destroy energy (Maxwell and Thomson were deeply religious[281]). Others viewed the world in terms of the romantic philosophy of an interconnected (but inexplicable) world.[282]

By about the turn of the century (1900), a number of key events had occurred – the first telephone, the electric light, the first motor car, the first flight, and so on. Germany invaded Belgium in 1914 and the First World

War had begun. Einstein had already marked himself as a political activist as early as the period of his professorship in Berlin. During the First World War, he participated in anti-war demonstrations, publicly discouraging people to refuse conscription, which would not have made him popular in Germany.[283]

In the last chapter, we also saw how, despite the war, Eddington had confirmed Einstein's theory of general relativity in the months after the demise of the German Empire at the war's end, which led to the Treaty of Versailles and the formation of the Weimar Republic. A new democratic Germany had emerged, burdened with the terms of an unfair treaty which sowed the seeds for a new war in the years to come.[284] Meanwhile, Eddington's confirmation had made Einstein an international celebrity and in 1921 Einstein travelled on his own tour to the United States. He travelled through Washington D.C., Chicago, Princeton, Harvard and many other places, including New York, where he was driven in the streets in a cavalcade to cheering crowds.[285] Amidst sold-out lectures he met President Warren G. Harding (who Marie Curie had also met in 1921 before successfully raising research funds). Einstein attended Charlie Chaplin's film *City Lights,* again to cheering crowds, prompting Chaplin to remark: "They cheer me because they understand me, and they cheer you because no one understands you." Chaplin and Einstein, who both shared left-wing views, became friends.[286]

In 1921, Einstein won the Nobel Prize in Physics for his paper on the photoelectric effect. Apparently, it was awarded belatedly in 1922 after the bizarre decision of the Nobel committee to withhold any award in 1921. Apparently, they regarded Einstein as a self-promoter but realised their decision was untenable in 1922. However, it seems that the Nobel committee has not corrected an even greater omission – they never awarded the Nobel Prize to Einstein for his Theory of General Relativity – one of the greatest theories in the history of science.[287] During this period, he received other prestigious awards such as the 1925 Copley Medal from the Royal Society, the Royal Astronomical Society Gold Medal for 1926 and the Max Planck Medal for 1929.[288] After his death, he appeared on the cover of *Time* magazine (14 June 1999) as topping the list of the top one hundred most influential people of the 20th century.[289]

After Germany was unable to meet the war reparation costs required under the Treaty of Versailles, the mark collapsed due to rampant inflation, and serious unemployment resulted. Hitler assumed power in this turbulent period from President Hindenburg, who was persuaded to accept Hitler as Chancellor, unaware of the dangerous 'phoenix' that lay hidden after Hitler's dormant years.[290] Hitler's ideology enshrined in his book *Mein Kampf* (*My Struggle*, published in July 1925)[291] was to become Germany's ruin. Einstein had fled Germany in 1932 because he had already felt the rise of anti-semitism in the country. With Hitler's rise, Einstein had foreseen the crisis point that was approaching, exemplified in a particular turning point in this period. On 9 November 1938, after Hitler learnt of the death of a German diplomat in Paris, assassinated over the issue of the wrongful deportation of Polish Jews to Poland, marauding gangs 'broke glass' ('*Kristall*') of Jewish shops throughout Germany in a night now known as *Kristallnacht*. Joseph Goebbels, who had become Hitler's Minister of Propaganda in 1933,[292] had announced that, although Hitler did not authorise the rampages, spontaneous outbursts should not be interfered with, sparking widespread violence and injustice against the German Jews.[293]

About a month after Hitler's appointment as Chancellor in 1933, Einstein's apartment was broken into by Nazi militia and he learnt that he was on Hitler's assassination list of prominent Jewish figures.[294] At the same time, a Berlin newspaper ran the headline: 'Good news from Einstein – he's not coming back'. Having visited Princeton in 1921, Einstein had commented: "[Princeton is]… young and fresh – a pipe as yet unsmoked." American educator Abraham Flexner had been establishing the Institute for Advanced Studies (est. 1933[295]) and was looking for prominent researchers.[296] Einstein drove a hard bargain in securing his new position as he also secured a position for his Jewish assistant Walther Mayer, who was also fleeing Nazi Germany. The Institute for Advanced Studies was a unique institution based on the educational principles of Plato for the higher forms of learning without laboratories, where the greatest scientists would gather to do research.[297] Thinking was paramount. Einstein would become one of the many Prizewinners of the Institute, including Bohr, Dirac, Pauli, Gell-Mann, Yang and many others.[298] In 1936, the Institute had bought 200 acres of land about half a mile south of the Princeton Campus with new offices planned but, by the time it was being built, Flexner was soon caught in a power struggle aimed at unseating him. By 1939, he had been succeeded by Frank Aydelotte.[299] Earlier on, when Einstein had received an invitation for he and Elsa to attend a dinner at the White House, Flexner had sent a letter in response: "Professor Einstein has come to Princeton for the purpose of carrying on his scientific work in seclusion… it is absolutely impossible to make any exception which would inevitably bring him into public notice."[300] Einstein had lost patience with Flexner with what many regarded as mismanagement and bad judgement. For example, ill-conceived appointments and the addition of economics to the Institute had prompted Einstein and others to meet with Flexner, urging him to consult them before further appointments could be made. Flexner did not agree.[301] Be that as it may, Einstein would stay at the Institute for the rest of his life.

In 1939, Hungarian physicist Leo Szilard advised Einstein that Hitler might create the atomic bomb. Einstein had replied that he had never thought of that, but it galvanised him into writing a letter to the US President Franklin D. Roosevelt. Szilard and another Hungarian physicist, Edward Teller (often seen as the father of the hydrogen bomb), drafted a letter and agreed that the letter should go to the president under Einstein's name. The letter, dated August 2 1939, went to the President, urging funding commence on nuclear energy and included the sentence: "Shall we put an end to the human race or shall mankind renounce war?" This was a remarkable letter for a pacificist such as Einstein to write and reflects the adage: 'If you wish peace you must prepare for war'. The next month after the letter, the president established an advisory committee on uranium, appointing Szilard and Teller but not Einstein to the committee. This was the first step in the United States developing the atomic bomb.[302]

He also supported the plight of the Jewish race in the shadows of the Holocaust and at one stage rewrote his Theory of Special Relativity, submitting it to auction in Kansas City. It was sold for an astonishing $6.5 million, which he donated to the war fund.[303] He also supported Zionism to the extent that due to public support in Israel, he was offered the presidency of Israel in 1952.[304] To the relief of the then prime minister of Israel, David Ben-Gurion, Einstein declined[305] and commented: "Equations are more important to me; because politics is for the present, but an equation is for eternity."[306]

In 1948, Einstein was admitted to hospital with acute abdominal pain which was diagnosed to arise from an aortic aneurysm in the major artery to the stomach. Einstein rejected an offer by medical experts for surgery, commenting, "I have done my fair share; it is time to go, I will do it elegantly." In April 1955, he died from that condition. He had worked on whilst ill and died with a draft speech on unified field theory he had proposed to give on Israeli Independence Day.[307]

Chapter Twenty Plate

Albert Einstein during a lecture in Vienna in 1921 and a photo taken through the ESA/NASA Hubble space telescope showing an example of gravitational lensing predicted by Einstein's theory of general relativity and later proven by Eddington's expedition. This photo shows light from a distant star, 5000 light years away, is bent by the curved space caused by the intervening white dwarf's gravity field through which the light has to travel (like light through a lense but the image of the star has shifted very slightly due to the bent path of its light).

TWENTY

EINSTEIN'S SCIENCE

Readers who wish to avoid mathematics can skip this chapter. (Readers who persist will be able to understand how Einstein derived $E = mc^2$.)

In Chapter 18, we saw that Einstein's great insight was to see that light and other transmissions of energy having a speed meant measurements of things like mass, speed and time could be distorted according to his Principle of Special Relativity. It was called *special relativity* because it involved non-accelerating objects. However, to see why mass, time and speed could be distorted, it is necessary to explain some of the mathematics involved.

Einstein originally set about using Newton's equations coupled with a specialised framework called Lorentz transformations to derive an equation that could determine the precise effect of distortions that would occur at 'light speeds'. Newton had used Galilean transformations, which assumed absolute space and instantaneous event speeds. (Galilean transformations involve transforming an object's position from x to a new value $x' = x - vt$, y to y', z to z', and t to t'.)[308] Einstein used a thought experiment involving an embankment and a railway carriage many times and varied the facts in his writing to show different concepts of this theory of special relativity.

The *Oxford Dictionary of Physics*[309] provides a key form of Einstein's thought experiment in this regard:

> *"[Einstein] invites the reader to imagine a very long train along an embankment with a constant velocity v in a given direction… Observers on the train use it as a rigid reference body, regarding all events with reference to the train. Einstein posed a simple question: Are two events which are simultaneous relative to the railway embankment [such as two lightning strikes at points A and B a distance apart]… also simultaneous relative to the observer on the train?" If an observer on the train is halfway between A and B (say, M') when lightning strikes at A and B, he will not see the strikes as simultaneous, whereas the observer on the embankment – also halfway between A and B – (say, M) does see the strikes as simultaneous. The observer on the train will see the strike at B first because the train is travelling towards B, but the explanation is puzzling.The* Oxford Dictionary *explains:"At first sight there may seem to be a problem here. If the observer at M' is 'hastening towards the beam of light from B', is this not equivalent to saying that the beam of light is travelling towards M' at a combined speed of* v + c *where c is the speed of light* in vacuo*? The resolution of this problem is the basis of special relativity. According to Einstein, the moving observer at M' must measure the speed of light* in vacuo *to be c… [since there is no way of distinguishing the embankment frame of reference from the train's frame of reference]… It is therefore the concept of time measurement that requires revision; that is, the time required for a particular event to occur with respect to the train cannot have the same duration as the same event when judged from the embankment."*[310]

This thought experiment leads to astonishing conclusions about the nature of reality. Time dilation and length contraction during high-speed motion are explained by Einstein in a question and answer format (see later in this chapter).

Thus, if an object moved from x' to x then as the position of the object x is defined as x = vt at any time then x'= x – vt. However, recall in Chapter 9 where we compare a tap running water at a constant speed showed a simple straight line graph to a real-life graph with a more complicated curve or function. So too here x' = x – vt is too simple. Einstein had pondered this simple Newtonian equation also but what did he do next? Physicist Leonard Susskind and author Art Friedman explain Einstein's mathematical steps: "Multiply the right side by a function of the velocity" x'

= (x – vt) f(v)… the function f(v) could be any function, but Einstein had one more trick up his sleeve: another symmetry… [of an object moving right or left]… That symmetry implies f(v) must not depend on whether v is positive or negative. There is a simple way to write any function that is the same for positive or negative v. The trick is to write it as a function of the square of the velocity v^2 (note …this is a slight variation of Einstein's approach). Thus instead of… [the last equation]… Einstein wrote x'= (x –vt) $f(v^2)$."[311] A similar derivation produces an equation for time t'= (t – vx) $f(v^2)$ where $f(v^2)$ is also a function of velocity linked to time governing the moving object.[312] Now the Lorentz transformations are defined in the *Oxford Dictionary of Physics* as follows:

$x' = \beta (x - vt)$

$y'=y$

$z'=z$

$$t' = \beta \left(t - \frac{vx}{c^2}\right)$$

Where v is the relative velocity of separation of 0 and O', c is the speed of light and $\beta = 1/\sqrt{(1 - v^2/c^2)}$.[313]

Now it can be seen that β is really $f(v^2)$, but these results have jumped a step. The quantity v that we were using is Newton's velocity without frame of reference, but all velocity should be a proportion of the speed of light because that is the maximum speed. Thus substitute v^2/c^2 for v in the equations. Now, when Newton's second law of motion[314] F= $d(mv)/t$ is subjected to some calculus but using v = v^2/c^2 an interesting result emerges:

$E = mc^2$

Using these equations, measurements of time, distance and even mass become distorted because when you apply this new framework for measuring quantities, these distortions produce curious effects. Einstein, in the book he co-authored with L. Infeld, described these effects in terms of a dialogue between an old physicist O and a modern physicist M (CS means 'co ordinate system(s)') explaining the difference between classical and Lorentz transformations:

M. Two observers… [each with their own clock]… in two different CS will assign not only different numbers to the position but also different numbers to the time at which this event happens.

O. This means that time is no longer an invariant. In the classical transformation it is always the same time in all CS. In the Lorentz transformation it changes and somehow behaves like the co ordinate in the old transformation. I wonder how it is with distance. According to classical mechanics a rigid rod preserves its length in motion or at rest. Is this true now?

M. It is not. In fact, it follows from the Lorentz transformation that a moving stick contracts in the direction of the motion and the contraction increases if the speed increases. The faster a stick moves, the shorter it appears…

O. This means that the rhythm of a moving clock and the length of a moving stick depend on the speed. But how?

M. The changes become more distinct as the speed increases. It follows from the Lorentz transformation that a stick would shrink to nothing if its speed were to reach that of light. Similarly the rhythm of a moving clock is slowed down, compared to the clocks it passes along the rod, and would come to a stop if the clock were to move with the speed of light, that is, if the clock is a 'good' one.

O. This seems to contradict all our experience. We know that a car does not become shorter when in motion and we also know that the driver can always compare his 'good' watch with those he passes on the way, finding that they agree fairly well, contrary to your statement.

M. This is certainly true. But these mechanical velocities are all very small compared with that of light, and it is, therefore, ridiculous to apply relativity to these phenomena."[315]

German mathematician Minkowski's four dimensions of space time, t, x, y, and z, were then adopted by Einstein for his special theory of relativity. In his famous paper, *Raum und Zeit* (*Space and Time*), he introduced this

4-dimensional geometry based on the Lorentz transformations used in the above derivation of $E = mc^2$.[316]

Einstein (and Infeld) then describe the quest for a unified theory of everything:

> *"The problem of formulating physical laws for every CS was solved by the general relativity theory; the previous theory, applying only to inertial systems, is called the special relativity theory. The two theories cannot contradict each other... But just as the inertial CS was previously the only one for which physical laws were formulated, so now it will form a special limiting case, as all CS moving arbitrarily, relative to each other, are permissible... our final aim is always a better understanding of reality."*[317]

Remember that the key point in the inertial systems is the study of frames that are at rest or travelling with constant velocity and are not accelerating. This magic insight only appeared when he considered the strange connection between accelerating frames and gravity, leading to his Theory of General Relativity.

The Equivalence Principle

Recall in Chapter 3 how Newton saw *weight* as measuring the gravitational pull of a body, whilst the *mass* measured the amount of matter. That is, *mass* and *weight* must therefore be precisely proportional to each other. Recall also physicist Ernest Mach's quote: "With regard to the concept of 'mass', it is to be observed that the formulation of Newton, which defines mass to be the quantity of matter of a body as measured by the product of its volume and density, is unfortunate. As we can only define density as the mass of a unit of volume, the circle is manifest."[318] We also saw how in the process of Newton's formulation of his Laws of Motion that he called his new concept of resistance *inertia* and concluded that forces must be proportional to the changes they cause to inertia of a body. This circularity was highlighted by Einstein (with Infeld) when he said that "the fundamental concept of mechanics, that of mass" disclosed a clue unnoticed for 300 years.[319] This clue was one of a number as to solving the mysteries of the universe.[320]

It is useful to follow Einstein's reasoning to derive his Theory of General Relativity as he explained in *The Evolution of Physics,* co-authored with

Infeld. Einstein saw that we can measure mass by applying the same force to two different bodies and measuring their different velocities or weighing them on a scale. He then observes that weighing them on a scale involves gravity, whereas *comparison* of velocities using equations does not. Einstein then concluded that mass determined using *comparison* was *inertial mass* and mass determined using *scales* was ***gravitational mass*** (despite being identical measurements).[321] Presumably, Einstein meant the 'weighing' results in a calculation to obtain *mass* and not *weight*. He then notes that, after Galileo's experiment showing bodies of all masses fall with equal acceleration, the force of gravity does not vary with the mass of the body because the body 'answers' gravity by falling at constant acceleration. He concludes that "the acceleration of a falling body increases in proportion to its gravitational mass and decreases in proportion to its inertial mass. Since all falling bodies have the same constant acceleration, the two masses must be equal." He also observes that "In contrast to electric and magnetic fields, the gravitational field exhibits a most remarkable property… Bodies which are moving under the sole influence of a gravitational field receive an acceleration, which does not in the least depend either on the material or the physical state of the body."[322]

All this led to Einstein's 'happiest thought in his life' (the image conjured by seeing men working on a roof who might fall and feel weightless), namely, his famous elevator thought experiment.[323] Einstein's chapter in his book *Relativity* entitled 'The Equality of Inertial Mass as an Argument for the General Postulate of Relativity' (Chapter 20) presents his famous thought experiment:[324]

> *"We imagine a large portion of space, so far removed from stars and other appreciable masses… to choose a Galileian reference – body for this part of space (world)… As reference-body let us imagine a spacious chest resembling a room with an observer inside who is equipped with apparatus. Gravitation naturally does not exist for this observer… a hook… [is fixed to the roof]… with rope attached, and now a being… begins pulling at this with a constant force… [and] …then begins to move 'upwards' with a uniformly accelerated motion… [the man]… is then standing in the chest in exactly the same way as anyone stands in the room of a house on earth. If he releases a body which he previously had in his hand, the acceleration of the chest will no longer be transmitted to this body, and for this reason the body will approach the floor of the chest with an accelerated relative*

motion... the man in the chest will thus come to the conclusion that he and the chest are in a gravitational field which is constant with regard to time."[325]

Einstein then observes that the man has quite naturally concluded he is in a gravitational field but then considers what we learn from knowing it is not in a gravitational field, yet its contents have behaved as though it was. He concludes:

"We have thus good grounds for extending the principle of relativity to include bodies of reference with respect to each other, and as a result we have gained a powerful argument for a generalised postulate of relativity... If... [the law of the equality of inertial and gravitational mass]... did not exist, the man in the accelerated chest would not be able to interpret the behaviour of the bodies around him... [to equate such behaviours]."[326]

The crucial point is that there is equivalence between gravity and an accelerated CS (ie. an accelerated frame of reference). He noted that gravity operated regardless of the mass of the object, and acceleration occurs regardless of mass, unlike what occurs with magnetic or electric fields. Also, unlike the force exerted by magnetic or electric fields, a mechanical force obeys **Force = Mass x Acceleration**, ie. a constant force causes an acceleration just like gravity does. Put another way, gravity exerts a constant force like a mechanical force. The incongruence (without the 'normalising' of inertial mass) is that a force of x newtons applied to a heavy body and light body will produce less acceleration for the heavy body, yet gravity will produce the same acceleration for both bodies. (The elevator and its contents were accelerated by the rope pulling as one body and the dropping of the object from the man's hand merely involved removal of the force, and so there was no incongruence there.) Einstein quantitatively solved this incongruence by altering Newton's F = inertial mass x acceleration and 'normalised' the equations by introducing new quantities, gravitational mass and intensity of the gravitational field in the following sequence:

Force = Gravitational mass x Intensity of the gravitational field

And combining the relations:

$$\text{Acceleration} = \frac{\textit{gravitational mass}}{\textit{inertial mass}} \times \text{Intensity of gravitational field}$$

He sums up this result:

> *"...the ratio of the gravitational to the inertial mass must likewise be the same for all bodies. By a suitable choice of units we can thus make this ratio equal to unity. We then have the following law. The gravitational mass of a body is equal to its inertial mass."*[327]

Thus, Einstein produced his principle of equivalence of gravitational mass to inertial mass. He then discusses the connection between them by referring again to the man in the chest: "A gravitational field exists for the man in the chest, despite the fact that there was no such field for the coordinate system first chosen. Now we might easily suppose that the existence of a gravitational field is always an *apparent* one." This passage appears in a difficult passage where Einstein is arguing against a misconception concerning existence of gravitational fields, but it appears to refer to a new concept he is introducing to extend the meaning of gravity to an apparent gravity or artificial gravity in more modern usage.

In that same chapter, he later refers to his railway carriage thought experiment and constructs a situation like the chest/elevator experiment where he imagines the carriage suddenly brakes and the passenger feels a jerk forward and has the passenger give an explanation: "My body of reference (the carriage) remains permanently at rest. With reference to it, however, there exists (during the period of application of the brakes) a gravitational field which is directed forwards and which is variable with respect to time."[328] Clearly, just like the artificial gravity in a rotating simulator for astronauts to simulate a multiple of g-forces, it seems Einstein uses the term 'gravitational force' in a similar way despite its different origin (not an attractive force between two bodies as in Newton's Law of Gravitation). He thus concludes in a later chapter referring to K (a Galilean system of coordinates away from other bodies such that the law of inertia – Newton's first law of motion[329] – holds using the mechanics of Galileo and Newton – an unaccelerated domain)[330] and K' , another domain or system of coordinates not so anchored, such as the chest or elevator) and says: "If we now refer to [domains in which gravitational fields are absent] to a reference-body K' possessing any kind of motion, then relative to K' there exists a gravitational field which is variable with respect to space and time… According to the general theory of relativity, the general law of the gravitational field must be satisfied for all gravitational fields obtainable in this way. Even though by no means all gravitational fields can be produced in this way, yet we may entertain the hope that the general law of gravitation will be derivable from all gravitational fields of a special kind. This hope has been realised in the most beautiful manner."[331] The meaning of this statement demonstrates key principles inherent in his general theory of relativity in its entirety, including (but not exhaustively) as expounded in his book *Relativity,* from which these quotes are taken.

Amidst these difficult concepts, we must now try and discern Einstein's reasoning in deriving his great theory. It is important to note that like Newton, Einstein did not discern the full nature and meaning of gravity, but he went further than Newton in discerning its behaviour and relation to reality and to determine new mathematical relations for the effects of gravity. Did the incongruence between a heavy body and light body producing the same acceleration under gravity but differing acceleration under force provide the spark for his general theory of relativity? It would seem not because in his original work it was the behaviour of light in the chest/elevator that created his great insight.

In *The Evolution of Physics,* Einstein revisits the chest experiment as a 'dilemma' between an outside observer and inside observer who must discern whether the elevator is in an inertial frame such as the earth's gravitational frame or in distant space under an accelerated force which mimics earth's gravitation: "There is, possibly, a way out of the ambiguity of two such different descriptions… Imagine that a light ray enters the elevator horizontally through a side window and reaches the opposite wall after a very short time. Again let us see how the path of the light would be predicted by the two observers." Einstein then discusses two possible viewpoints of the outsider and the insider. He surmises that an insider would predict a straight beam of light crosses the elevator, whilst the outsider would predict a bent beam of light due to the movement. He surmises that the insider would say, "A beam of light is weightless and, therefore, it will not be affected by the gravitational field," to which Einstein would reply, "This cannot be right! A beam of light carries energy and energy has mass. But every inertial mass is attracted by the gravitational field as inertial and gravitational masses are equivalent. A beam of light will bend in a gravitational field exactly as a body would if thrown horizontally with a velocity equal to that of light."[332]

Does this mean that photons, ie. light, have a tiny mass accounting for light bending in curved space? Physicists assert light has no mass but light does behave like a particle, and presumably this includes behaving *as if* it has mass. If so, the measurement of such behaviour derives from $E = mc^2$. Einstein clearly relies on $E = mc^2$ or rearranging $M = E/c^2$ which would make the mass in a light beam very tiny but clearly enough to be bent either in a gravitational field on earth or a 'gravitational field' arising from an accelerated frame in deep space. However, is the answer merely that if light travels through a gravitational field such as the sun's field then space is curved due to that field and therefore the light travels along that curved path? This is not what Einstein implies as he refers to a body being thrown. Thus, to combine that principle with curved space causing light to bend, one then has to state that a photon with a tiny mass must also travel 'in its lane' along the curved path *and* veer across that curvature 'to the next lane' due to the sun's gravity. This means both attraction and space curvature cause two curved trajectories to combine – if photons have mass. Roger Penrose comments whether light has mass whilst discussing photons (as bosons): "...the photon appears to be singled out... by being a *massless* particle. Indeed, the mass of the photon, if non-zero, would certainly have to be less than 10^{-20} of an electron mass..."[333] Presumably, this is well below experimental limits and so the issue remains open.

Einstein thus had available a number of insights or clues, as he called them, as to how to frame a general theory of relativity:

- The absence of the ether proven by Michelson and Morley.
- Mach's Principle that the inertial mass of objects depends on the rest of mass of objects in the universe.[334]
- The phenomenon of 'Frame Dragging', discovered by two Austrian physicists in 1918, Josef Lense and Hans Thirring, which seemed to confirm Mach's Principle. The phenomenon involved spinning objects apparently dragging time and space around the objects.[335]
- A rotating disc which has the Lorentz transformation applied to it

distorts under the Lorentz contraction effect to become a dish,[336] which effect appears to have been at work with the Frame Dragging phenomenon.

- The fact that light was proven to have a maximum speed and travelled in straight lines allowed thought experiments with this constancy of the speed of light cutting across all types of motion.
- The need to explain Newton's laws of gravitation – Force = m_1m_2G/d^2 displaying great *action at a distance* governed by their centres of mass – eg. two spheres would have distance d measured from their centres (ie. centres of mass).
- The fact that the application of Newton's laws to the orbit of Mercury produced results that varied from what was expected, indicating an anomaly.
- The Special Theory of Relativity which carried the insight of distortions of time, mass and length in a flat Cartesian plane caused by the constancy of the speed of light.
- $E= mc^2$ resulting from the application of Lorentz's mathematics to the Special Theory of Relativity.
- The Equivalence Principle – The gravitational mass of a body is equal to its inertial mass, which meant the incongruence of a heavy body and light body producing the same acceleration under gravity but differing acceleration under force (and adaptation of Newton's Laws to this principle as explained earlier).
- Not only this incongruence between weight and mass but the fact that the gravitational field applied to an object in that field caused an acceleration of 9.8 M/s^2 (=g)[337] for both a heavy body and light body in contrast to electric or magnetic fields.
- A further incongruence was that Newton's Law of Gravitation involved the force of gravity to act instantaneously between objects, yet nothing can travel faster than the speed of light – even force. Thus, nothing can travel faster between two points in that field.
- That light will bend in both the gravitational field of a large body like the earth and in the elevator in space.

After the 'happiest thought in his life', what 'dots were joined' by Einstein in his puzzle of reality? Clearly, many of the above factors were at play in his mind, but what is striking is the behaviour of light across the elevator. First, recall the second last point above that any force such as gravity cannot travel faster than the speed of light. Secondly, if light bends (and it must according to Special Relativity), then just as something must 'give' in the train carriage and embankment thought experiment (ie. time dilation and length contraction, etc.) so too in the elevator experiment something must 'give' if light bends, and this must be space distorting to account for the bent light beam. And if space distorts in the elevator then due to the Equivalence Principle, so too will space distort around any gravitational field caused by presence of mass. How does space distort? It must preserve integrity sufficient to carry light and objects and so space distorting to curved space seems logical. The instances to test the theory were thus observing the orbit of mercury against the background of stars (during an eclipse as performed by Eddington) or to measure distortion of spectral lines from stars with the same spectral lines from the same materials on earth. Light passing through distorted (ie. curved) space is tested for a discrepancy between the shortest possible (ie, straight) path and a curved path. Further instances of testing general relativity include gravitational lensing (where the gravity of intermediate celestial bodies may bend starlight over vast distances – the effect was observed in 1979 when the Double Quasar was observed in the constellation Ursa Major[338]) and gravitational waves expected from giant accelerating celestial bodies.[339]

The synthesis of this and all his work was his conclusion that space was curved and amidst this insight his theory of general relativity emerged. In short, Einstein's own description of his principle of general relativity reads:

> ***"All bodies of reference K, K', etc. are equivalent for the description of natural phenomena (formulation of the general laws of nature), whatever may be their state of motion."***[340]

Thus, his special theory of relativity with its mathematics of Lorentz transformations applying in a 2D framework for all non-accelerating moving bodies (in a Euclidean system[341]) is then subsumed into a larger general theory, *prima facie*, in three dimensions to account for curved space (ie. a non-Euclidean system[342]). However, recall that time was distorted as well and so Minkowski's concept of 4D space became the most appropriate system to apply to adopt general relativity. In addition to Minkowski's system, a new coordinate mathematics was required and in fact the great mathematician Gauss had developed his concept of a

Gaussian system of coordinates. As space was curved, he had to develop new mathematics and the Gaussian system of parallel ***u*** curves as one axis with ***v*** curves as the other axis allowed points in curved space to be manipulated in equations.[343] However, as the Minkowski space is but a special instance of the vast possible instances of moving bodies,[344] a deeper system of mathematics is required for general relativity which Einstein found with the Riemann geometry which proved ideal for such a task.[345] Central to Riemann's mathematics is the Riemann Curvature Tensor.[346] With Special Relativity, we can imagine that a vector which merely describes magnitude and direction of the force linearly (ie. in two dimensions on a Cartesian plane) is sufficient. With four dimensions of x, y, z and t (length, height, width and time), we could perhaps handle this by merely adding a third dimension to the Cartesian plane and then laboriously having successive snapshots of this 3D system for time (or utilising what are known as phase space systems). However, these four dimensions must be represented in curved space, which requires a new system.

Thus, with General Relativity, different systems (manifolds) such as those encoding 2D vectors become 4D vectors with more encoded information. With relativity, such a system or manifold was adopted, which uses tensors instead of vectors and expands the length, height, width and time coordinates to *towards/away, left/right, up/down* and *past/future*. Vectors are simple tensors but tensors in relativity encode more information, such as differing values and directions. These tensors, adapted for general relativity, became known as Einstein Tensors.[347] Clearly, the mathematics and geometry of general relativity becomes far more complicated. In an example of 'mathematical convergent evolution',[348] the great mathematician David Hilbert had been developing a parallel line of mathematics to derive some equations for the path of least resistance of a travelling object through space and time with the same objective as Einstein's General Relativity. This is now known as the Einstein-Hilbert Action.[349]

Einstein had created a new reality with his theory of general relativity. Yet his contribution to science had not ended because even Einstein's new reality was being challenged by the new science of quantum mechanics which he helped establish. However, these realities were to clash, as we will see in the next chapter.

TWENTY-ONE

EINSTEIN'S REALITY

Einstein had transformed our ideas of space and time. However, the change in *reality* is so great that it is important to try and grasp what it means because the implications are difficult to comprehend. Mathematician Ian Stewart wrote (see page 294):

Chapter Twenty-One Plate

Upper: Artist's impression of the curved nature of space warped by gravity fields of the two planets shown as envisaged by Einstein pursuant to his general theory of relativity showing a new reality to that envisaged by Newton.

Source: Author: Henze, NASA

http://www.nasa.gov/centers/goddard/universe/gwave.html

http://www.nasa.gov/centers/goddard/images/content/146978main_gwave_lg5.jpg

Lower: Rubber sheet model of Einstein's curved space time showing how 2 objects such as planets warp space time

Author: Tokamac

"Instead of a hypothetical force that causes a particle's path to curve, Einstein substituted a space time that is already curved, and whose curvature affects the path of a moving particle. No action at a distance is needed: space time is curved because that's what stars do to it, and orbiting bodies respond to nearby curvature. What we and Newton refer to as gravity, and think of as a force, is actually the curvature of space time."[350]

Physicist Carlo Rovelli, in his book *Reality is Not What it Seems,* describes the paradigm shift to reality in vivid terms:

"And it's here that Einstein's extraordinary stroke of genius occurs, one of the greatest flights in the history of human thinking: what if the gravitational field turned out to be Newton's mysterious space? What if Newton's space was nothing more than a gravitational field? This extremely simple, beautiful, brilliant idea is the theory of general relativity. The world is not made up of space + particles + electromagnetic field + gravitational field. The world is made up of particles + fields, and nothing else; there is no need to add space as an extra ingredient."[351]

Einstein himself saw space differently in two ways:

(i) With respect to the Special Theory of Relativity, he said "...'now' loses for the spatially extended world its objective meaning. It is because space and time must be regarded as a four-dimensional continuum that is objectively unresolvable."[352]

(ii) With respect to the General Theory of Relativity under *The concept of space in the general theory of relativity,* he said: "This theory arose primarily from the endeavour to understand the equality of inertial and gravitational mass. We start out from an inertial system S_1 whose space is, from the physical point of view, empty"[353] [ie. no matter, no fields]. After summarising space under special relativity and the Riemann mathematics (see last chapter), he concludes with this vital reasoning: "...space (space-time) has an existence independent of matter or field. In order to be able to describe what fills up space and is dependent on co ordinates, space-time... must be thought of as existing, for otherwise the description of '*that which fills up space' would have no meaning*' ... On the basis of the general theory of relativity, on the other hand, space as opposed to 'what fills space', which is dependent on co ordinates, has no separate existence... if we imagine a gravitational field... to be removed there does not remain... [space[354]]... but absolutely nothing, and also no topological space... Space-time does not claim existence on its own, but only as a structural quality of the field."

Thus, Descartes was not so far from the truth when he believed he must exclude the existence of empty space. The notion indeed appears absurd, as long as physical reality is seen exclusively in ponderable bodies. It requires the field as the representative of reality, in combination with the general principle of relativity, to show the true kernel of Descartes' idea; there exists no space 'empty of field'."[355] [my italics].

It would seem that Einstein's education under *Bildung* with its blending of philosophy with science is well demonstrated here and indeed our discussion of Newton's space vs Leibniz's space versus Descartes's space in Part I is well justified in this context also.

However, this concept of space seems hard to swallow because we still hear the word 'space' frequently in scientific theories, so the term has not disappeared. We hear of the quest for the mysterious dark energy which acts as a type of filler to balance out cosmological equations governing the expanding universe. It would seem that, compared to such a quest, a search for the medium that separates the stars and the planets, namely space, is more immediate! Or is it merely a reclassification that what we used to call space is a spatial field called gravity or is a new term required? (At least 'grace' as a combination of space and gravity provides the following pun!)

However, there is perhaps one 'saving grace' for space. In the above direct quotes from Einstein's book *Relativity*, highlighted in italics, the reader may have noticed a little footnote ([1]) after the phrase *'that which fills up space' would have no meaning*[1], which was omitted in the above quote but which is now reproduced:

> ***"If we consider that which fills space (e.g. the field) to be removed, there still remains the metric space [governing objects in the Special Theory of Relativity], which would also determine the inertial behaviour of a test body introduced into it."***[356]

In this regard, we have already referred to the insightful physics adage more than once in this book: 'The map is not the territory',[357] meaning that 'maps' of theories are abstractions and are not the fabric of reality. Einstein may be referring to metric space as that fabric of reality here. Consistently with this insight, Rovelli has not adopted any new term for *space* and indeed in his other book – *Seven Brief Lessons on Physics* – his fifth lesson is on 'Grains of Space'. In that chapter, in discussing a theory called loop quantum gravity, he describes the 'structure of reality' as: "General relativity has taught us that space is not an inert box… [as Newton had thought]… but something dynamic: a kind of immense, mobile snail-shell in which we are contained – one which can be compressed and twisted. Quantum mechanics, on the other hand, has taught us that every field of this kind is 'made of quanta' and has a fine, granular structure. It follows that physical space is also 'made of quanta'."[358]

This brings us to quantum mechanics, in which Einstein played a key role. Although Einstein had revolutionised physics, the theories of relativity had still preserved *classical physics.* Thus, even though light bent past the sun resulted from a cause and effect framework, it could be precisely measured and read – a certain result is given. Whilst Einstein's great theories were taking shape, a major anomaly seemed to appear in classical physics. In physics folklore, the story goes that in 1900, two physicists, Lord Rayleigh and James Jeans, were experimenting with black bodies (surfaces that absorb all radiation and then re-emit radiation) and attempting to analyse the energy spectrum to find the relationship between frequency of vibrations and temperature. However, the classical laws of thermodynamics predicted an energy distribution that made no sense when compared to experiment.[359] The inexplicable results were called 'ultraviolet catastrophe'. Although the story continues with Max Planck solving this impasse by introducing the quantisation of light, an article in *Physics World* (December 2000) states: "The story is a myth, closer to fairy tale than historical truth. Quantum theory did not owe its origin to any failure of classical physics, but instead to Planck's profound insight in thermodynamics." Instead, the article claims that originally the experiment of Wilhelm Wien showed the spectra of radiation from a black body did not agree with experimental data and that "faced with this grave anomaly, Planck looked for a solution, during the course of this he was forced to introduce the notion of energy quanta." They continue to explain that he was reluctantly inspired by Boltzmann's probabilistic notion of entropy (later to be enshrined in Boltzmann's famous equation S=klogW, where S is entropy (degree of order/disorder),

W is molecular disorder and k is Boltzmann's constant. This dichotomy was interesting because Planck initially felt the solution would be based on a classical and not probabilistic solution. Drawing from a third source of physics history, "[Planck] took the Rayleigh and Wien laws and in 1900 constructed a single equation from them which agreed exactly with the experimental results."[360] The result was that Planck presented his landmark paper in 1900 to the German Physical Society, stating that he regarded the energy "as made up of a completely determinate number of finite equal parts, and for this purpose I use the constant of nature h= 6.55×10^{-27} (erg sec),"[361] which became known as Planck's constant in the equation $E_{1-} E_2 = hv$, where v is frequency and the energy of oscillating atoms drop from energy state E_1 to E_2.[362] Apparently, he told his son that "he had made a discovery which would prove comparable in importance to those of Newton."[363]

In 1905, apart from originating Special Relativity and solving Brownian motion, Einstein submitted a doctoral dissertation to the University of Zurich calculating the size of molecules.[364] In 1906, he also used quantum relations to deduce that matter existed in a 3-dimensional lattice.[365] Another puzzling effect was the photo electric effect. When light hit certain metals, electrons would be given off which could not be explained under classical physics. Einstein's theory was that light exhibited both wave-like and particle-like behaviour, leading to the famous wave – particle duality of light – now a foundation of physics for which Einstein received the Nobel Prize. He had used Planck's quantisation of light to devise his theory. Professor of Mathematical Physics J.C. Polkinghorne explains: "Below a certain critical frequency no electrons were ejected, however intense the radiation. Einstein could explain this in quantum terms. Light was behaving as a swarm of particles, a broadside of torpedoes. Only those electrons which were struck by one of these projectiles would be affected. Whether such an electron was ejected… depended on how much energy this projectile had. According to Planck's equation… the quantity of energy depended directly on the frequency… From these sorts of consideration it became clear that light is made up of lots of particles. It was natural to call them photons."[366] We saw in the last chapter Einstein's intriguing statement to the effect that a beam of light carries energy and energy is equivalent to mass making photons behave as if they have mass. This raised the issue of whether photons have mass or not. Moreover, the energy

E of a photon fits nicely into the equation E = hv, where h is Planck's constant and v is frequency (particles of light photons were observed in experiments by British physicist Arthur Compton[367]). If 'mass' is viewed as a mirage of waves behaving like a particle then perhaps the issue might be resolved by viewing the nature of mass as wavelike.

Around a decade later, Einstein would add two more discoveries affecting the quantum world. First, in 1915, he discovered the Einstein-de Haas effect with Dutch physicist W.J. de Haas when a landmark experiment showed that the spin of electrons aligned in the same direction as the magnetic field surrounding them. Thus, the magnetic material would rotate showing aligned electron spins in a ferromagnet can be converted to mechanical angular momentum.[368] His other quantum discovery occurred in 1917 when Einstein discovered the basis for the LASER, namely, that pumping extra light into atoms stimulated emissions of pure powerful light – thus the acronym, Light Amplification through the Stimulated Emission of Radiation.[369]

By introducing the wave particle duality, Maxwell's Equations – a foundation of classical physics – were shaken to their foundations because smooth wave equations might not obey discrete particle behaviour now discovered in light. (Maxwell's Equations described the behaviour of electromagnetic waves, and after Fresnel's 1817 theory of light showed the wavelike nature of light, it seemed inevitable that light would share many properties and also be governed by Maxwell's Equations.)[370] Now, we saw in the Marie Curie chapters that Rutherford had been discerning the correct model for the atom and after years of experimentation had proposed the 'solar system' model of the atom of a nucleus surrounded by orbiting electrons. However, there was a serious objection to the model because Maxwell's theory predicted that electrons would send off waves of all frequencies (unlike true atoms) and would spiral into the nucleus.[371]

Professor Polkinghorne continues to describe the next step in this intriguing saga in vivid terms:

> ***"The first important step was made by an amateur, a Swiss schoolmaster called Balmer. A lot of people hanker after making a big scientific discovery in their spare time. Any reasonably well-known professional scientist will receive***

from time to time letters written by well-meaning people who indicate, usually in rather guarded terms, that they have in their possession the solution to the riddle of the universe and they just need a little help in polishing it up or propagating it. Dealing with them is a rather sad business. They do not realise that to make a discovery of some magnitude requires not only great ability and a bit of luck but also considerable time and effort in mastering a demanding discipline… It is good therefore to be able to record at least one such effort that scored a stunning success. Balmer had been thinking about the spectral lines from hydrogen… In 1885 Balmer discovered a striking numerological relationship between the frequencies of the most prominent lines. In a slightly rewritten form due to Rydberg the frequencies v are given by the formula:

$$v_n = cR\left[\frac{1}{2^2} - \frac{1}{n^2}\right]$$

Where n takes integral values 3,4… c is the velocity of light and R is a constant called the Rydberg. For a long time no one knew what to make of this; it appeared just a curiosity, but obviously an intriguing one."[372]

J.J. Thomson had demonstrated electrons were deployed in the atom and for a long time the atom was thought of as a type of plum pudding with electrons scattered about like raisins.[373] Rutherford's experiments showed that the plum pudding model was not viable and that there was a very compact nucleus and the 'solar system' model of the atom prevailed.

Niels Bohr was to become an important colleague of Einstein in the development of quantum mechanics. The authors of one physics text describe Bohr's landmark breakthrough in quantum history:

"[To save the crippling objections to Rutherford's model of the atom]… Bohr, in 1913, introduced a very drastic theory. Having accepted the Rutherford model, his first step was to renounce Maxwell. The second was to forbid all electron orbits but a select group. And the third was to allow jumps from one orbit to another, provided the energy differences were taken care of by single photons. The theory was more a theory of electrons moving around the nucleus than of the atom as a whole… it is only possible for an electron to move in an orbit for which its orbital angular momentum L is quantized by the relationship $L = nh/2\pi$ where n = 1,2,3… and contrary to Maxwell's theory, an electron moving in an allowed orbit does not radiate electromagnetic energy."[374]

Einstein's relativity was used by physicist Arnold Sommerfield to improve Bohr's theory in 1915, but further improvements were to come.[375]

And this is where Balmer's equation was transformed by Bohr. After calculating the mathematics of the orbits L, he was able to calculate the energy of the orbits and this turned out to be:

$$E_n = -\left[\frac{me^4}{2\hbar^2} x \frac{1}{n^2}\right]$$

Where m is the mass and e is the electric charge of the electron. Thus, when Bohr looked at the equation for electrons jumping from orbit to orbit, Balmer's equation was suddenly explicable so that loss of energy or using Planck's equation $E_{1-} E_2$= hv (note $\hbar = h/2\pi$ where $\hbar$ is the Dirac constant):

$$Loss\ of\ Energy\ v_n = -\frac{me^4}{2\hbar^2}\left[\frac{1}{2^2} - \frac{1}{n^2}\right] = h.v_n$$

(By combining Planck's equation $E_{1-} E_2$= hv or e = hv with E = mc^2, we get hv = E = mc^2, a precise frequency v is found directly proportional to its mass, ie v = *m x c^2/h* because *c^2/h* is a universal constant – note the use of v for frequency is a little confusing as v is normally velocity and ω is normally frequency).[376]

In 1925, Wolfgang Pauli was acutely aware of anomalies with Bohr's theory, one of which was why all electrons were not in the innermost shell of the atom.[377] He then proposed his famous Pauli Exclusion Principle. Extending Bohr's three quantum numbers to four quantum numbers, he labelled different orbits as n,k,j and m for those quantum numbers defining energy levels and stated his exclusion principle:

> ***"There never exists two or more equivalent electrons in an atom which agree in all quantum numbers n, j, k and m. If there exists in the atom an electron for which these quantum numbers have definite values, this state is 'occupied'."***[378]

However, Pauli did not understand why his exclusion principle worked but three years later British physicist Paul Dirac applied Einstein's Special Relativity to Pauli's equations to formulate a system of what we use today as quantum mechanics.[379] Professor Polkinghorne describes it as follows:

> ***"By consistently applying quantum mechanics to Maxwell's theory of the electromagnetic field he constructed the first known specimen of a quantum field theory.This provided an example of a well-understood formalism which if interrogated in a particle-like way gave particle behaviour and if interrogated in a wave-like way gave wave behaviour."***[380]

In 1925, Einstein had given praise to a French prince, Louis de Broglie, for his 1923 paper which postulated that electrons did not behave like tiny particles circling the nucleus but like waves, as if following an invisible torus.[381] In fact, both Einstein's and de Broglie's concepts of wave particle duality can be seen to draw upon an 1860s notion called the Hamilton-Jacobi theory of treating motion based on waves rather than particles.[382] Erwin Schrödinger, who would become Professor of Physics at Berlin University in 1927, heard of Einstein's support for the principle and within a few months produced his famous Schrödinger equation. (In fact, a modification to the Hamilton-Jacobi equation leads to the Schrödinger equation.[383]) Author of *Quantum Physics,* Professor Michael Raymer[384], refers to Schrödinger's equation as beautiful and inspiring to physicists and reproduces the equation, which you don't need to understand but simply admire:

$$i\hbar \frac{\partial \psi}{\partial t} = -\frac{\hbar^2}{2m} \nabla^2 \psi + V\psi \text{ "}$$

Professor Polkinghorne provides a simplified form for this equation as follows:

$$i\hbar \frac{\partial}{\partial t} |\psi> = H |\psi>$$

He explains: "The left hand side is iħ times the rate of change with time of a state vector. The right hand side equates with the effect of an operator, the Hamiltonian, which is simply the observable corresponding to the energy of the system under consideration…The proper symbol for a state vector is | >. It is like a half bracket and in fact is called a ket vector. (The so called bra vector is also used and written < |)… This bra-ket terminology was invented by Dirac."[385] Professor Polkinghorne saw Dirac as "the greatest British theoretical physicist since Maxwell and one of the great figures of twentieth-century culture".[386] Dirac was Lucasian Professor of Mathematics for thirty-seven years (a position held by Newton and Hawking)[387] and won the Nobel Prize with Schrödinger for developing the Schrödinger equation to take account of relativity which we record here (in the spirit of Professor Raymer's approach) for its 'beauty' also (!) from the *Oxford Dictionary of Physics* (without an explanation):

$$i\alpha\nabla\psi + \frac{mc}{\hbar}\beta\psi = \frac{i}{c}\frac{\partial\psi}{\partial t}$$

Where m is the mass of a free particle, c the speed of light, t the time and ħ is the Dirac Constant. The wave function is ψ and α and β are square matrices and $i = \sqrt{-1}$."[388]

Max Born, a friend of Einstein, and a professor at German University, Göttingen, observed that the de Broglie waves in Schrödinger's equation were probability waves.[389,390] This meant that the electron could not be precisely located here or there but that its probability of being here or there was the best measurement that could be made. From this major insight, one of Max Born's pupils, Werner Heisenberg, emerged with a new theory – Heisenberg's famous Uncertainty Principle. Heisenberg recognised that to measure the location and momentum of a particle required bombarding it with a stream of electrons, but this created a two-fold problem.

First, high energy short wavelength photons would affect the particle, distorting the reading. Second, using lower energy photons would give less accurate readings because of their longer wavelength. In theoretical and experimental terms, this conundrum meant that, by reason of the act of observation and measurement, precise knowledge of both the position and momentum of the particle is unknowable. If a precise position of one aspect of the particle is known then the other aspect cannot be precisely known. Heisenberg then decided that the only way to deal with this conundrum was to bring in a system which utilised Boolean algebra where, in simple terms, if state A is known then this would mean not state B. He would do this merely by observing inputs and outputs to the particle under observation and the symmetries involved. In 1927, he was thus able to devise a matrix system involving symmetry which involved vectors and states in systems of linear equations housed in matrices which are ideal for this type of system, including the bras and kets devised by Dirac. The Heisenberg Uncertainty Principle is enshrined in the equation:

$$\Delta x \Delta p \geq \frac{h}{4\pi}(1) \qquad (1)$$

"Where Δx is the uncertainty of the x co ordinate of the particle, Δp is the uncertainty in the x-component of the particle's momentum and h is the Planck constant."[392]

Polkinghorne called these years 1925–1926 'marvellous years', when "many theoretical physicists moved in to apply the new mechanics to a host of significant problems. Speaking of that exploitive period Dirac once said to me: 'It was a time when second-rate men did first-rate work."[393] (It seems it was also a time when first-rate human beings unwittingly made second-rate comments.)

During his discovery, Heisenberg commented: "It was almost three o'clock in the morning… I was far too excited to sleep." Although Born accepted Heisenberg's new system, Einstein did not because what Heisenberg had discovered disclosed a major rift between classical physics and quantum physics.[394]

Bohr had postulated that electrons revolved in fixed discrete orbits – planet-like yet unlike the planets obeying Newtonian physics. Bohr's theory involved *discrete* (quantum) hops or jumps between orbits which de Broglie and Born had determined to be probabilistic.

Einstein had said of Bohr's theory: "The weakness of the theory lies… in the fact… that it leaves the duration and direction of the elementary processes to chance." In another cryptic comment, Einstein said: "The real joke presented to us here by the eternal riddle setter has not yet been understood."[395] Bohr's foundation for quantum physics (giving way to further developments) and Heisenberg's uncertainty principle eventually prompted Einstein's famous statement "God does not play dice with the universe." This dichotomy went to the core of Einstein's reality. He was a great admirer of Spinoza[396] – again, his philosophical foundation under the German education system of *Bildung* blending science with philosophy shone through. Einstein had said "[Spinoza]… was convinced of the causal dependence of all phenomena, at a time when the success accompanying the effort to achieve... [that knowledge]... was still quite modest." Einstein is said to believe in Spinoza's (cosmic) God. Author David Bodanis comments on Einstein's beliefs: "He didn't believe in the tenets of revealed religion... but that is far from meaning he was not religious. He thought that being an atheist was presumptuous, and he was awed at the intelligence manifested in natural laws. 'This feeling is the guiding principle of [a scientist's] life and work', he wrote, 'in so far as he succeeds in keeping himself from the shackles of selfish desire." So the very core of Einstein's intellectual and spiritual life depended on the premise that all underlying reality was clear, exact, understandable. He was not going to believe that the universe was fundamentally unknowable".[397] However, the depth of Einstein's intellect should not be measured by this apparent outright rejection of quantum uncertainty because in his 1917 paper 'On the Quantum Theory of Radiation' he postulated that gamma rays were governed by a statistical law.[398]

Recall Stephen Hawking's comment reproduced in Chapter 12 about Heisenberg's Uncertainty Principle – "The doctrine of scientific determinism was strongly resisted by many people, who felt that it infringed God's freedom to intervene in the world, but it remained the standard assumption of science until the early years of this century..."[399]

Both Einstein's and Hawking's comments point to a philosophical void which I have tried to fill in previous chapters by drawing together concepts of determinism, order, chaos and the inability of mathematics to solve key infinities that strike at the heart of the origin of matter and life. I will now summarise these concepts in the context of Einstein's thought.

In Chapters 5 and 13, we saw Einstein's discomfort with inserting the cosmological constant (Λ) into his famous field equation in his theory of General Relativity[400] because to him the 'universe always has order'.[401] However, he considered the insertion of lambda (Λ) 'the greatest blunder in my life'.[402] The conclusion to Chapter 13 – 'Is a Perfect World Possible?' – is that due to the mathematical and logical necessity of *universal chaos* (a term from Chapter 13 referring to the synthesis of chaos, turbulence, irrationals, infinities, quantum shredding and the Infinite Wall Principle) was that there is a deeper uncertainty principle inherent in all of reality which destroys the notion of a perfect world which obeys deterministic laws. This turns the standard assumptions underlying both standard scientific determinism and religious determinism on their heads. Heisenberg overturned scientific determinism and religious determinism can only be overturned by the rigour of philosophers. If religious determinism is overturned then what is the new hypothesis? A plausible explanation is twofold:

(a) Heisenberg uncertainty is an essential feature of the world to preserve free will – God does play dice with the universe to protect individuality – we are not puppets of heaven because this is hard wired into reality – God's choice – and it may also be a feature such as 'lubrication' of discontinuities or other insolubilities of our universe solved by quantum 'cheats' in the form of quantum jumps, foam and fluctuations.

(b) To create an ordered universe meets the challenge of inherent unsolvability arising from universal chaos which cannot be solved by deterministic processes – only a stochastic process like quantum events can overcome the indeterminate causality in events to achieve the objective of an ordered universe or evolution to achieve ordered complex life (but defects in this order are a mathematical necessity – the cosmological constant being one instance). Creation can produce empirical solutions to otherwise unsolvable 'creation problems' despite the side effect of some residual bad outcomes – thus Einstein may have had this insight later when he made his famous quote: "God does not care about our mathematical difficulties. He integrates empirically."

As stated in Chapter 13, it may be that Einstein became aware of these inherent limits to reality being deterministic and continuously solvable when he said: "What interests me is whether God had any choice in the way He made the world." But Heisenberg's Uncertainty marked the beginning of an intellectual struggle with Bohr and Heisenberg. Schrödinger seemed to share Einstein's reticence for the new theory and challenged Heisenberg in discussions over the theory. Einstein even said to his friend Michele Besso regarding Heisenberg's elaborate matrix system on inputs and outputs that it was 'a veritable witches' multiplication table... "Exceedingly clever and [yet] because of its great complexity safe against refutation as being incorrect."[403] That was almost an allegation of obfuscation.

The rift came to a head at a scientific conference in Brussels in October 1927. Author David Bodanis describes the mood at the conference regarding Heisenberg's February 1927 paper: "[The scientists]… like Einstein, were impressed that his early computations had had such success in accounting for how electrons responded to blasts of light, but weren't convinced that reality could be so unclear, so vaguely glued together, that at the most detailed level we really had to accept uncertainty forever."[404] And after Bohr spoke in favour of Heisenberg: "'The uncertainty principle proved that these subatomic goings-on were unknowable', Bohr insisted. [405]…[and one of the attendees Paul Ehrenfest recalled] '…Einstein was like a chess player. He was a perpetual motion machine, intent on breaking through uncertainty' but there was also Bohr, who 'out of a cloud of philosophical smoke' would lean forward, musing and musing until he came up with the tools that could undermine Einstein's new examples [ie. thought experiments]." The conference was deemed a draw between classical physics and Heisenberg's uncertainty.[406]

It should also be noted that at this same conference (or indeed just after it) Einstein was approached by a scientist and catholic priest Father Lemaître, who asked Einstein whether he had read a paper submitted to a Belgian journal by Lemaître where he had analysed various values of Λ in Einstein's field equation and that the most interesting result was when Λ was set to zero. (By inserting Λ, Einstein had changed his field equation from $G=8\pi\gamma T$ to $G=8\pi\gamma T + \Lambda g$ because it made "possible a near-static distribution of matter, as required by the small velocities of the stars… [although it was]… gravely detrimental to the formal beauty of the theory"[407].) Lemaître had urged Einstein to consider current scientific work that showed the universe was expanding, which would bring the equation back to its original form $G=8\pi\gamma T$ (because Λg obviously becomes zero). This was the trigger for Einstein to realise that the insertion of Λ in his field equation was 'his biggest blunder'.[408] However, in effect, the insertion of Λ proved correct after Edwin Hubble showed the universe was expanding,[409] and so it was not Einstein's greatest mistake because Λ has indeed been restored to his field equation and now possibly regarded as dark energy. The field equation did need 'tweaking' after all.[410]

The rift was not over, for about three years later in October 1930 a similar scientific conference anticipated an intellectual tussle between Einstein and

Bohr over Heisenberg's Uncertainty Principle. Again, author David Bodanis describes Einstein's challenge to Bohr in the form of a thought experiment: "He told Bohr to imagine a box that had a fine cloud of radiation – think of it as a cloud of light particles, or photons – floating inside it. There's a tiny shutter in one wall, controlled by a very precise clock. The whole apparatus is on a scale so it can be weighed. When the clock strikes a particular time, the shutter opens, one photon is let out, and then the shutter closes. The box is weighed before and after, and that way it's obvious how much mass has been lost. Doing this, we know how much energy that lost photon carries: the scale tells us (because mass and energy are equivalent). We also know what time it is when the photon flies out: the clock tells us. This was something that should never happen if Heisenberg's uncertainty principle were true (see Heisenberg uncertainty equation earlier in this chapter). Since the clock has no connection with the scale – unlike a tyre pressure gauge, where measurement interferes with accuracy – Heisenberg's argument is ruined. Certainty is possible. The classical world of cause and effect is saved... Einstein's thought experiment overwhelmed Bohr."[411]

However, the following morning, after Bohr had time to consult with others and ponder the problem, a stunning riposte was constructed by Bohr to rebut Einstein, again related by David Bodanis: "When the shutter opens and the photon flies out, the mass of the box goes down. But the weight of the box is being measured. That means it has to be on a scale. When the photon flies out, the scale rises up – very little, but at least a bit. That means it's ever so slightly higher in the earth's gravitational field. By Einstein's own theory of relativity, time is seen to operate at different rates in a stronger versus a weaker gravitational field... the uncertainty in the weighing, because of that tiny gravitational shift, was just enough to match exactly what's predicted by Heisenberg's uncertainty principle."[412]

Einstein had been outfoxed by his own theory, producing a triumph for the quantum mechanics demonstrating its departure from classical physics. "Einstein never again attended such a meeting; never again attempted to refute Bohr or Heisenberg in public debate."[413] It seemed that God did play dice with the universe. This also had implications for Schrödinger's equation because whilst the equation governed the behaviour of a particle continuously, this ceased upon wave collapse and discontinuities are not easily digestible by Schrödinger's equation. Mathematicians can cope with this but the probabilistic element involved in the measurement step places challenges to explaining physical reality.[414]

However, Einstein did mount another challenge, which became known as the EPR paradox. 'EPR' referred to Einstein-Podolsky-Rosen, who together published another thought experiment. This thought experiment anticipated the famous Bell's Inequality Principle published by physicist John Bell in 1964 (later confirmed by many experiments). The key principle of the EPR paradox was that if two particles are connected but then separated then once the properties of one particle are measured then the outcome of this measurement determines the state of the other particle. This was predicted by quantum theory and Einstein called it 'Spooky action at a distance'. Presumably, Einstein could not believe it would prove to be true. Einstein and his colleagues maintained that somehow classical physics still applied and that it was just the fact that hidden factors – hidden variables– had not been found. This hypothesis is now known as the 'hidden variable theory'. The link between the two separated particles became known as 'quantum entanglement'. Bell's Inequality examined the concept of Einstein's 'local realism', namely that objects have essential features that when measured will return those features. Bell devised experiments to test whether measurements on pairs of particles which become separated would return results to detect shared features dependent by each particle on the other, eg. the polarity feature where each particle has the opposite polarity to the other. Such experiments would show that the local realism worldview was contrary to quantum mechanics.[415] John Bell put the problem simply in terms of a professor colleague of his:

> ***"Dr Bertlman likes to wear two socks of different colours. Which colour he will have on a given foot on a given day is quite unpredictable. But when you see… that the first sock is pink you can be already sure that the second sock is not pink…"***[416]

Thus, to continue this allegory, if Dr Bertlman separated his legs into two lead-lined tubes then 'spooky action at a distance' would continue to mean that measuring one foot (by observation) would determine the colour of the other sock!

Bohr-Heisenberg's Copenhagen Interpretation required that you can extract a maximal amount of data from two particles. However, Bell's inequality states that the amount of data that can be extracted from when particles are separated is less than that 'quantum' maximal amount of data when non-local particles are measured separately (noting that any 'conspiratorial' messaging between the particles – hidden variables – cannot travel faster than the speed of light). Yet in a landmark experiment by Alain Aspect in 1982 in Paris, Bell's inequality was proven to be broken, showing there is still full correlation of maximal data as predicted by quantum theory, which seems to refute Einstein- Podolsky-Rosen's hidden variable theory and local realism.[417] Yet hidden variable theories survive in two forms. First, those hidden states are merely inaccessible and would, if known, provide full predictability. It is as if the data we receive is merely a cross section of another dimension, which if accessible could yield the equations for these hidden variables.[418] Second, the hidden variables are inaccessible because matter is quantum mechanical in nature and thus the variables are not detectable.[419]

The rift still remains but the Copenhagen Interpretation, namely, the Bohr-Heisenberg system of probabilistic analysis between different states of particles, now dominates in favour of hidden variable theories such as those held by Einstein, Podolsky and Rosen. Theoretical physicist Lee Smolin has written a book, *Einstein's Unfinished Revolution –The Search For What Lies Beyond the Quantum*,[420] where he describes Einstein, de Broglie and Schrödinger as realists and Bohr and Heisenberg as 'anti-realists'.[421] He quotes Bohr: "An independent reality in the ordinary physical sense can… neither be ascribed to the phenomena nor to the agencies of observation…" and how Bohr, Dirac, Pauli and Jordan – as quantum revolutionaries – formed a 'powerful network of academic power brokers' to cement quantum mechanics as the new paradigm.[422] This was 'paradigm building in action' in the sense of Thomas Kuhn's scientific paradigms.[423] Lee Smolin's book outlines the continued search for a true reality-based theory that can match the predictability that quantum mechanics provides. Two of the theories he examines are, first, David Bohm's pilot wave theory originally proposed by de Broglie, which involves a 'wave-particle' theory and, second, a wave collapse theory which involves only waves.[424] He describes the wave collapse theory as championed by Sir Roger Penrose. Penrose was seeking to 'relativise the quantum' rather than quantise gravity by introducing time as a factor and dispensing with superposition and linearity in favour of non-linear quantum states which undergo wave collapse as part of a 'real physical process'. The new process would combine

Schrödinger's equation with quantum mechanical wave collapse.[425] A key insight of Lee Smolin is that any hidden variable theory must overcome the conflict with special relativity that Bell's non-locality ('spooky action at a distance') entails or 'how do particles share features instantaneously if nothing can travel faster than the speed of light?'[426]

Einstein, Curie, Bohr, Dirac, Schrödinger, Heisenberg and all the others formed an elite group of scientists peeling back each layer of reality which emerged gradually throughout the 20th century and became known in the 1970's as the Standard Model of Particle Physics. It involved a table of elementary particles classified into a collection of quarks, leptons and mediators governed or mediated by the four fundamental forces of the universe – strong and weak forces, the electromagnetic force and gravity. However, the Standard Model has yet to unify these four fundamental forces into a 'Grand Unified Theory' (GUT). For example, a new particle (or energy quantum) has been postulated for the Standard Model called the *graviton* to mediate gravity interactions. However, it is only conjecture and has never been detected by the Large Hadron Collider (LHC) in Geneva, the major testing facility for detecting elementary particles.

Einstein's reality was his earnest wonder at the cosmos and in 1933 the physicist priest Georges Lemaître said to him: "The evolution of the universe can be likened to a display of fireworks that has just ended: some few wisps, ashes and smoke. Standing in the cooled, cylinder, we see the fading of suns, and try to recall the vanished brilliance of the origin of the worlds." Einstein described this as "the most beautiful and satisfying interpretation of creation I have listened to."[427]

Einstein described this reverence by scientists for the cosmos: "[Their] religious feeling takes the form of a rapturous amazement at the harmony of natural law, which reveals an intelligence of such superiority that, compared with it, all the systematic thinking and acting of human beings is an utterly insignificant reflection… [whoever did not have this sense of wonder]…i s as good as dead, and his eyes are dimmed."[428]

Charles Darwin echoed a similar sentiment when he spoke similar words, saying:

"I am inclined to look at everything as resulting from designed laws, with the details, whether good or bad, left to the working out of what we may call chance. Not that this notion at all satisfies me. I feel most deeply that the whole subject is too profound for human intellect. A dog might as well speculate on the mind of Newton."[429]

NASA
LECTURE
SERIES

YEARS

NASA

HAWKING

Chapter Twenty-Two Plate

Stephen Hawking being presented by his daughter Lucy Hawking at the lecture he gave for NASA's 50th anniversary.

Source: Author: NASA/ Paul Alers

https://www.nasa.gov/50th/NASA_lecture_series/hawking.html

TWENTY-TWO

STEPHEN HAWKING

Stephen Hawking was born 8 January 1942 on the 300th anniversary of Galileo's death (and close to the anniversary of Newton's birth[430]) into an interesting family who lived in North London.[431] Although his father was a doctor, he came from a line of Yorkshire farmers (like Newton's family – farmers do spawn innovation!). His great-grandfather, John Hawking, had gone bankrupt by reason of the purchase of too many farms. Indeed, his son had tried to rescue his father and had also gone bankrupt. It seems Stephen had grown up in an atmosphere of austerity after his family endured the hardships of the Second World War. For example, his father, long after he was able to afford power bills from a doctor's income, used to walk about their home with his dressing gown over his clothes instead of turning the heaters on.[432]

Like Einstein, Stephen Hawking's life has been portrayed in movies (in at least two feature films) because of the spectacular but very challenging life he led, which placed considerable strain on his loved ones. It is difficult to succeed as a physicist per se, but he was diagnosed with a life-threatening disability just at the critical phase of gaining his qualifications. Yet, as stated in the Preface, Hawking was described as the 'most brilliant scientist of our time'. In the book *The Great Scientists,* the authors state: "...while Einstein's relativity played a key part in both the Big Bang and Black Hole theories, the other great revolutionary idea, quantum physics, seemed almost to have been sidelined as irrelevant to cosmology... It was Stephen Hawking's brilliant insight that brought the Big Bang and black holes – relativity and quantum physics – all together to give an extraordinary theoretical picture of quantum forces at work."[433] In fact, Einstein's relativity predicted black holes in prescribing the curvature of space-time and the existence of singularities – the ultimate sinks of the universe, but it was Hawking that took this further by combining what he termed "our search for an understanding of the universe from our theory of the extraordinarily vast to our theory of the extraordinarily tiny" (the latter being quantum mechanics).[434]

(We explain what has been described as the 'Rosetta Stone of physics' Hawking accomplished in the next chapter). One of Stephen Hawking's contemporaries, Leonard Susskind, who wrote of him in his book *The Black Hole War: My Battle with Stephen Hawking to Make the World Safe for Quantum Mechanics*, recognised his achievements when he said: "Physicists incessantly question whether Hawking belongs among the greatest physicists of all time and where he ranks in the hierarchy. In response to those who doubt Hawking's greatness, I will only suggest that they go back and read his 1975 paper 'Particle Production by Black Holes'." At the beginning of his book, Susskind quotes Stephen Hawking: "What is it that breathes fire into the equations and makes a universe for them to describe?" a quote displaying Hawking's philosophical approach to physics.[435]

Stephen had entered Oxford at age seventeen and commenced natural science, which included a physics course which "was arranged in a way that made it particularly easy to avoid work."[436] He later admitted to loafing because he had only done about one hour a day of work over three years and said: "I'm not proud of this lack of work, but at the time I shared my attitude with most of my fellow students. We affected an air of complete boredom..." He also wrote of another curious attitude at Oxford concerning so-called 'grey men' which he explained: "You were supposed to either be brilliant without effort or accept your limitations and get a fourth-class degree. To work hard to get a better class of degree was regarded as the mark of a 'grey man', the worst epithet in the Oxford vocabulary."[437] Yet again, Stephen's frank appraisal of himself renders his success in gaining a 'first-class' degree as a 'grey area', for he describes his performance in his final exams in theoretical physics as "borderline between the first and second class degrees, and I had to be interviewed by the examiners to determine which I should get."[438] In fact, Stephen admitted to the examiners that if he only received a second-class degree he would stay at Oxford (and perhaps he was surprised when he received a first- class degree, allowing him to go to Cambridge).[439] Although he had hoped to study cosmology with the famous astronomer Sir Fred Hoyle, he was disappointed when he could only study under Dennis Sciama (who later championed Stephen's work).

A few months after arriving at Cambridge to study cosmology, Stephen was diagnosed with amyotrophic lateral sclerosis (ALS) yet had fallen in love with Jane Wilde as she then was. But it is the human side of Stephen's life that leaves the public in awe. No one describes this better than Jane Hawking herself, who supported Stephen during the critical years of extraordinary endurance (with her own endurance). After learning from her friend that Stephen had been diagnosed with ALS and given two years to live, Jane had displayed a rare moral courage because although she had only just met him, she persevered with her interest in him. Committing to Stephen was a difficult decision for Jane because she writes of a key conversation with some of her friends who advised her, "If he needs you, you must do it," referring later to a decision to marry Hawking as the right thing to do at the time.[440]

Jane also wrote of his philosophical viewpoint on cosmology in her book *Travelling to Infinity: My Life with Stephen:* "Like his parents, Stephen had no hesitation in declaring himself an atheist, despite the strong Methodist background of his Yorkshire grandparents. It was understandable that, as a cosmologist examining the laws that governed the universe, he could not allow his calculations to be muddled by a confessed belief in a creator God, quite apart from the confusion his illness might be creating in his mind."[441]

Prior to his diagnosis, he had been carefree and, as seen above, physics did not seem to motivate him greatly, but after facing possible death within two years, he recalls dreaming that he was going to be executed.[442] He described his feelings: "I suddenly realised there were a lot of worthwhile things I could do, if I were reprieved." On his 75th birthday, he referred to this period in his speech and said: "While there's life, there's hope." Jane and Stephen were married on 14 July 1965 and were together for twenty-five years, producing three children.[443]

As Stephen began with new determination, he had rejected Dennis Sciama's suggestion to study astrophysics because he was determined to do cosmology. He had commenced studying general relativity and tried to get a feel for the subject. He also had to learn maths in more depth and to aid this process he decided to tutor maths at Gonville and Caius College.[444] He attended maths lectures, given by the Master of Pembroke College, Sir William Hodge, with another emerging physics intellect, Brandon Carter, (an Australian researcher famous for his strong/weak anthropic principles).[445]

Dennis Sciama did introduce him to an interesting theory at the time called Wheeler-Feynman electrodynamics, which stated that electricity and magnetism were time symmetric.[446] Stephen described the implications of the theory: "For Wheeler-Feynman electrodynamics to work, it was necessary that all the light travelling out... [from a theoretical lamp]... should be absorbed by other matter in the universe. This should happen in a steady state universe... but not in a Big Bang universe... This was supposed to explain the arrow of time, the reason disorder... [ie. entropy]... increases and why we remember the past but not the future". (Recall also in Chapter 21 where Einstein had realised that the universe was expanding rather than being static and had removed the cosmological constant from his field equation.[447]) Hoyle and another researcher, Narlikar, had modelled both expanding universes and steady state ones yet stuck with the steady state model. This set the scene for an early clash between scientific titans, which is well evidenced by Jane Hawking's account:

> *"Professor Fred Hoyle, who had rejected Stephen's postgraduate research application, was at the time pioneering the use of television to popularize science to great effect... In advance of publication, Hoyle's latest paper, expounding further aspects of the theory of the steady-state universe... was presented to a distinguished gathering of scientists at the Royal Society... [When Stephen asked a question]... He, a very junior research student who as yet had no academic research of any note to his credit, struggled to his feet and proceeded to tell Hoyle and his students as well as the rest of the audience that the calculations in the presentation were wrong. The audience*

was stunned, and Hoyle was ruffled by the piece of effrontery. 'How do you know?' he asked, quite sure that Stephen's grounds for disputing his new research could easily be dismissed. He was not expecting Stephen's response. 'I've worked it out,' he replied, and then added, 'in my head.' As a result of the intervention, Stephen began to be noticed in scientific circles." [448]

Stephen's own account included:

"...and in the question period I said that the influence of all the matter in the universe would make his masses infinite. Hoyle asked why I said that, and I replied that I had calculated it. Everyone thought I meant that I done it in my head during the lecture, but in fact I had been sharing an office with Narlikar and had seen a draft of the paper ahead of time, which had allowed me to do the calculations before the meeting." [449]

In my own experience in running court cases in Australia as a lawyer, it is rare for witnesses to agree on the exact words used at a meeting, let alone a turbulent one, but what emerges is the gist of the meeting, which appears clear. Stephen commented on the aftermath:

"Hoyle was furious. He was trying to set up his own institute, and threatened to join the brain drain to America if he didn't raise the money. He thought I had been put up to sabotage his plans. However, he got his institute, and later gave me a job, so apparently didn't harbour a grudge against me." [450]

Further battles awaited Stephen, as will appear later.

Stephen's insight in 1965 that led to his proposing the Big Bang theory as a singularity arose after Brandon Carter told him of a lecture Roger Penrose gave at King's College London in January 1965.[451] Stephen's intrigue grew regarding the mathematics involving a collapsing star, which hinged on whether it was spherical or not. Penrose had shown the star would collapse whether spherical or not. This led to Stephen, Roger Penrose and Bob Geroch researching space-time, black holes, the universe, etc. to develop a theory of causal structure in relativity and an essay which brought Stephen the Adams Prize at Cambridge in 1966 and formed the basis for his book *The Large Scale Structure of Space-Time* in 1973.[452] In his 2013 book *My Brief History*, Stephen comments about this whole area of quantum mechanics linking to relativity: "Anyway, it is impossible to be rigorous in quantum physics, because the whole field is on very shaky ground".[454]

By this time, his lecturer Dennis Sciama had proved a useful ally. In fact, Leonard Susskind remembers a lecture given by Dennis Sciama in New York on black holes just before Stephen's black hole discovery was known. Although Dennis referred to Stephen as 'his student Hawking', he referred to the source of these discoveries as 'Stephen says this' and 'Stephen says that'. Also, Stephen's collaboration with Roger Penrose saw increasing international scientific recognition and bore fruit. Again, Stephen describes his ascent:

> *"...I had gotten engaged to a very nice girl, Jane Wilde. But in order to get married, I needed a job, and in order to get a job, I needed a PhD. In 1965 I read about Penrose's theorem that any body undergoing gravitational collapse must eventually form a singularity. I soon realized that if one reversed the direction of time in Penrose's theorem, so that the collapse became an expansion, the conditions of his theorem would still hold, provided the universe were roughly like a Friedman model... Penrose's theorem had shown that any collapsing star must end in a singularity, the time-reversed argument showed that any Friedman-like expanding universe must have begun with a singularity... The final result was a joint paper by Penrose and myself in 1970, which at last proved that there must have been a big bang singularity provided that general relativity is correct and the universe contains as much matter as we observe. There was a lot of opposition to our work, especially from the Russians because of their Marxist belief in scientific determinism and partly from people who felt that the whole idea of singularities was repugnant and spoiled the beauty of Einstein's theory. However, one cannot really argue with a mathematical theorem. So in the end our work became generally accepted and nowadays nearly everyone assumes the universe started with a big bang singularity."*[455]

Thus, the famous Big Bang theory in its modern form[456] emerged from this landmark paper yet using the term that Fred Hoyle had coined 'Big Bang'. (Digressing, a similar irony occurred with the famous Australian battle at Tobruk where the Australian forces held out the mighty Panzer division of General Rommel – the Germans had made fun calling the Australians the 'rats of Tobruk', but this was heralded as a badge of honour in Australian newspapers and ever since has been adopted and proudly proclaimed by Australians in the history books.)

Carl Sagan concluded his Introduction to Stephen's *A Brief History of Time* with "a universe with no edge in space, no beginning in time, and nothing for a creator to do", which seems contrary to the concept of a point of creation such as the Big Bang, where both time and space were created along with the laws of the universe in an instant. Stephen explains this contrary position referring to his joint paper with Penrose: "It is perhaps ironic that, having changed my mind, I am now trying to convince other physicists that there was in fact no singularity at the beginning of the universe... it can disappear once quantum effects are taken into account." This surprising back flip will be examined in Chapter 25.

In the period 1965 to 1970, Stephen's focus had been on relativity, black holes, singularities and the fact that the Big Bang was implied by this work.In 1974, Stephen was admitted as one of the youngest fellows of the Royal Society as his reputation grew. Yet he was now confined to a wheelchair and his voice was unintelligible to everyone except close family and friends.[457]

In the week his name was being published as a Fellow of the Royal Society, he received an attractive offer to join the staff at the California Institute of Technology (Caltech), Pasadena, USA. Jane Hawking describes the offer: "The offer was lavish in the extreme. Quite apart from a salary on an American scale, it included a large, fully furnished house rent free, the use of a car and all possible aids and appurtenances, including an electric powered wheelchair to allow Stephen maximum independence... [and physiotherapy, medical care and schooling for the children]... Stephen's students, Bernard Carr and Peter De'Ath were also invited to accompany him."[458]

However, this dream offer was tempered by a quiet pressure reflected by a remark made at a dinner party told to Jane some years later as she explains: "...Stephen's likely fate was indicated in a remark delivered with consummate indifference by a senior don. 'As long as Stephen Hawking pulls his weight, he can stay in this university, but as soon as he ceases to do that, he will have to go...'."[459]

However, as the logistic difficulties in looking after Stephen's daily regime of arduous activities sunk in, the prospect was daunting to Jane, as she explained: "Only a miracle could resolve the problems we faced. However, that Easter a miracle of an idea floated into my mind like a thistledown seed gliding to earth. It lightened my step and removed my anxieties at the impracticability of well-meaning attempts from the other side of the world to offer us a welcome change of scene. The idea was quite simple: we should invite Stephen's students to live with us in our large Californian house. We could offer them free accommodation in return for help with the mechanics of lifting, dressing and bathing. This was all the more essential since Stephen was no longer able to feed himself at all and needed a constant watchful eye... Stephen's first reaction was automatic rejection, but when he had time to think about it and realised the fate of the Californian venture might hang on his decision, he changed his mind."[460] Bernard Carr and Peter De'Ath graciously accepted the offer as a good arrangement for all.

The allure of Cambridge drew the Hawkings back in 1975 (of their own volition[461]) but during that time Hawking had mingled with the greats of physics and Nobel Prize winners such as Richard Feynman and Murray Gell-Mann and even received a medal from the Pope – the Pope Pius XI Gold Medal for science – peculiar as Stephen was an atheist. However, apart from the discussions of particle symmetry, the symmetry of their success was reflected in a comment made by Ruth Hughes, a Caltech voluntary organiser, as Jane Hawking explains: "...that she had first seen Stephen in the Athenaeum, the Caltech Faculty Club, and while everyone else was praising his courage and brilliance... she said to herself that there must be someone equally courageous behind him or he simply would not be there. Nobody had ever said anything like that to me before and it quite threw me off my stride. Later when Stephen was awarded the Papal medal, Ruth presented me with a pearl brooch because, she said, I should be given something too."[462]

In 1982, Stephen's finances were such that it seemed a good idea to build on his scientific success by writing a popular science book. Stephen's reputation had grown considerably. For example, in about 1973, the BBC had aired a documentary where Stephen featured on the topic of the Origins of the Universe.[463] His daughter's school fees were a worry, but he also had the urge to explain to everyone "how far I felt we had come in our understanding of the universe... [and]... how we might be near finding a complete theory that would describe the universe and everything in it".[464] He had told a literary agent, Al Zuckerman, that he would like to write a science book that would sell in airport bookstores.[465] Zuckerman told him there was no chance of that and, after Stephen gave him a first draft of his new book, Zuckerman secured a number of publishers offering to publish it. After accepting the offer from Bantam books, Stephen worked with publisher Peter Guzzardi, who was intensely interested in making Stephen's technical writing accessible to the masses. As Stephen explains: "[He] made me rewrite the book so that it would be understandable, to non-scientists such as himself. Each time I sent him a rewritten chapter, he sent back a long list of objections and questions he wanted me to clarify. At times I thought the process would never end. But he was right: it is a much better book as a result."[466]

In 1985, Stephen had finished the book and visited the particle accelerator, CERN, in Geneva, Switzerland. Unfortunately, after some breathing problems, he was admitted to hospital.[467] Jane Hawking describes being asked into a 'grey side room' by a doctor at the hospital in Geneva: "At first I thought he simply wanted to verify the facts of Stephen's exceptional existence... Having ascertained various details about his phenomenal longevity, and his self-management, the doctor came abruptly to the point. The question was whether the staff should disconnect the ventilator while Stephen was in a drugged state, or should try to bring him round from the anaesthetic. I was shocked. Switching off the life supply was unthinkable. What an ignominious end to such a heroic fight for life, what a denial of everything that I, too, had fought for! My reply was quick and ready... 'Stephen must live. You must bring him round from the anaesthetic.'"[468] Stephen was brought back but it turned out that a tracheotomy was necessary due to his windpipe being blocked by pneumonia. (The tracheotomy was required to wean him off the respirator and bypass the infected area of his throat.)[469] Jane helped nurse him back by reading such literature to him as Borges's *Libro de Arena* (*The Book of Sand*), which included a short story *The Other*, concerning 'identity, time travel, dreams and prediction, of history repeating itself and knowing the future', seen by Jane as an 'antidote to physics'.[470] When Stephen was fit to fly, they took him by a 'small red jet' to Cambridge Airport and then to the intensive care unit at Addenbrooke Hospital (a teaching hospital in Cambridge).[471]

Stephen could only communicate by blinking his eyes and so, after various solutions were considered, he was offered computer software developed by Wal Woltosz to enable him to synthesise a voice so as to communicate. This became the world renowned 'trademark' voice of Stephen Hawking.[472] Jane Hawking describes working on his manuscript during Stephen's failing health: "I myself had read it and marked in red the passages where the science was incomprehensible, and the publishers pointed out that every equation would halve its sales… [Apparently Eddington was given similar advice!]… In his present circumstances, it was unlikely that Stephen would be able to effect the fundamental changes required. I approached one of Stephen's former students, Brian Whitt, to enlist his help with the rewriting."[473]

After such a gruelling task to complete the book in his condition, it seemed the book was destined for greatness because after Bantam Books published his famous book *A Brief History of Time* in April 1987, it went on to sell over 10 million copies[474] and became the bestselling science book of all time. John Farndon summed it up in *The Great Scientists:* "Nobody quite knows why it did so well, but perhaps many people felt this strange, brilliant man might just reveal some of the infinite truths about our universe. It seemed important to be in the know. In the final chapter, Hawking talks about the nature of God".[475] The book elevated him to international celebrity status followed by the creation of further documentaries, series, TV appearances and two feature films about his life. Sadly, his marriage to Jane Hawking of twenty-five years broke up before he commenced a relationship with his nurse, Elaine.[476] He received many awards, with the most notable being the Copley Medal from the Royal Society in 2006 and the Presidential Medal of Freedom in 2009. He was appointed Lucasian Professor of Mathematics in 1979, a post previously held by icons such as Paul Dirac and Isaac Newton.

Not only was his life an inspiration for all, but his book and work served to popularise science greatly in the 20th and 21st centuries, leaving us with his enduring legacy. At the end of *A Brief History of Time,* Hawking talks about the 'theory of everything' with these final words:

> ***"However, if we do discover a complete theory… [of everything]… it should in time be understandable in broad principle by everyone, not just a few scientists. Then we shall all, philosophers, scientists, and just ordinary people, be able to take part in the discussion of the question of why it is that we and the universe exist. If we find the answer to that, it would be***

the ultimate triumph of human reason – for then we would know the mind of God." [477]

STARMUS

Chapter Twenty-Three Plate

Stephen Hawking at Starmus 2016, Tenerife, Canary Islands, Spain.
(Ref. The Stephen Hawking Medal for Science Communication 2017)

TWENTY-THREE

HAWKING'S WORLD

Both Einstein's and Hawking's lives were disrupted by war. Although Hawking's family lived in Highgate, North London, he was born in Oxford to escape the Blitz.[478] During the war, the British and Germans had agreed that neither side would bomb the other's prize educational institutions (namely, Cambridge and Oxford for the British and Göttingen and Heidelberg for the Germans).[479] Highgate was an area where scientific and academic people lived at the time. His family had lived in a 'tall, narrow Victorian house' despite the common view that London was likely to be flattened by the Blitz. Hawking comments: "In fact, a V-2 rocket landed a few houses from ours."[480] We saw how the *Bildung* education system in Germany helped build Einstein's great intellect. What built the mind of Hawking? The great philosophers of England in the 20th century (such as Bertrand Russell, Alfred North Whitehead and Wittgenstein) exemplified the materialist, logical, scientific and mathematical approach to the big questions. Surprisingly, the analysis of language (for which Wittgenstein is famous) dominated English philosophy and was strongly influenced by the Vienna school (a group of Austrian intellectuals active in the early twentieth century).[481] Philosophers such as Moritz Schlick and Rudolf Carnap were members of the Vienna school that introduced logical positivism (a new version of the positivism that we saw in Chapter 15, 'Curie's World', and Chapter 19, 'Einstein's World', which has also been described as scientific empiricism).[482] One English philosopher and proponent of logical positivism was A.J. Ayer, who said: "The philosopher, an analyst, is not directly concerned with the physical properties of things. He is concerned only with the way in which we speak about them. In other words, the propositions of philosophy are not factual, but linguistic in character."[483] A.J. Ayer also used the concept of 'Linguistic Turn' created by US Philosopher Richard Rorty, which essentially saw language as a system which 'constructs its own reality'.[484] Philosopher Julián Marías describes the English emphasis on language (still influential during Hawking's early life): ".the trend which has the most followers is the one which we can call 'linguistic analysis' in which almost all present-day British thinkers… [ie. mid-1960s Britons]… participate, although to very different extents."[485]

One philosopher described logical positivism as a principle "that some apparently intelligible questions are really nonsense, and that therefore there is no point looking for answers to them. For the logical positivists, most questions of religion and metaphysics were to fall on the nonsense side of the line and most scientific questions on the sense side. The pressing question, then, is how to draw the line."[486] Carnap wrote *The Logical Syntax of Language* and the *Logical Foundations of Probability* about how scientific statements are backed by evidence. One of his key quotes was "The metaphysician believes that he travels in territory in which truth and falsehood are at stake. In reality, however, he has not asserted anything, but only expressed something, like an artist."[487] Clearly, he did not foresee the metaphysical-based hypotheses of Hoyle anticipating the origin of Carbon-2 using the (metaphysical) anthropic principle or Gellman/Swieg's (metaphysical) symmetries predicting quarks. Moreover, the entire body of string theory, which we will examine in Chapter 26, might even be nonsense to Carnap. It seems that by the time Hawking was born, critics had uncovered the weakness of logical positivism, namely, where does one draw the line between sense and nonsense? If there is a legacy for Carnap, it is the view that any philosophical claim must be precise enough to be tested against evidence.[488]

Still, it seems that *logical positivism* – or, at least in its 19th century guise, just *positivism* – still existed in Hawking's prime, as explained by Jane Hawking:

"The suggestion of the presence of a Creator-God was an awkward obstacle for an atheistic scientist whose aim was to reduce the origins of the universe to a unified package of scientific laws, expressed in equations and symbols... Strangely to the happy band of the initiated, the equations were said to reveal a miraculous, breathtaking mathematical beauty... In the face of dogmatic rational arguments, there was no point in raising questions of spirituality and religious faith... questions which ran completely counter to the selfish reality of genetic theory. Issues of morality, conscience, appreciation of the arts, were best kept out of the arena lest they too became victims of the positivist approach."[489]

In his book *The Grand Design*, Hawking and his co-author, Leonard Mlodinow, raised some of the big questions, namely, 'How can we understand the world in which we find ourselves? How does the universe behave? What is the nature of reality? Where did all this come from? Did the universe need a creator?' and said: "Traditionally these are questions for philosophy, but philosophy is dead. Philosophy has not kept up with modern developments in science, particularly physics. Scientists have become the bearers of the torch of discovery in our quest for knowledge."[490]

It seems there is some truth in Julían Marías's comment quoted in Chapter 15 that after science has exhausted its search for the metaphysical, philosophy, being empty, becomes a 'theory of science' (which is one way to view the meaning of Hawking's execution of philosophy).[491]

Albert Einstein once said: "The man of science is a poor philosopher."[492] and so he may have questioned the distinguished authors' qualifications to banish philosophy to the sands of time.

News of the death of philosophy (like other things) is greatly exaggerated. To take immediate examples of philosophers keeping up with science, one need look no further than two philosophers of science who wrote books on very advanced topics touching quantum mechanics: *Quantum Mechanics and the Philosophy of Alfred North Whitehead* by philosopher Michael Epperson[493] and *Philosophy of Science* by philosopher of science Steven French.[494]

Of great relevance to this book is the great philosopher of the 20th century, Wittgenstein, who described Newton's reality:

"Newtonian mechanics, for example, imposes a unified form on the description of the world. Let us imagine a white surface with irregular black spots on it. We then say that whatever kind of picture these make, I can always approximate as closely as I wish to the description of it by covering the surface with a sufficiently fine square mesh, and then saying of every square whether it is black or white. In this way I shall have imposed a unified form on the description of the surface. The form is optional, since I could have achieved the same result by using a net with a triangular or hexagonal mesh. Possibly the use of the triangular mesh would have made the description simpler: that is to say, it might be that we could describe the surface more accurately with a coarse triangular mesh than with a fine square mesh (or conversely), and so on. The different nets correspond to the different systems for describing the world. Mechanics determines one description of the world… [based on]… the axioms of mechanics."[495]

Thus, in Chapter 21, 'Einstein's Reality', and throughout this book, I have mentioned the physics adage: 'The map is not the territory' and in this regard, Wittgenstein comments further: "[the axioms of mechanics supply]… the bricks for building the edifice of science… [ie. Newton's reality. However]… the possibility of describing the world by means of Newtonian mechanics tells us nothing about the world: but what does tell us something about it is the precise way in which it is possible to describe it by these means."[496] We saw the genius of Newton in the mathematisation of gravity, but he stopped short in saying his mechanics (his 'map') was gravity (the 'territory), ie. he did not produce a hypothesis of what gravity is. We also saw Leibniz apply the 'net' of theses to produce his masterpiece of reality – theodicy. Wittgenstein's net woven around reality is a useful way of classifying such complex systems as the space or manifold of the Calabi-Yau space explained in Chapter 26.

Hawking's new quasi-philosophical approach (for want of a better term) was summarised well in Stephen's famous book *A Brief History of Time* in cosmologist Carl Sagan's introduction to the book when he wrote about Stephen's doubts on the *need* for a creator God (to use the same phraseology he used to answer Larry King in the Preface):

"This is also a book about God… or perhaps about the absence of God. The word God fills these pages. Hawking embarks on a quest to answer Einstein's famous question about whether God had any choice in creating the universe. Hawking is attempting, as he explicitly states, to understand the mind of God.

> *And this makes all the more unexpected the conclusion of the effort, at least so far: a universe with no edge in space, no beginning in time, and nothing for a creator to do."*[497]

This last statement is based on what is called the Hartle-Hawking 'no-boundary' proposal (which we will discuss in Chapter 25) and postulates that one of the quantum foam fluctuations in chaotic space time became the universe, ie. it was created from 'nothing'. Hawking also said: "Because there is such a law as gravity, the universe can and will create itself from nothing. Spontaneous creation is the reason there is something rather than nothing, why the universe exists, why we exist."[498]

Philosopher of science John Leslie responds to this proposal:

> *"No matter how you describe the universe – as having existed forever, or as having originated from a point outside space-time or else in space, but not in time, or as starting off so quantum – fuzzily that there was no definite point at which it started, or as having a total energy that is zero – the people who see a problem in the sheer existence of Something Rather than Nothing will be little inclined to agree that the problem has been solved."*[499]

Clearly, something must have been fluctuating first to produce the universe from 'nothing' and quantum foam space is not 'nothing'. In fact, in string theory, 'nothing' is important because it takes the form of a 'vacuum state' with all particles removed from it, which physicist Andrei Linde explained: "The properties of the vacuum state determine what its particles will look like and what the physics of their interactions will be if it were populated."[500] It turns out there are a vast number of potential vacuum states.[501] Obviously, 'nothing' comes with a vast number of configurations and so it seems there was something for a creator to do, ie. create the space with those configurations so that the 'no-boundary' principle could start to work.

It was Hawking who, in the present age, elevated the issue of God in public discussion despite his atheism. Emeritus Professor of Mathematics (Oxford), John Lennox, refers to both Hawking and biologist Richard Dawkins as 'new atheists' in the context of the science versus religion debate. In past chapters, we saw the fading of God since the time of Newton and Leibniz ('religion had left the laboratory') through Voltaire's great doubts about

God after the Lisbon earthquake of 1755, which must have prompted him to write his famous book *Candide*, challenging Leibniz's theodicy (see Chapters 11 and 13). Some believe that the Lisbon earthquake was the real catalyst towards atheism that had grown since the Scientific Revolution. In 1927, Bertrand Russell gave a lecture on why he was not a Christian during which he said: "There is no reason why the world could not have come into being without a cause; nor, on the other hand, is there is any reason why it should not have always existed. There is no reason to suppose the world had a beginning at all."

Russell gave that speech in 1927, which might be considered the peak of logical positivism (scientific empiricism) and the Vienna circle's activities before its decline. Science and atheism were now 'joined at the hip' to the extent that materialism or positivism was (and still is) a dominant paradigm of science. Yet since 1927 the Big Bang theory, fine tuning and advances in biology, including the 'black hole' in origin-of- life science we saw in Chapter 5, may have placed cracks in this paradigm.[502] Clearly, Russell would have been surprised by the Big Bang theory becoming scientifically established because it ran against his beliefs and may have altered his position (although he held other objections to faith). As Hawking remarked: "Many people do not like the idea that time has a beginning, probably because it smacks of divine intervention."[503] It cannot be denied that the Big Bang is quite extraordinary (even apart from things like the Big Bang's entropy paradox[504]). John Leslie identified a serious metaphysical flaw in any theory of the cosmos without proper hypothesis. In viewing a synthesis of today's science, it is hard to see a better cosmological hypothesis than the 'Cosmic God Hypothesis' when one considers that a singularity of infinite space-time curvature with the entire mass of the universe was compressed to a point and then inflated to something as vast as our universe. Of course, God is distasteful as an explanation in science because it is seen as 'giving up' scientific enquiry. The answer is that there will always be room for scientific enquiry despite the efforts of Einstein's 'riddle setter'. The goal of science might then remain as Newton saw it, namely, to uncover the workings of the creator despite the positivists' 'no Gods' scientific paradigm.

In discussing some of the topics in this book and the 'cosmic faith' of Newton, a friend said to me, "It is easy with our present knowledge to be an atheist," to which I replied, "What do you know that Newton did not know that changes your beliefs?" to which my friend was silent. If there is no body of knowledge for that change in belief then the change might follow a cultural paradigm akin to Thomas Kuhn's scientific paradigms.[505] Whatever spiritual paradigm prevails in science, the triumph of the human spirit is exemplified by Stephen Hawking's mastering of his disability to become the world's most famous physicist. His success was recognised when he was famously asked to open the 2012 Paralympic Games with fanfare including a 'Big Bang' display. During the ceremony he spoke these beautiful words:

> *"The Paralympic Games is also about transforming our perception of the world. We are all different, there is no such thing as a standard or run-of-the-mill human being, but we share the same human spirit. What is important is that we have the ability to create. This creativity can take many forms, from physical achievement to theoretical physics. However difficult life may seem there is always something you can do, and succeed at. The Games provide an opportunity for athletes to excel, to stretch themselves and become outstanding in their field. So let us together celebrate excellence, friendship and respect. Good luck to you all."*[506]

As author, celebrity, scientist and even as philosopher, he came to be considered by many as the greatest scientist of our time. He was once asked the meaning of life on television and in response referred to the wonder of advanced primates pondering their place in the universe. In a National Geographic production aired on cable channels in 2017 – *Genius by Stephen Hawking* – an episode in the series asked 'Why are we here?' An extract from Hawking's conclusions appears below:

> *"Since the Big Bang itself the universe has been governed by the laws of nature... the universe provides stable conditions allowing complex creatures to evolve but when we look closer we discover that the universe appears to split into all possible universes all the time and that would lead to a pretty remarkable realisation. We may be tiny and feeble but in a very real way it exists just for you because the universe you see is the one you have chosen out of all the possible universes and that is why you are here. So no matter how bad things get, I always say don't look at your feet but look up at the stars."*

Chapter Twenty-Four Plate

"This artist's concept illustrates a supermassive black hole with millions to billions times the mass of our sun. Supermassive black holes are enormously dense objects buried at the hearts of galaxies (smaller black holes also exist throughout galaxies). In this illustration, the supermassive black hole at the center is surrounded by matter flowing onto the black hole in what is termed an accretion disk. This disk forms as the dust and gas in the galaxy falls onto the hole, attracted by its gravity."

TWENTY-FOUR

HAWKING'S SCIENCE

Hawking took the singularity in a black hole – where the laws of physics break down and mass shrinks to an infinitesimal point – and in a brilliant display of lateral thinking postulated that the beginning of the universe was 'a black hole in reverse'. By doing this, he transformed cosmology. In other words, Hawking and Penrose in their 1965 paper linked black hole singularities to the 'ultimate singularity', namely, the Big Bang. In this chapter, we will see how he built his theory of black hole science by linking Einstein's general relativity to quantum mechanics – a paradigm shift in science because the connection between the micro world of quantum field theory had not been linked to the macro world. The fascination with the mathematics of black holes and the life cycles of stars had emerged against some of the key historical insights of scientists since 1783, when a Cambridge scientist, John Mitchell, published a paper regarding 'dark stars'. He postulated that a star with large mass would have such a strong gravitational field that light could not escape. This was an innovative idea because Einstein's theories of relativity showing that light could be affected by gravity had not yet been thought of. Simon Laplace had also included it in the first and second edition of his book *The System of the World* (but left it out of later editions). Hawking comments: "Perhaps he decided that it was a crazy idea."[507] After this, an early science of black holes developed. New Zealander Roy Kerr developed equations governing rotating black holes based on Einstein's relativity.[508] In 1917, Karl Schwarzchild had also applied relativity to black holes to predict the shape and minimum radius of a star before inevitable gravitational collapse transformed it into a black hole.[509] Roger Penrose and John Wheeler postulated that it would be spherical, but others argued it could not be perfectly spherical and would thus form what is known as a 'naked singularity'.[510] Hawking and Penrose between 1965 and 1970 had shown that there "must be a singularity of infinite density and space-time curvature within a black hole."[511] As the laws of nature broke down inside the singularity, an observer outside the black hole would be unable to see this break down because light could not escape the black hole. Thus, Roger Penrose postulated that "God abhors a naked singularity," known as the cosmic censorship hypothesis or, in other words, singularities like this only occur in regions of uncertainty such as black holes. Such regions are the exception and so hidden from view, as it were.[512]

The life cycle of a star occurs after a large quantity of gas collapses into its centre, causing the atoms to collide with each other more frequently as they crowd together, causing the gas to heat up from such increased frequencies. The gas, which is mostly hydrogen, fuses to form helium releasing massive heat, causing the star to shine. A balance between the escaping heat and the capturing gravity then makes the star stable, with the star continuing to shine. In 1939, Robert Oppenheimer, who had led the Manhattan project in the development of the atomic bomb in the USA during World War II, showed that a star that had run out of nuclear fuel could not support itself against gravity.[513] It followed that eventually the star begins to run out of fuel, causing its balance to become a one-sided 'capture' by gravity, shrinking the star and then, depending on its mass, a black hole may result.[514]

The behaviour of light near the star is indicated by 'light cones' (light is modelled in a cone of light according to Einstein's laws of relativity) and these light cones thus reflect the changes in space and time. Hawking explains: "As the star contracts, the gravitational field at its surface gets stronger and the light cones get bent inward more... [and when]... the gravitational field at the surface becomes so strong that the light cones are bent inward so much that light can no longer escape... neither can anything else; everything is dragged back by the gravitational field... So one has a set of events, a region of space-time, from which it is not possible to escape to reach a distant observer. This region is what we now call a black hole. Its boundary is called an event horizon, and *it coincides with the paths of light rays that just fail to escape from the black hole*"[515] [my italics]. The significance of the words in italics is that they were the lynchpin to Hawking's later success which came in a burst of ideas one night as he pondered matter or radiation drawn into a black hole.[516] He explains: "...one evening in November... [of 1970]... shortly after the birth of my daughter, Lucy, I started to think about black holes as I was getting into bed... Suddenly I realized that the paths of these light rays... ['that just fail to get away from the black hole']... could never approach one another... But if these light rays were swallowed up by the black hole, then they could not have been on the boundary of the black hole. So the paths of the light rays in the event horizon had always to be moving parallel to, or away from, each other... [this meant that]... the area of the event horizon might stay the same or increase but it could never decrease... this nondecreasing property of the event horizon's area placed an important restriction on the possible behaviour of black holes. I was so excited with my discovery that I did not get much sleep that night. The next day I rang up Roger Penrose. He agreed with me..."[517] and he explains that Penrose probably realised this also but was working with a 'slightly different definition of a black hole'. Apart from the beginnings of a comprehensive theory of black holes, it also pointed to the operation of the second law of thermodynamics which involves 'entropy'. Entropy is a measure of the disorder in a system (and in physics systems, disorder tends to increase). The second law can be expressed as "the entropy of a closed system increases with time"[518] and "when two systems are joined together, the entropy of the combined system is greater than the sum of the entropies of the individual systems."[519] Hawking saw that this non-decreasing behaviour of the event horizon area was similar behaviour to that of entropy. However, a puzzling insight occurred to him, namely, that the second law of thermodynamics would be violated because if a body of gas with high entropy dropped into a black hole then (counter-intuitively) the entropy of the gas would drop, *not increase*.[520]

Hawking's puzzling insight was built on by Jacob Bekenstein, a Princeton University research student who then said that the event horizon area of a black hole measured the entropy of the black hole (despite *volume* being expected as the parameter, not area).[521] This was an apparent resolution of the theoretical violation of the second law but Hawking noticed a 'fatal flaw', namely, that if the black hole had entropy, it would have a temperature which meant it would emit radiation despite the wisdom at that time that 'nothing can escape the event horizon of a black hole'. Hawking later said that he was irritated that Bekenstein had 'misused' his discovery concerning the increase of area of event horizons[522] and in retaliation wrote a paper in 1972 ruling out the apparent similarity between entropy and the area of the event horizon. After a visit to Moscow and discussions with two Soviet scientists who put the case that particles were emitted from rotating black holes based on Heisenberg's uncertainty principle, Hawking decided to review the position with a different mathematical approach. He then went on to develop a stunning theory which linked general relativity to quantum mechanics. In his own words: "How is it possible that a black hole appears to emit particles when we know nothing can escape from within its event horizon? The answer, quantum theory tells us, is that particles do not come from within the black hole but from the 'empty space' just outside the black hole's event horizon!... What we think of as 'empty' space cannot be completely empty because that would mean that all the fields, such as the gravitational and electromagnetic fields, would have to be exactly zero. However, the value of a field and its rate of change with time are like position and velocity of a particle: the... [Heisenberg] ... uncertainty principle implies that the more accurately one knows one of these quantities, the less accurately one can know the other... There must be a certain minimum amount of uncertainty, or quantum fluctuations... [thus virtual pairs of matter particles such as electrons or quarks will make it possible]... for the virtual particle with negative energy to fall into the black hole and become a real particle or an antiparticle... or having positive energy, it might escape from the vicinity of the black hole as a real particle or antiparticle... To an observer at a distance, it will appear to have been emitted from the black hole."[523] Hawking's theory then involves conservation of energy and entropy. As positive energy is emitted from the black hole, negative energy is absorbed into it, which results in the black hole losing mass (energy is proportional to mass by Einstein's equation $E=mc^2$). "As the black hole loses mass the area of the event horizon gets smaller but this decrease in the entropy of the black hole is more than compensated by the entropy of the emitted radiation."[524] This was in contrast to his initial brainstorm involving matter and radiation drawn into the black hole where the horizon cannot decrease.

His discovery in 1974 of this radiation became known as Hawking Radiation, and when his paper was published in the journal *Nature*, postulating that the black hole would eventually evaporate as pure radiation and explode, it was hailed as one of the classic works of cosmology.[525] He postulated that "it would disappear completely in a tremendous final burst of emission, equivalent to the explosion of millions of H-bombs".[526] Further insights have even been postulated based on superstring theory,[527] which we will examine later.

Also in 1974, Professor Leonard Susskind, now a physics professor at Stanford University, attended a lecture by Dennis Sciama. Recall that Stephen had been left with Dennis Sciama as his lecturer after being disappointed that he had not been able to study under Fred Hoyle. During that lecture in New York, Professor Susskind first heard of Stephen Hawking as a 'rising star in the world of general relativity'.[528] During that lecture, not only did Professor Susskind hear about the mysterious radiation from black holes but also that they could evaporate; both concepts unheard of. This implied that this radiation, being thermal and with no correlation to temperature, disappeared, violating the principles of thermodynamics involving entropy. Professor Susskind wrote a book on what followed which he called *The Black Hole War: My Battle with Stephen Hawking to Make the World Safe for Quantum Mechanics,* which we referred to in Chapter 22. Professor Susskind comments that "Hawking's calculation showing how black holes evaporate was more than a brilliant tour de force. I believe that in time when its repercussions are fully understood, physicists will recognise it as the beginning of a great scientific revolution… it will touch on the deepest issues: the nature of space and time, the meaning of elementary particles, and the mysteries of the origin of the universe."[529]

The problem concerned the other implications of Hawking's theory, but by 1981 Professor Susskind and his colleague Nobel Prize winner Gerard 't Hooft[530] had 'picked up the gauntlet thrown down by Hawking', namely, that any information sucked into a black hole is lost forever, violating the conservation of information principle arising from quantum mechanics (called unitarity).[531] This became known as the Information Paradox. It led to a number of 'scientific *confrontations*' and the reader is referred to Professor Susskind's book for full details.[532] One example is reminiscent of the contests between Einstein and Bohr/Heisenberg where thought experiment trumped thought experiment. In about 1990, Professor Susskind met Hawking at a conference in a ski resort at Aspen, Colorado and put a thought experiment to Hawking and others present. He asked everyone to imagine tiny Xerox machines copying information (eg. a written document) just outside the horizon of a black hole, producing two identical copies which both drop into the black hole. The first copy drops into the black hole and is destroyed at the singularity. The second copy is radiated back via Hawking radiation. Thus, the postulate is that observers can decode it and "see every bit of information returned in the Hawking radiation" and thus Hawking's assertion that information is destroyed is wrong.[533] Susskind described Hawking's reaction: "But Stephen said nothing. Slumped in his wheelchair, he had a wide smile on his face. It was clear that he knew something that Sidney... [one of the other scientists]... did not." The thought experiment was flawed because there was no such thing as a perfect copying machine because it would violate Heisenberg's uncertainty principle (the state of particles inputted may not be identical to particles outputted). After an hour defending his position, Susskind then heard Hawking's real reaction from Hawking's synthesised voice: "'So now you agree with me!' There was a mischievous twinkle in his eyes."

At a conference in Santa Barbara in 1993, Hawking addressed about one hundred attendees – mainly physicists eager to hear the latest Hawking revelation on the Information Paradox. The speakers included Hawking, Gerard 't Hooft and Susskind. After presenting their opposing positions, a vote was taken and it seemed that the pro-Susskind-'t Hooft position (that information is conserved and is emitted by the Hawking radiation) won by a vote of thirty-nine votes to twenty-five votes, excluding other minor options.[534] However, Susskind's description of Hawking is worth a mention: "Stephen sat slumped in his wheelchair, head too heavy to hold straight, while the rest of us waited in hushed expectation... His electronic voice delivered a pre-recorded message... Despite its robotic sound, his voice was rich with personality... Stephen's face conveys magnetism and charisma that few men have... The most unusual thing about Stephen's lectures is the question-and-answer period that follows... [after a question]… the auditorium becomes deadly quiet. One hundred acolytes turn into mute monks in a bizarrely silent cathedral. Stephen is composing his answer. The method by which he communicates with the outside world is amazing. He cannot speak or lift a hand to do sign language… He has neither the strength nor the coordination to type on a keyboard… He has a small computer screen attached to the arm of his wheelchair, and a series of electronic words and letters flash across the screen, more or less in continuous succession… Meanwhile, as the oracle is composing his answer, the room is as silent as a crypt… What is it about Stephen that commands the rapt attention that a holy man, about to reveal the deepest secrets of God and the universe, might receive?"[535]

During the years that the information paradox dispute simmered, Professor Susskind reached the moment of truth in a surprising way.[536] He decided to analyse black holes in terms of string theory in contrast to Hawking's approach using classical theory (although Hawking had used quantum mechanics on his key breakthroughs). The position Hawking had taken was that a black hole is a 'black hole' for information and we really can't know what is inside it. Susskind thought that string theory could probe deeper into the mysteries of the black hole and set about constructing a conceptual framework that could reveal more about the black hole and more particularly about its entropy. First, he constructed a model using M-theory, which includes string theory and branes. We will be analysing string theory in detail in Chapter 26 but it will suffice here to know that all microscopic particles and bonds between them are postulated to consist of a range of 'branes' (from mem*branes*), and branes include strings as their basic unit. A point particle is a D0-brane and a D1-brane is a string. Strings are a basic unit, somewhat like the original concept of an atom that Democritus had envisaged (as we saw in Chapter 4), but when you add energy it increases their mass.[537] D-branes are joined to strings at both ends of the string as the fabric of reality. Second, Susskind conceptualised a black hole as a ball of string. Indeed, string theory postulates that everything is made from strings or branes. Thirdly, Susskind makes a series of assumptions as follows:

(i) Entropy is proportional to mass in a general sense because if entropy is equated to information content then each twist and turn of a string or each link or 'bit' is a piece of information – call it a 'bit'.

(ii) More analysis tightens up the relation. He takes the Schwarzchild radius equation which shows that the area of the black hole surface (the event horizon) is proportional to the square of the Schwarzchild radius and links this to the radius being proportional to mass – thus, he alters his relation in (i) to 'Entropy is proportional to mass squared' or $S{\sim}M^2$.

(iii) He then introduces gravity into this relation and a thought model of the relation. He asks you to imagine you can dial the size of the black hole by making it bigger or smaller by dialling the gravity up or down. He notices a curious feature of this model as the equations 'tweak' the black holes larger or smaller, ie. entropy does not change.

(iv) Susskind also uses string theory to model what occurs at the event horizon and it seems that the 'bits' (ie. information) contained in the twists and turns of the strings at the event horizon result in some loops of the strings protruding from the event horizon (making it a 'hairy black hole' contrary to previous theory) – like Hawking, Susskind applies quantum mechanics and it seems that strings have a probability distribution where some percentage of the time loops break off. Now, these loops represent particles that can be emitted from the black hole, consistent with Hawking radiation. So Susskind's model is taking shape.

(v) At this point, he is 'stuck' because he needs a more precise relation than $S{\sim}M^2$, but Harvard professor Cumrun Vafa[538] advises him that the best black hole to use for his modelling purposes is an *extremal* black hole. This is a black hole where there is a perfect balance between gravitational attraction and electric repulsion.

(vi) In fact, an Indian physicist, Ashoke Sen, had completed a paper introducing some mathematics for extremal black holes which proved helpful to Susskind.[539] Sen had developed his own thought model where instead of imagining black holes as a ball of string in string theory (with its four dimensions and six curled-up dimensions), he imagines the black hole as a cylinder where the same length of string is wound perfectly around that cylinder. The 'bits' entropy (ie. information) that was in the twists and turns of the strings is now in the form of 'wiggles' in the string. (Note that strings have vibrational modes and so the wiggles are quite consistent with string theory.) Sen demonstrated that the entropy was proportional to the stretched event horizon of the black hole.

(vii) The final breakthrough then came in the form of applying D-branes to the event horizon itself. Professor Vafa and his colleague physicist Andy Strominger had more mathematics to enrich the model. The initial problem was that the event horizon appeared to be a mixture of classical and quantum mechanics. The classical approach to the event horizon combined with the string approach to the branes (strings + D-branes) representing the entropy or bits of information is then modelled using the quantum fluctuations in the 'pregnant' wiggles (ie. pregnant with information), allowing for emission of Hawking radiation. The calculations were found to be in precise agreement with the Hawking equations that have stood for twenty years.[540]

Moreover, another two physicists, Curt Callan and Juan Maldacena, had independently prepared their own paper using open strings and branes to arrive at the same conclusion.[541] Why was their calculation significant? The fact that they used quantum mechanics via a different route implied that the canonical rule of unitarity, ie. that information cannot be lost during quantum events, cannot be violated. Thus, information is not lost when an object drops into a black hole and must be emitted with the Hawking radiation. Susskind also refers to a paper of Ed Witten appearing two months after Maldecena's paper entitled 'Anti-De Sitter Space and Holography' and comments: "Whatever else Maldacena and Witten had done, they had proved beyond any shadow of a doubt that information would never be lost behind a black hole horizon."[542]

In July 2004, Hawking announced that he had lost a famous bet he had made with his colleague Kip Thorne concerning the Information Paradox. The quantum fluctuations that facilitate Hawking radiation would allow information to leave the black hole preserving the rules of quantum mechanics.[543] Hawking describes the concession himself: "The paradox had been argued for thirty years, without much progress, until I found what I think is its resolution. Information is not lost, but it is not returned in a useful way. It is like burning an encyclopaedia: the information contained in the encyclopedia is not technically lost if one keeps all the smoke and ashes, but it is very hard to read."[544] (Perhaps it would be impossible if chaos intervened, as we saw in Chapter 13.)

In one of the BBC Reith lectures – 'Black holes ain't as black as they are painted', broadcast on 2 February 2016, Hawking restated his solution to the paradox: "Finally, I found what I think is the answer. It depends on the idea of Richard Feynman that instead of one single history there are many different possible histories, each with its own probability. In this case, there are two kinds of history. In one, there is a black hole, into which particles can fall; in the other, there is no black hole. The point is that from the outside, one can't be certain whether there is a black hole or not... This possibility is enough to preserve the information, but the information is not returned in any useful way." Again, the concept aligns with the indecipherability of chaos – we know the information is there but knowledge of its content is unknowable, contrary to Laplace's demon's powers in his deterministic Clockwork Universe (see Chapter 12).

Unlike the thought experiment battle between Einstein and Bohr/Heisenberg, the black hole war raged for about thirty years, revealing today's 'weapons of choice' for how physicists do battle. What it reveals is quite extraordinary because the battle is largely fought in the (yet-to-be-confirmed-by experiment) metaphysical realm – possibly the realm of philosophy of science – because the theoretical hypothesis of string theory is applied to another theoretical hypothesis – the physics of black holes. The outcome is a synthesis of systematised thought distilled from these hypotheses to describe a new emerging reality – Hawking's reality of fusing quantum mechanics, information, entropy and relativity.

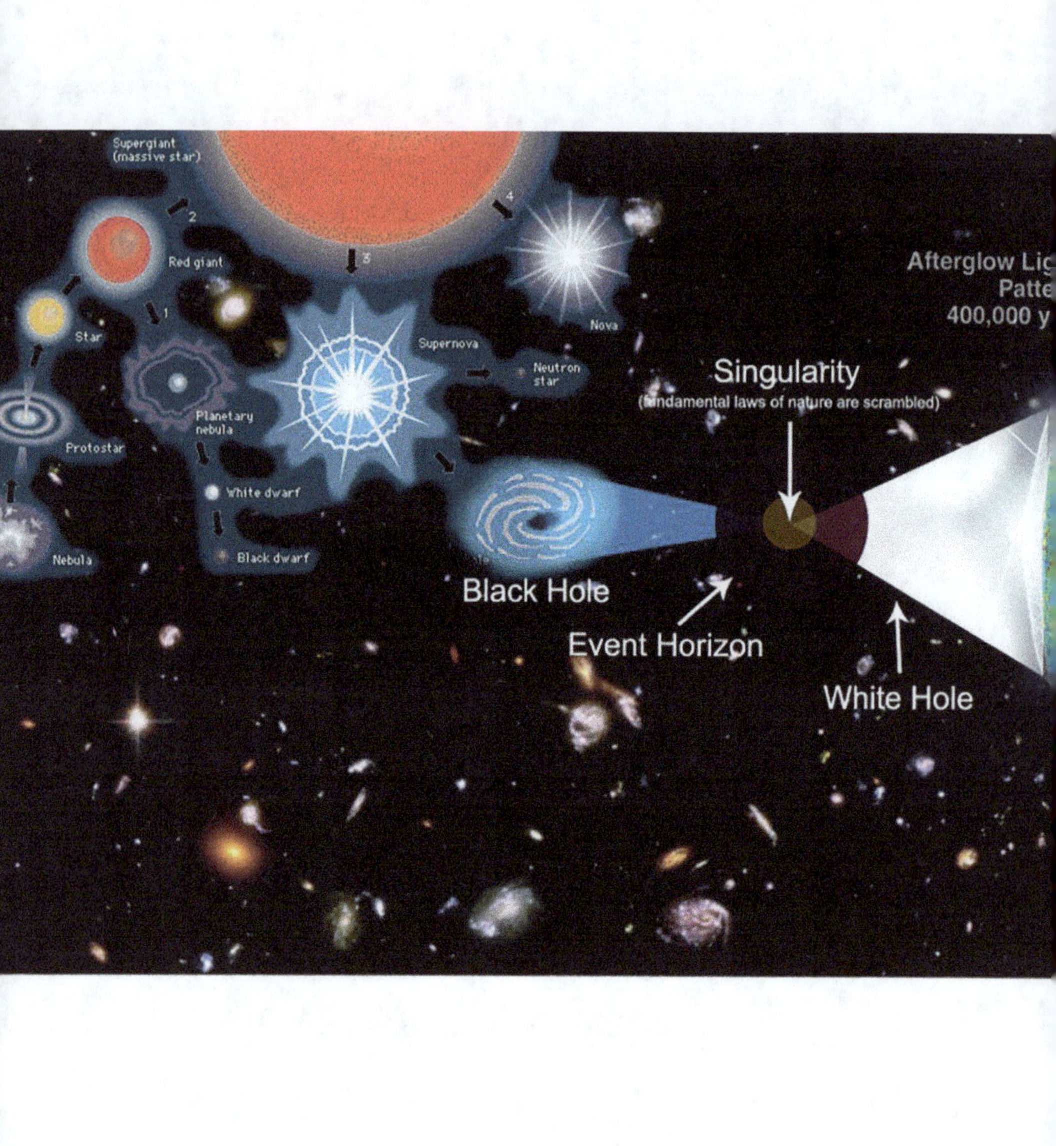

Supergiant
(massive star)
2
4
3
Red giant
1
Star
Supernova
Nova
Neutron
star
Planetary
nebula
Protostar
White dwarf
Nebula
Black dwarf
Afterglow Lig
Patte
400,000 y
Singularity
(fundamental laws of nature are scrambled)
Black Hole
Event Horizon
White Hole

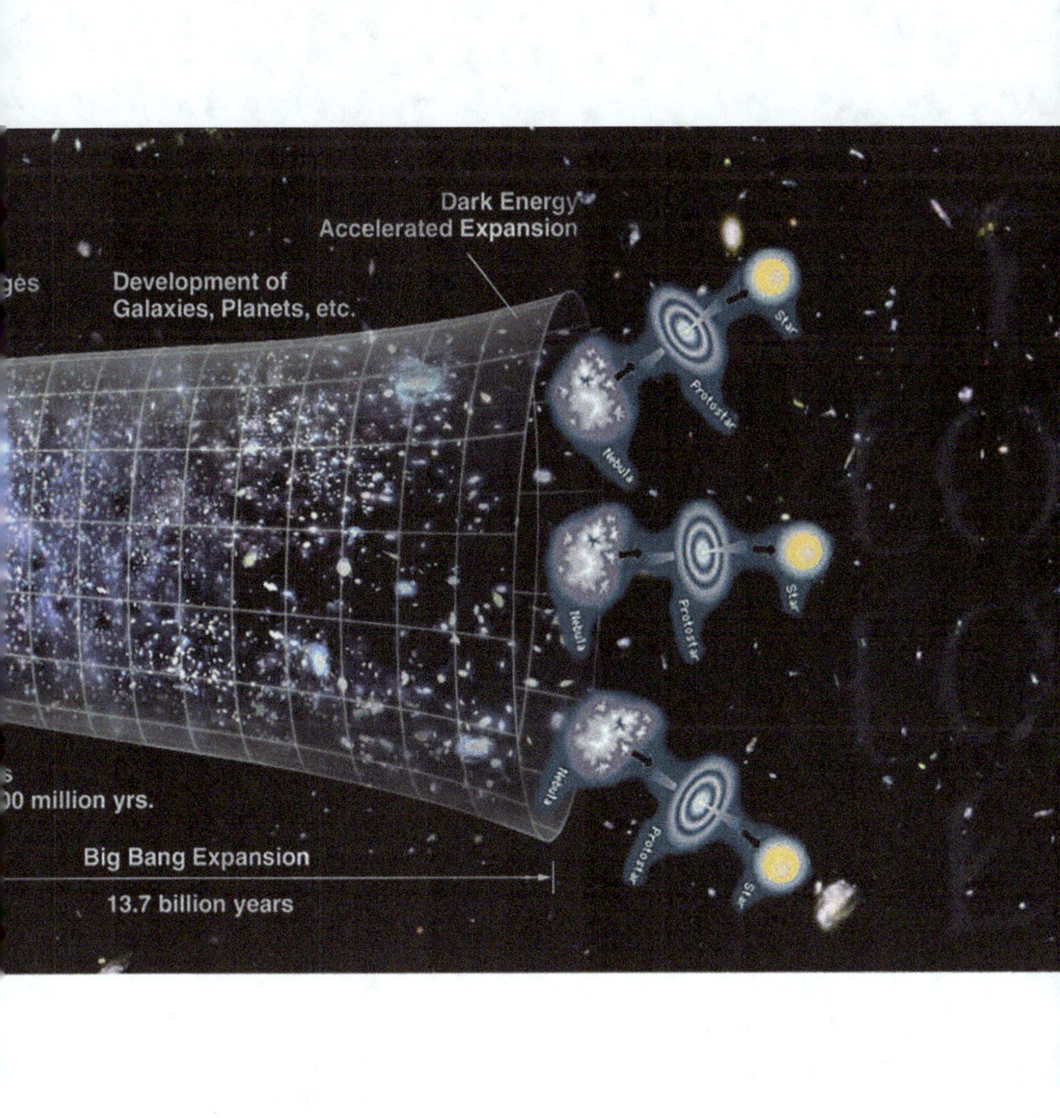

Dark Energy
Accelerated Expansion
Development of
Galaxies, Planets, etc.
Nebula
Protostar
Star
Big Bang Expansion
13.7 billion years

Chapter Twenty-Five Plate

Stephen Hawking's reality encompassed black holes, big bang theory, an expanding universe, the strings and branes of M-theory and their cosmic implications –the multiverse. Andrei Linde's theory of chaotic inflation of the universe also implied a megaverse/multiverse with island universes. Our universe could have been one of these island universes.

TWENTY-FIVE

HAWKING'S REALITY

In 1981, Stephen Hawking (who had been awarded the Papal medal for science some years earlier) had just attended a conference on cosmology organised by the Jesuit order of the Catholic Church. After he had later met the Pope, he said: "I was glad then that he did not know the subject of the talk I had just given at the conference – the possibility that space-time was finite but had no boundary, which means that it had no beginning, no moment of creation."[545] Hawking explained that the Catholic Church had 'seized' the Big Bang theory in 1951 as it demonstrated creation, and was therefore the work of God.[546] (The Big Bang theory was originally proposed by Jesuit scientist George Lemaître in the 1930s, but in 1948, scientists Herman Bondi, Thomas Gold and Fred Hoyle had proposed the Steady State theory, which Newton, in effect, had accepted over 200 years earlier.[547])

Recalling Hawking was a co-founder of the Big Bang theory (in its modern form), his change of heart is surprising: "It is perhaps ironic that, having changed my mind, I am now trying to convince other physicists that there was in fact no singularity at the beginning of the universe… it can disappear once quantum effects are taken into account." Why the back flip?

In his landmark book *A Brief History of Time,* he summarises his change of mind in Chapter 8, 'The Origin and Fate of the Universe'.[548] After referring to the 'generally accepted' Friedman model of an expanding universe known as the 'hot big bang model', he underscores a key scientific correlation. Recall first the discovery by Nobel Prize winners Penzias and Wilson in 1965 that a background radiation occurred throughout the universe, indicating that this was the after-effect of an event such as the Big Bang. Secondly, this discovery confirmed a famous paper in 1948 by scientists George Gamow and Ralph Alpha predicting that radiation in the form of photons from a very early hot phase of the universe should still be evident today, namely, a few degrees above absolute zero (-273 C°).[549] This prediction was confirmed by the Penzias/Wilson discovery. Clearly, Hawking's analysis shows that any alternative theory must explain the same radiation. Indeed, Hawking concedes: "It is, moreover, very difficult to explain in any other way why there should be so much helium in the universe. We are therefore fairly confident that we have the right picture, at least back to about one second after the Big Bang."[550] The reason for this appears to be that helium needs to be forged at very hot temperatures and this would have occurred only in the first few hours of this 'Hot Big Bang' and subsequent 'hot' events like clouds coalescing with nuclear fusion producing more helium from hydrogen; otherwise, one cannot account for the quantity of helium in the universe.[551]

Hawking then tackles this 'hot universe that cooled' model with some important questions:

1. Why was the early universe so hot?

2. Why is the universe so uniform on a large scale, such as having about the same microwave background temperature everywhere? According to the theory of relativity, "if light cannot get from one region to another, no other information can… Unless for some unexplained reason they happened to start out with the same temperature" (such as would occur with the Big Bang theory).

3. Why did the universe start out with so nearly the critical rate of expansion that separates models that recollapse from those that keep expanding forever, so that now ten thousand million years later, it is still expanding at nearly the critical rate?

4. What is the origin of local irregularities or density fluctuations in the universe?[552]

Hawking then appears to accept the theory to within an instant *after* the Big Bang, referring first to the mathematical impossibility of a singularity at the point of a Big Bang (as the density of matter and curvature of space-time becomes infinite[553] because mathematics can't really deal with infinity[554]). Second, on a more philosophical note, "These laws may have been originally decreed by God, but it appears he has left the universe to evolve according to them and does not now intervene in it. But how did he choose the initial state or configuration of the universe? What were the boundary conditions at the beginning of time?"[555]

After he dismisses God making conditions that are inexplicable (at least to scientists), he suggests chaotic boundary conditions which would imply "either that the universe is spatially infinite or that there are infinitely many universes."[556] He then sees the outcomes in the universe we see as random and thus a lucky fluke that we happen to live in a smooth part of the universe fit for life (or in one of the many universes fit for life).

You might recognise this as the anthropic principle we saw in Chapter 5 (or in one of its many forms). It is interesting that the importance of helium in the Big Bang model aligns with the insight of Hawking's opponent to the Big Bang, Fred Hoyle. In Chapter 5, we saw the anthropic principle as 'science in action' regarding the essential building block of life carbon-12. Hoyle's prediction was that a mechanism of stable resonance relating to 3 helium-4 nuclei fusing at 100 million degrees inside a red giant existed to produce carbon-12. "It was an astonishing triumph, and remains the only successful prediction based on the anthropic argument ahead of experiment – that is, *'I exist, therefore it must exist'.*"[557]

Hawking also brings in the fine tuning hypothesis we saw in Chapter 5 and raises arguments for and against it. He outlines the strong version of the anthropic principle, namely, that the universe appears bio-friendly and fine-tuned for life and considers whether this is a fluke. Before considering his analysis, we must revisit the mind-bending definitions we saw in Chapter 5 of both the weak and strong version of the anthropic principle as propounded by its founder Brandon Carter (a colleague of Hawking), according to the *Stanford Encyclopaedia of Philosophy* – fine-tuning:

Weak Anthropic Principle

"[W]e must be prepared to take account of the fact that our location in the universe is *necessarily* privileged to the extent of being compatible with our existence as observers (Carter 1974: 293, emphasis due to Carter)."[558]

Strong Anthropic Principle

[T]he Universe (and hence the fundamental parameters on which it depends) must be such as to admit within it the creation of observers within it at some stage (Carter 1974: 294).[559]

The ***Oxford Dictionary of Philosophy*** states: "The strong anthropic principle postulates many different universes with different structures, fundamental constants, and histories… to sceptics the strong principle offends against Ockham's razor" (that the simpler explanation should always be preferred).[560]

Hawking has a glib answer to the general question raised by the anthropic principle (in its many forms) 'Why is the universe the way we see it?', to which he answers: "If it had been different, we would not be here!"[561] Yet he seems to balance the issues 'finely' (!): "The remarkable fact is that the values of these numbers seem to have been very finely adjusted to make possible the development of life" (these 'numbers' being the thirty or so essential constants referred to by Professor Paul Davies – see Chapter 5).[562]

> ***"For example, if the electric charge of the electron had been only slightly different, stars either would have been unable to burn hydrogen and helium, or else they would have exploded… Most sets of values would give rise to universes that, although they might be very beautiful, would contain no one able to wonder at that beauty. One can take this either as evidence of a divine purpose in creation and the choice of laws of science or as support for the strong anthropic principle".***[563]

Hawking then seems to weigh the strong anthropic principle: "There are a number of objections… to the strong anthropic principle[564] … (referring to 'multiverses').

First, in what sense can all these universes be said to exist? If they are really separate from each other, what happens to another universe can have no observable consequences in our own universe. We should therefore use the principle of economy and cut them out of the theory".[565] (Recall that he postulated the *possibility* of the multiverse in the context of a big bang with chaotic boundary conditions implying "either that the universe is spatially infinite or that there are infinitely many universes".[566])

This sweeping assessment by Hawking is surprising because on the one hand he dismantles the strong anthropic principle by also rejecting multiverse theory, but on the other hand he acknowledges how compelling fine tuning is if this is the only world we have. Clearly, the glib refutation involving the 'selection effect' (namely, having a bias because we are already here) is not sufficient to refute the theory. (It is interesting that Leibniz's best of all worlds theodicy – see Chapter 11 – relies on creation consistent with an anthropic principle based on multiple worlds that *might be*, but the strong anthropic supporters rely on multiple worlds that *are* – yet both are really based on mathematical not physical realities.) Curiously, atheists rely on the strong anthropic principle including the multiverse to show that the laws of the universe are not fine-tuned because they merely appear as an inevitable outcome of all possible sets of laws out of all possible sets of universes. (In his co-authored 2011 book, *The Grand Design*, Hawking and co-author Leonard Mlodinow support the existence of multiverses on the basis of what is known as M-Theory despite the quote above, for reasons we will return to in the next chapter.[567])

Indeed, the founder of the theory, Everett, conceived of the theory to show an alternative to wave collapse under the Copenhagen Interpretation, but this theory implies many parallel universes branching off with the *same laws* to explain the multiple outcomes of momentum and position of a particle (unless they now propose a multi-Schrödinger's Equation!).

Based on Hawking's principle of economy to 'cut [multiverses] out of the theory',[568] he then considers the other basis for the strong anthropic principle, namely, whether different laws can apply to different regions of the universe. He dismisses this also because otherwise we could not travel freely within the universe. He does not seem comfortable with multiverses or the anthropic principle[569] yet he does employ the weak version of the anthropic principle a number of times in his 1988 book.[570] Despite this,

his more recent views support M-theory which *does* imply multiverses, requiring us to delve deeper into his rationale!

Hawking's reasoning appears to verge on no longer accepting the big bang model. His treatise in *A Brief History of Time* revolves around aspects of the design of the universe when he refers to the problem of random initial conditions at the Big Bang evolving into the smooth universe we see, and indeed sees this model as lacking in divine design. "It would be very difficult to explain why the universe should have begun in just this way, except as the act of a God who intended to create beings like us." It almost appears as if Hawking is trying to fine-tune God out of the equation. (However, the reader might wish to reconsider this approach with deeper analysis of the rationale that a cosmic God might have in the light of Chapter 13 in which the opposing hypothesis is presented, namely, that God needed a chaotic universe for creation to work at all.)

Hawking then comments: "One would feel happier about the anthropic principle, at least in its weak version, if one could show that quite a number of different initial configurations for the universe would have evolved to produce a universe like the one we observe."[571] He refers to the famous theory of scientist Alan Guth of the inflationary universe which involved the theory of a very fast chaotic expansion of the universe. This theory was consistent with the cosmological constant[572] in Einstein's field equation[573] and was consistent with the critical expansion rate applicable to the universe now. Like bubbles of steam in boiling water, the bubbles of 'universe' would expand and join up as the phase transitions proceeded to the present day.

Hawking remembers a Moscow conference he attended given by a then young Russian Andrei Linde, who proposed a new chaotic inflationary model and Hawking pronounced: "In my personal opinion, the new inflationary model is now dead as a scientific theory, although a lot of people do not seem to have heard of its demise and are still writing papers as though it is still viable. A better model, called the chaotic inflationary model, was put forward by Linde in 1983. In this there is no phase transition or supercooling. Instead, there is a spin 0 field, which, because of quantum fluctuations, would have large values in some regions of the early universe. The energy of the field in those regions would behave like a cosmological constant… One of these regions would become what we now see as the observable universe... This model... does not depend on a dubious phase transition, and it can moreover give a reasonable size for the fluctuations in the temperature of the microwave background that agrees with observation... [t]he work on inflationary models... [show] that the initial state of the part of the universe that we inhabit did not have to be chosen with great care."[574]

Again, Hawking almost displays a didactic (atheist) activism, saying that if the universe could begin chaotically, this removes God from his diametrically opposed models of God's 'perfect world' Big Bang versus a 'scientific' Big Bang. As stated in Chapter 13, a perfect world is not mathematically possible and indeed creation of both physical matter and biological matter cannot be created without chaos as the soil against which numerical iteration – evolution of both types of matter – is required because analytical creation is insufficient. Indeed, as stated above, Hawking himself dismisses this diametrical approach when he says: "Must we turn to the anthropic principle for an explanation? Was it all just a lucky chance? That would be a counsel of despair, a negation of all our hopes of understanding the underlying order of the universe."[575]

An interesting comparison that may assist in understanding sentiments such as this was Hawking's comment on the discovery of the Higg's Boson when he said: "Physics would be far more interesting if it had not been found."[576]

Hawking's exposition of the universe in his *A Brief History of Time* and Big Bang theory marches on in his chapter 'The Origin and Fate of the Universe'. Remember also that although this is his view in 1988, it echoes his final views (considered below). He observes also that predicting how the universe began involves assuming laws of the universe "that hold at the beginning of time" but all the known laws of the universe break down at a "point of infinite density and infinite curvature of space-time" – namely the Big Bang singularity that he and Roger Penrose first proposed in their landmark 1965 paper.[577]

He goes on to explain that this conundrum can be saved by a mathematical nicety – the use of imaginary numbers. To explain imaginary numbers, compare them to what are called radical numbers such as square roots. We do not know exactly what the square root of a number is (eg. the square root of 2), so we have extended the number system by adding square roots such as $\sqrt{2}$ even though an exact value for $\sqrt{2}$ does not exist. So too, even though there is no way of calculating what $\sqrt{-1}$ means, we can still say that when you square $\sqrt{-1}$ it equals -1. $\sqrt{-1}$ is defined as i (or j in engineering). It has been found invaluable for discovering hidden solutions to polynomial equations. For example, $x^5 = 1$ seems to have only one solution that is $x=1$, but in fact there are four more solutions if imaginary numbers are added. Fascinating also is the fact that if you graph imaginary numbers on a graph (called an Argand diagram), you find that the five solutions for $x^5 = 1$ form five points which if joined form a perfect pentagon (perhaps the universe is shaped like a pentagon!). More seriously, imaginary numbers help solve problems. Just as in statistics where x^2 becomes a better measure for some probability distributions, so too adding imaginary numbers into equations can become a type of red-green light mechanism converting the result to plus or minus (once you have a variable squared to get a result and that variable is i then $i^2 = -1$ applied to the result can make it plus or minus).

Space-time mathematics using imaginary numbers has been named Euclidean space-time. The adding of i removes time from space-time. It is a little like legislating time from the space-time equations. It is hard to see how Hawking could have been comfortable with that having regard to the arrow of time, entropy and the second law of thermodynamics anchored so well to time having a direction. In any event, he explains what is known as the Hartle-Hawking proposal: "In the classical theory of gravity, which is based on space-time, there are only two possible ways the universe can behave: either it has existed for an infinite time, or else it had a beginning at a singularity at some finite time in the past. In the quantum theory of gravity, on the other hand, a third possibility arises. Because one is using Euclidean space-times, in which the time direction is on the same footing as directions in space, it is possible for space-time to be finite in extent and yet to have no singularities that formed a boundary or edge: if you sail off into the sunset, you don't fall off the edge or run into a singularity... [thus]... the quantum theory of gravity has opened up a new possibility, in which there would be no boundary to space-time and so there would be no need to specify the behaviour at the boundary. There would be no singularities at which the laws of science broke down and no edge of space-time at which one would have to appeal to God or some new law to set the boundary conditions for space-time… [this 'no-boundary universe']… would be completely self contained and not affected by anything outside itself. It would neither be created nor destroyed. It would just BE."[578]

There are occasions when a purely theoretical mathematical-based theory such as the 'no-boundary universe' has preceded a discovery such as the quark predicted by Gellman /Zweig or the Higgs Boson predicted by Professor Peter Higgs, incidentally, despite Hawking predicting that it would be impossible to detect due to microscopic black holes.[579] Further conflicts between Hawking and Professor Peter Higgs spilled over into the media.[580] Roger Penrose, after analysing the 'Hartle-Hawking no-boundary' proposal, states:

> *"There is a fundamental problem that I have with any proposal (inflation or the Hartle-Hawking… [no-boundary] …proposal) that attempts to address the problem of space-time singularities within an apparently* time asymmetric *physics.*[581] *There is no time asymmetry in inflationary physics and, as far as I can make out, there is none in the Hartle-Hawking proposal either, so this proposal should be applied also to the final singularities of collapse (in black holes, or in the Big Crunch if there is one) as well as to the Big Bang. Hawking (1982) has argued that it can be, but in a decidedly exotic way, the space in the neighbourhood of a final singularity being 'closed off without boundary', by taking the universe all the way back to the Big Bang, the 'Euclideanization'… [ie. using imaginary numbers]… being applied only there… I have to say that I have great difficulties with this argument – and indeed any argument where there is no explicit time asymmetry in the physical laws themselves. (In Hawking's 'exotic' argument… it would appear that there is still a 'boundary' at the final singularity of collapse, even though there has been a smooth boundary-free closing off only 'on the other side' of the space time.)"*[582]

The problem seems to be one of topology and whether entropy (or time) is uniform throughout that topological space. During Roger Penrose's argument against the 'no-boundary' proposal, he states that applying imaginary time should not stop at just the Big Bang but should apply to black holes. A black hole could be seen as its own manifold (ie. its own system), but we know it is not a 'closed universe unto itself'. Yet it is true that within the black hole, the 'spherical'[583] event horizon has no boundaries (analogous to the tiny 'nose' of the Big Bang), but the 'other side' is just the universe (to use Penrose's term). It seems to be a case where the mathematics of the 'no-boundary' proposal works but matching reality is more difficult.

In any event, Hawking seems to largely adhere to the Big Bang theory he helped create subject to his quest to refine the theory. The joint milestone of Hawking and Penrose following their 1965 paper, namely, the Big Bang theory, is still based on solid foundations for very good reasons, as Penrose explains in his comment: "Referring to a sub chapter in his book… [*27.13 Our extraordinarily special Big Bang*]… In 27.13, I drew attention to the extraordinarily special state in which the universe appears to have started out. The main way in which this state was special, and which gave it its absurdly low entropy, was a very precise spatial isotropy and homogeneity, so that the universe's space time geometry is (still) in remarkably close accord with one of the standard FLRW models…"[584] (FLRW[585] refers to a group of standard Big Bang expanding universe models.) He goes on to point out that the Big Bang may not have produced a purely symmetrical model but theories that quantum effects which cascade to become larger effects do not account for any alleged deviation from a symmetrical expanding universe after the Big Bang. Furthermore, a new theory in addition to gravitational and quantum mechanics is required, namely, a system of state reduction that will be consistent with the experimental evidence from such data as is now available from COBE, BOOMERanG and WMAP is required.[586]

About ten years after Hawking's *A Brief History of Time,* astronomers in Australia and the USA concluded that the universe is not only expanding but accelerating. This was based on observing more than one supernova. This implied a positive cosmological constant but more importantly to the Hartle-Hawking 'no-boundary' proposal, it implied that a key parameter, spatial curvature – the quantity K[587] – was such that K=0 but in the Hartle-Hawking proposal it required K>0 . Apparently, even a tiny positive K is enough to 'tweak' the Hartle-Hawking theory into shape.[588] However, more recently, doubts persist, such as expressed in these comments by the authors of a 2013 paper: *Hartle-Hawking no-boundary proposal in dRGT massive gravity: Making inflation exponentially more probable,* by Misao Sasaki, Dong-han Yeom and Ying Li Zhang[589], who said: "However, it is known that the Hartle-Hawking no-boundary proposal is in severe conflict with the realization of successful inflation (3)." (Reference (3) refers to A. Vilenkin, Phys. Rev. D 37, 888 (1988).)

There is perhaps one deeper problem to the 'no-boundary' proposal, namely, that it postulates that the universe can create itself out of nothing. We saw John Leslie's comment on this proposal in Chapter 23. Philosopher Michael Epperson further describes the Hartle-Hawking proposal in more technical language but, in effect, a type of 'cart before the horse' where "a vacuous space-time is purported to evolve quantum mechanically from a void of pure potentiality – potentially somehow abstracted from actuality. Such a void, often termed a 'quantum vacuum' or 'quantum foam' is a fundamental incoherent construction, given the concept of actuality is necessarily presupposed by the concept of potentiality, such that the latter cannot be abstracted from the former." (Recall also Hawking's comment – see Chapter 24 – that empty space is not completely empty; indeed, string theory implies many vacuum states.) In other words, any system of quantum mechanics where particles, fields, waves, etc. appear from nothing presupposes the space (and scientific laws) out of which it appears, and just as Hawking mentioned there was no time before the universe, the same argument can be applied to say there was no space (and scientific laws) before the universe. The enigma remains.

Perhaps the missing unknown system of state reduction identified by Penrose (in addition to gravitational and quantum mechanics) could be string theory or its close relation, M-theory? This will be one of the themes in the next chapter.

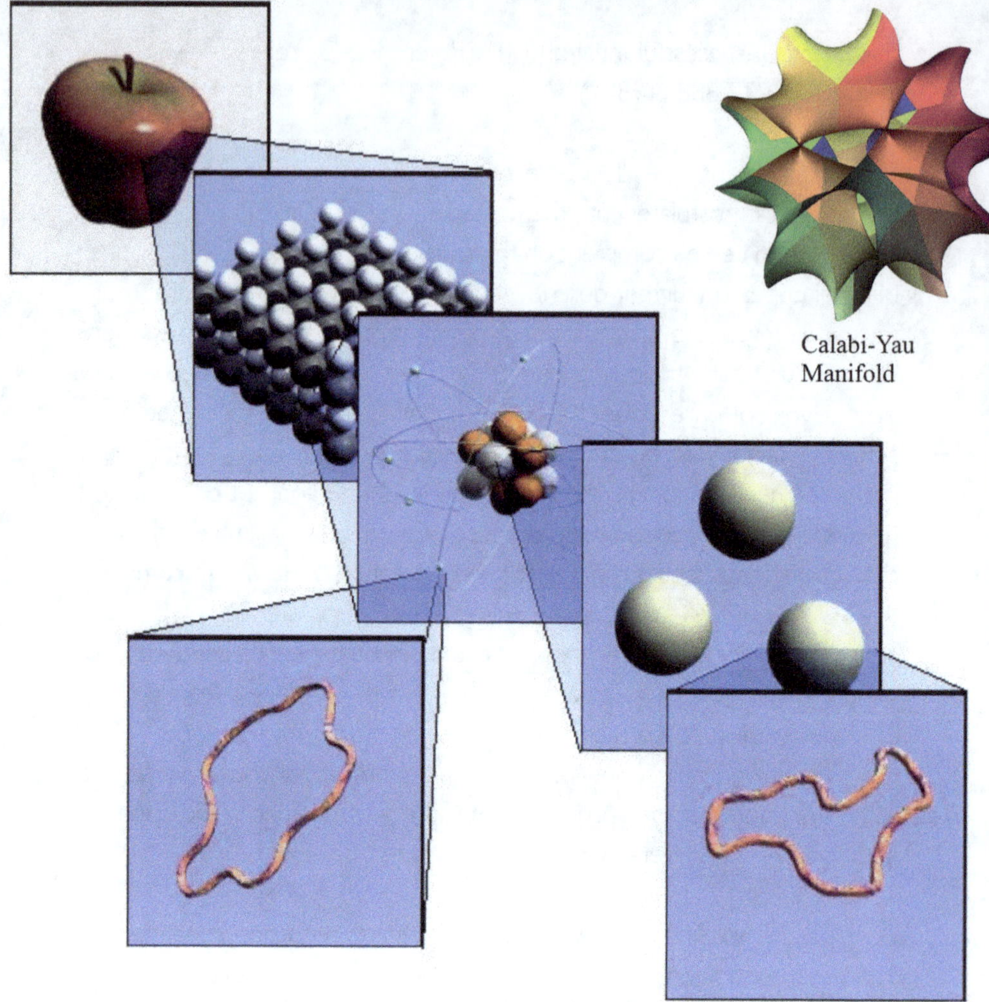

Chapter Twenty-Six Plate

Since Newton's apple revealed the reality of gravity, hypotheses of reality have now grown to encompass such things as M-theory (strings/branes). One of the spaces/manifolds that is said to be capable of housing the extra dimensions beyond our own 3D + 1 dimensions is the mathematical Calabi-Yau manifold shown above (top right).

Author for Apple to String particles diagram: Nina Hernitschek

Source for Calabi-Yau Manifold :Author: or Andrew J. Hanson, Indiana University (Ref."A construction for computer visualization of certain complex curves," Notices of the Amer. Math. Soc. 41 (9): 1156-1163, (November/December 1994).)

TWENTY-SIX

NEWTON'S REALITY AFTER HAWKING

Physicist Carlo Rovelli refers to the three great pillars of reality as time, gravity and quantum mechanics and that "Time sits at the centre of problems," which can unify these pillars. "...Using quantum mechanics Hawking successfully demonstrated that black holes are always 'hot'... The heat of black holes is a quantum effect upon an object, the black hole, which is gravitational in nature. It is the individual quanta of space, the elementary grains of space, the vibrating 'molecules' that heat the surface of black holes and generate black hole heat. This phenomenon involves three sides of the problem: quantum mechanics, general relativity and thermal science. The heat of black holes is like the Rosetta Stone of physics, written in a combination of three languages – quantum, gravitational and thermodynamic – still awaiting decipherment in order to reveal the true nature of time".[591]

In the last chapter, we explored Hawking's modified Big Bang theory including the 'no-boundary' principle, his ambivalent views on multiple universes, God, fine-tuning and the anthropic principle. He also relied upon the intriguing M-theory to analyse the mysteries of the universe. We thus need to deepen our analysis to probe the 'mind of Hawking', which can only be discerned from his later books. First, in **The Grand Design,** he and his co-author state:

> *"Model-dependent realism can provide a framework to discuss questions such as: if the world was created a finite time ago, what happened before that? An early Christian philosopher St Augustine (354–430) said that the answer was not that God was preparing hell for people who ask such questions, but that time was a property of the world that God created and that time did not exist before creation… that is one possible model, which is favoured by… [those supporting Genesis as literally true]… One can also have a different model in which time continues back 13.7 billion years ago to the big bang. The model that explains the most about our present observations, including the historical and geological evidence; is the best interpretation we have of the past… Still neither model can be said to be more real than the other."*[592]

Second, in the book published after his death in 2018, *Brief Answers to the Big Questions,* Hawking addresses the question 'How did it all begin?' in a chapter of the same name and relates the sequence of theories, as set out earlier, beginning with his famous 1965 paper with Roger Penrose showing that the universe had a beginning.[593] Again, he introduces the concept of imaginary time and how Feynman's mathematical techniques (recall his 'sum of histories' for the path of any particle) can be made to work with imaginary time. "Scientists are now working to combine Einstein's general theory of relativity and Feynman's idea of multiple histories into a complete unified theory that will describe everything that happens in the universe."[594] With the 'no-boundary' condition, he states there is a 'superabundance' of histories but we must pick out the possible history that includes human beings, eg. a history that has galaxies (again relying on the anthropic principle).[595] He draws in the 'sanity check' to any theory of the origin of the universe, namely, the background radiation discovered by Penzias and Wilson and, after referring to a 1982 paper on quantum fluctuations that would occur during an inflationary universe, draws a direct parallel with black holes aligning with his original joint concept with Roger Penrose: "This idea is basically the same mechanism as so-called Hawking radiation from a black-hole horizon… except that now it comes from a cosmological horizon, the surface that divided the universe between the parts that we can see and the parts that we cannot observe."[596] The lynchpin for answering 'How did it all begin?' comes at the end of that chapter where he concludes:

> ***"The beginning of the universe itself in the Hot Big Bang is the ultimate high-energy laboratory for testing M-theory, and our ideas about the building blocks of space-time and matter. Different theories leave behind different fingerprints in the current structure of the universe, so astrophysical data can give us clues about the unification of all the forces of nature. So there may well be other universes, but unfortunately we will never be able to explore them."***[597] He and his co-author also say: ***"The laws of M-theory therefore allow for different universes with different apparent laws, depending on how the internal space is curled. M-theory has solutions that allow for many different internal spaces, perhaps as many as 10^{100}, which means allows for 10^{100} different universes each with its own laws".***[598]

Hawking thus appears to rest his hope to explain the origin of the universe upon M-Theory yet seems ambivalent about his support for multiverses. What is M-theory that it can lead to multiverses and how can it assist us in finding some new system to understand the origin of the universe? (Recall Roger Penrose's view cited at the end of the last chapter that gravitational and quantum mechanics were not sufficient to explain the 'extraordinarily special' nature of the Big Bang with my comment that perhaps string theory could be the answer.) Recall also that Hawking rules out multiverses (using the principle of economy to 'cut them out of the theory'[599]) as they are not observable when arriving at the 'no-boundary' origin of the universe theory. However, later he appears to accept their possibility by relying on M-theory, which may imply multiverses. Recall also that he may not have been comfortable with the weak anthropic principle, let alone the strong anthropic principle, implying he did not support the implication that fine-tuning is explained by multiverses. If it is not explained by multiverses then what is it explained by? These are questions that we need to answer as we probe deeper into Hawking's Reality.

It is at this stage that we need to immerse ourselves into some very strange territory indeed. M-theory is, according to Roger Penrose, related to string theory which dominates the "physics departments and physics institutes right across the globe..."[600] and so we must strive to explain it despite its mind-boggling complexity. To allay the fears of those who are not quantum physicists or mathematicians, a candid quote from mathematician Edward Frenkel is called for. Frenkel writes of the difficulty of explaining the mathematics of string theory to his father: "All this stuff is quite heavy: we've got Hitchen moduli spaces, mirror symmetry, A-branes, B-branes, automorphic sheaves...[601] One can get a headache just trying to keep track of all of them. Believe me, even among specialists, very few people know the nuts and bolts of all elements of this construction".[602] The reader, encouraged by those comments, will now commence a slow winding descent into the depths of M-theory.

As M-theory originates through superstring theory, superstrings must be explained first. Strings have been described as "the basic building blocks of reality... [ie.]... vibrating, one dimensional loops of energy that quiver in 10 or more dimensions to strum out the elementary particles and fundamental forces of nature."[603] So how did string theory originate? String theory had its origins in Regge theory – the study of the behaviour of binding or scattering of a particle such as an electron. Curiously, the theory involved the particle moving as a function of angular momentum not only as a multiple of ħ (Planck's constant) but can take on complex values (ie. imaginary numbers) governed by a combined exponential and gamma function.[604] We saw imaginary numbers in the last chapter and so it is mysterious that real-world behaviour can be governed by the mathematics of 'impossible values'. Reality is strange! (Recall also that Feynman's sum of histories involves imaginary time.)

The origin of string theory began when one researcher, Gabriele Veneziano, trying to decipher Regge trajectories, conceived a theory involving self-consistent scattering amplitudes which became the first string theory. He was analysing hadrons which consist of protons, neutrons and pions bound by strong resonances called the strong forces. After a systematic relationship was discovered between spin and mass of hadrons, Veneziano proposed a model 'describing the quantised motion of a string'.[605] This saw the bonds between newly discovered quarks like strings of elastic. (Quarks and anti quarks are building blocks of protons, neutrons and pions.) This was a paradigm shift for reality because prior to this, particles were considered as points and now they could be considered as elastic strings. However, a more compelling theory explaining the strong interactions (involving quarks and gluons) arrived with the successful gauge theory called quantum chromodynamics (QCD). QCD involved a system of strong inter-quark forces linked by gluons – the quanta of the strong force fields holding quarks together and involving three types of charge dubbed 'colours' where gluons and quarks are defined with colours. Gluons can thus be seen as strings also. Symmetries in the form of gauge groups became a fundamental principle in the evolution of string theory, but the problem that emerged with QCD was that despite workable symmetries it could not be reconciled, ie. unified, with the electro weak subatomic forces, let alone gravity. String theory later evolved to include 'supersymmetry', which included the graviton in its core principles, thus presenting a possible Grand Unified Theory of the four forces (strong, weak, electromagnetic and gravity) to complete the Standard Model.

The Large Hadron Collider near Geneva, Switzerland is called that because, as mentioned, hadrons are basic particles such as protons, neutrons and pions. So, taking the proton or neutron (both are called nucleons as they are located in the nucleus of the atom), these nucleons are viewed as strings in string theory (ie. M-theory) which stretch, vibrate and rotate like a string. Other particles such as mesons (seen as sub hadrons known as sub-atomic quarks – anti quark pairs[606] – note also that pions, mentioned earlier, are a type of meson) are modelled as forces in string theory. Nucleons have three quarks. Some hadrons are closed strings such as 'gluons', which are seen as bonds/links inside quarks,[607] and three quarks are seen as the fundamental units of hadrons.[608] The symmetry of these particles and many others in the 'zoo' of quantum particles form the mathematical system for such things as the (fundamental) quarks predicted by Nobel Prize winners Gell-Mann and Zweig.

Then a 'super system' or 'supersymmetry' developed, linking two great categories of particles, bosons and fermions. Whereas symmetry involved particles such as bosons and fermions, supersymmetry involved such things as matching every boson to a fermion partner. (Fermions are elementary particles such as electrons or quarks which become selectrons or squarks.[609]) This supersymmetry is then used to form a unified theory of fundamental interactions between objects modelled as tiny strings called superstrings of a size of 10^{-35} m but with very high energies of 10^{19} GeV, well above the capability of modern particle accelerators, meaning they cannot be observed (yet).[610] Hawking himself then explains that basic objects are not particles but strings: "A particle occupies one point in space at each instant of time... A string, on the other hand, occupies a line in space at each moment of time. So its history in space-time is a two- dimensional surface called a world sheet."[611] Superstrings involve these 'string particles' and display supersymmetries between fundamental particles such as bosons and fermions matched as described to become selectrons or squarks.

In the formulation of a new speculative theory, the initial problem was that there were initially an unwieldy twenty-six dimensions for bosons and ten dimensions for fermions. Both bosons and fermions are elementary particles such as atoms or nuclei, but bosons have integral spin and fermions have ½ spin. Two researchers, who were proponents of earlier string theory involving fermions in 10 dimensions, noticed that the model seemed to imply a particle with 0 mass and spin 2 which described the graviton.[612] Their theory became known as the Schwarz-Schwerk string theory.[613] The first four dimensions could be dealt with using the four dimensions of space and time. Hawking explained this problem of dimensions: "String theories, however, have a bigger problem: they seem to be consistent only if space-time has either ten or twenty-six dimensions, instead of the usual four! Of course, extra space-time dimensions are a commonplace of science fiction..."[614] The remaining dimensions were dealt with using a technique called compactification, adopted from the Kaluza-Klein theory devised in the 1920s. Kaluza and Klein's work was followed by Einstein, due to their goal of unifying electromagnetism and gravity.[615] They had claimed the 5th dimension was too small to be observed – "curled up or compactified, like origami folded from the fabric of reality".[616] The string theorists followed this technique with the extra six dimensions 'curled up' as they were not observable either.

The advantage of the string theory was that the mathematics necessarily implied the existence of the graviton – the spin 2 particle that solved the problem of how to quantise gravity thus unifying general relativity with quantum mechanics.[617] Thus, the concept of 'super gravity' became part of the theory by linking the spin 2 graviton particle with the other 'super particles'.[618] The theory was neat not only because it unified quantum mechanics with gravity but because it solved other problems such as the inherent infinities in quantum modelling (solved by the least preferable option of 'renormalisation'), the chirality problem of symmetries (point theories at the time involved eleven dimensions, and ten dimensions fitted symmetrical models better) and it became an 'anomaly free' theory.[619] Hawking describes how – despite a landmark paper in 1974 by two scientists mentioned earlier, Joël Schwerk and John Schwarz – it lost appeal until 1984 when it was revived for two reasons: "[First]… people were not really making much progress toward showing that super gravity was finite or that it could explain the kinds of particles that we observe… [and, secondly, the publication of a paper by John Schwarz[620] and Mike Green of Queen Mary College, London]… showed that string theory might be able to explain the existence of particles that have a built-in left handedness… (with the upgraded string theory called heterotic string theory).[621] Independently, particle physicist Leonard Susskind, then at Yeshiva University, New York, had also been working on some mathematics to describe the strong force binding protons and neutrons and realised that modelling particles on vibrating loops of string which oscillate and separate was viable.[622] Susskind (a pioneer of string theory) was elated until he found out that Schwarz and Green had hit on the same idea but with wider application. This led to a sensation in the physics world in about 1984–85 with articles in *Nature, Science, Time* and the *New York Times.*[623] Nobel Prize winner Murray Gell-Mann referred to string theory as: "The theory of everything – gravity, weak, strong, and electromagnetic interactions plus a lot of things all together – a completely unified theory of nature."[624] Nobel Prize winner Stephen Weinberg said: "I dropped everything I was doing, including several books I was working on, and started learning everything I could about superstring theory."[625]

Yet this euphoria was tempered by worried physicists such as Sheldon Glashow and Paul Ginsparg who wrote in *Physics Today* an article entitled 'Desperately Seeking Superstrings', including this quote: "Years of intense effort by dozens of the best and the brightest have yielded not one verifiable prediction… [from superstring theory]… The theory depends for its existence upon magical coincidences, miraculous cancellations, and relations between unrelated (and possibly undiscovered) fields of mathematics. Are these properties reasons to accept the reality of superstrings?"[626] Another commentator referred to string theory as 'recreational mathematical theology.'[627]

Ed Witten, a pioneer of superstring theory, also said: "String theory was invented essentially by accident in a long sequence of events, starting with the Veneziano model that was formulated in 1968... it was invented by a lucky accident." Witten saw string theory as twenty-first or twenty-second-century mathematics and it "should not have been invented until our knowledge of some of the areas that are prerequisite for string theory had developed to the point that it was possible for us to have the right concept of what it was all about."[628] Ed Witten was considered one of the 'greatest living theoretical physicists' and has been a leading proponent in a type of 'math unification initiative' known as the Langlands program.[629] He is also the first physicist to win the Fields medal – the 'Nobel Prize of Mathematics'.[630] Mathematician Edward Frenkel has written a book – *Love and Mathematics* – about the Langlands program, which assisted the development of string theory.[631] Frenkel also details his fruitful cooperation with Ed Witten including a joint paper supporting string theory on the exotic topic of branes existing on tori ('toruses') in Hitchen's moduli spaces(!)[632] These moduli spaces span *Riemann surfaces* (eg. a simple function such as y = logx is single valued – you plug in a value for x and you get one value for y easily represented as a curve on a 2D x-y graph. However, if you use imaginary numbers, the complex function gives multiple values. For example, the complex function w=logz results in w having multiple values, which is not so simple to graph. Thus, Riemann invented a way to show multiple values on a 3D graph eg. w-logz results in a spiral ramp. In string theory, Riemann surfaces/manifolds take on even more exotic shapes), but with functions graphed on these manifolds in string theory they have something in mind far more exotic than a spiral ramp.

Imagine a sculpture of an octopus dancing where you can't really discern that it is an octopus.[633] That is the type of manifold (eg. the Calabi-Yau moduli spaces) that has been proposed as a workable manifold for string theory – namely, to serve as the co ordinate system for the six curled-up dimensions. Edward Frenkel describes the origin of the obscure Hitchen moduli spaces: "[British Mathematician and Oxford Professor, Nigel Hitchen's] insight... was that equations could be written on any curved Riemann surface, such as the surface of a donut or a pretzel, as well. Physicists missed this... But Hitchen saw that, mathematically, solutions on these surfaces were quite rich. He introduced his moduli space M(X,G) as the space of solutions on a Riemann surface X (in the case of the gauge group G)..." (Presumably, G could be variables representing a group of quantum particles or strings that need solutions to describe their behaviour.) "He found it was a remarkable manifold, in particular, it possessed the 'hyper-Kähler metric'..."[634] (The Calabi-Yau manifolds are examples of Kähler manifolds which Roger Penrose explains are manifolds with a Riemann surface co ordinate system but not 1-dimensional, rather 'complex-3-manifolds'.[635]) Although Calabi-Yau spaces can potentially appear in thousands of different configurations, this is seen as an advantage to model particles because their oscillating rotations slice through varying amplitudes that the Calabi-Yau space is flexible enough to represent (eg. representing spinors[636], which are quantum particles similar to a vector but which change sign on each rotation as would be expected by the vectors in Schodinger's equation using the Dirac Vectors).[637]

If the reader showed the last passage to anyone, they would be likely to scratch their head unless they are a quantum physicist or mathematician, but it is hoped that mathematician Edward Frenkel's comments quoted earlier in this chapter have allayed any 'string phobia' somewhat as it seems few fully understand it all. Frenkel also comments on the mutually beneficial relationship between physics and mathematics: "The interaction between math and physics is a two-way process... At different times one of them may take the lead in developing a particular idea, only to yield to the other subject as focus shifts."[638] John Bagger of the Institute for Advanced Study (Princeton University), a colleague of Ed Witten at the time, said: "Superstring theory may tell us why the universe looks the way it looks. Why is it so large? Why isn't the universe the size of a marble...? In other words, how does the theory know what's called the vacuum structure...? This is the old question of the cosmological constant. Why is the cosmological constant so small?"[639] (The cosmological constant is estimated to be one part in 10^{120}.) In fact, the cosmological constant question was thought to be the only important question that superstring theory had not answered.[640]

This brings us back to M-theory. The *Oxford Dictionary of Physics* does not contain a separate definition for M-theory but mentions it under 'Superstring Theory' in the following way: "All the viable superstring theories and also super gravity are related by duality, with the unified theory underpinning this being an 11-dimensional theory called M-Theory. Both the entropy of black holes and Hawking radiation have been explained in terms of superstring theory... There is much more to superstring theory than strings alone, as the theory has membranes ('branes') and also solutions that are extended objects in more than two space dimensions."[641]

This duality was the key breakthrough that transformed string theory to become M-theory and in fact it has been suggested that string theory in all its forms be now called M-theory. Author and Professor of Physics Steven S. Gubser (Princeton University) explains it as the duality between the Type IIA string theory and M-theory being created by modelling the string interactions very strongly and enlarging the eleventh dimension to almost its own manifold or co ordinate system and then linking the two theories.[642] Michio Kaku also explains this transformation:

> *"[The string theories]... are based on different symmetries with exotic names like E(8)xE(8) and O(32)... In 11 dimensions... there are alternate super theories based on membranes as well as point particles (called supergravity). In lower dimensions, there is moreover a whole zoo of super theories based on membranes in different dimensions... For the p-dimensional case, [someone] dubbed them p-branes (pronounced "pea brains")... [Physicist, Ed Witten and others showed]... that there was a duality between 10 dimensional Type IIa strings and 11 dimensional supergravity! The non-peturbative region ofType IIa strings, which was previously a forbidden region, was revealed to be governed by 11 dimensional supergravity theory, with one dimension curled up... I remember saying to myself, 'But that's impossible!' All of a sudden, we realised that perhaps the real 'home' of string theory was not 10 dimensions, but possibly 11, and that the theory wasn't fundamentally a string theory at all!... Lurking in the 11th dimension was an entirely new theory which could reduce down to 11 dimensional supergravity as well as 10 dimensional string theory and p-brane theory."*[643]

Professor Gubser provides a good example of string theory applied to black holes: one of the family of membranes or 'branes' forming the fabric between strings is the D0-brane (a point) with layers in the millions to become a black hole event horizon, whilst the D3-brane is layered inside the black hole linked by strings/gluons (gluons are the colour charge bonds between quarks).[644] Curved black hole geometry is equivalent to the gluon gauge symmetry from quantum mechanics. This curved geometry is ten dimensional in contrast to the 3 space and 1 time dimension governing gluons.[645] The horizon increases in temperature by reason of the vibrational energy of the D3-branes, allowing the energy of the strongly interacting gluons to be measured (noting that measurement in a black hole is theoretically impossible because nothing can escape except Hawking radiation).[646] (Recall in Chapter 24 how Professor Susskind used string theory to analyse the entropy of a black hole.) Now two of the great dualities in string theory are first, supergravity – theories (the quantised gravity is linked to other fields via the graviton) where low energy dynamics of superstrings of the lowest vibrational modes unify with the weak gravitational force. The second duality is what Professor Gruber calls "long, straight strings and long straight branes," which he describes as "zero temperature black holes in supergravity."[647]

Michio Kaku also describes Ed Witten's pivotal role in string theory:

> ***"Not knowing what to call it ,Witten has dubbed it M-theory… In one stroke, M-theory has solved many of the embarrassing features of the theory, such as why we have 5 superstring theories and from Witten himself, 'I think there are still a couple more superstring revolutions in our future, at least and one of its other founders, Schwarz, states of M-theory – 'Whether it is based on something geometrical (like super membranes) or something completely different is still not known. In any case, finding it would be a landmark in human intellectual history.'"***[648]

How does M-theory imply many worlds or the multiverse then? In the last chapter, recall how Hawking attended a conference by Russian Andrei Linde. Linde had solved a problem with Guth's inflationary universe problem by proposing a 'chaotic beginning' Big Bang with a slowing down of inflation and a 'graceful exit' to the inflationary state (instead of Guth's use of quantum tunnelling which was problematic).[649] Chaos creates fractal patterns and the universe presents as the ultimate fractal in experimental results such as data maps captured by the COBE and WMAP satellites showing imprints of quantum fluctuations in the early universe.[650] The WMAP data picture of the early universe was based on the microwave background radiation and gave an age for the universe of 13.7 billion years (now generally adopted) and also indicated that 70% of the universe consists of dark matter.[651] Importantly, Linde explained: "Galaxies are children of random quantum fluctuations produced during the first 10^{-35} seconds after the birth of the universe" or the 'graceful' exit via the slowdown in chaotic inflation.[652]

At the same time, Linde swung his focus from the early universe to the end process happening now in cosmic time (our time) and realised that this dissipative fractal process would probably spawn 'islands' of space-time in pockets that he called 'island universes' where inflation ends or as he explained: "As a result the universe becomes a multiverse, an eternally growing fractal consisting of exponentially many exponentially large parts... These parts are so large that for all practical purposes they look like separate universes." Now separate to Linde's work, the string theory emerging from Schwarz, Green and Susskind indicated a geometry of dimensions where "the theory allowed for many combinations of physical laws and constants," which implied a staggering 10^{500} possible solutions.[653] This was tempered by Susskind in describing the result: "Mathematical existence is not the same as physical existence. Discovering that string theory had 10^{500} possible solutions explains nothing about our world unless we also understand how the corresponding environments came into being."[654] Thus, in the Preface, Hawking said to Larry King: "Gravity is a consequence of M-theory which is the only possible unified theory – it is like saying 'why is 2 + 2 = 4?'." The only problem with that comparison is that gravity is governed by relativity and with time dilation 2+2 does not always equal 4 seconds. Moreover, whereas Newton was careful to construct or *abstract* his gravitational equation from observation without describing its essence or origin, Hawking reverses this stance by supporting laws of a theory to generate gravity.

Parallel to this (mathematical) possibility, Linde saw his fractal island universes as potentially huge with differing 'properties' which some might wrongly interpret as differing 'laws' or perhaps, just as a singularity implies a breakdown in the laws of physics some change could take place in these island universes. Linde described these differing properties as follows: "It's not that the laws of physics are different in each universe, but their realisations... [are different]... An analogy is the relationship between liquid water and ice. They're both H_20 but realized differently."[655]

Everett's many-worlds hypothesis is quite different from this multiverse concept. At present, both are only possible mappings under a mathematical function but they are quite different. If the laws of the multiverses varied like a zone in the universe, such as a singularity in a black hole, it is hard to see a coherent theory between those multiverses and our universe in the same way we can still theorise between the singularity in a black hole and the outside universe. In other words, the laws of the universe (and daughter universes) are 'grand fathered' in the sense that the universe still governs those daughter universes (as described by Linde), which is quite different to the mathematical model of 10^{500} different solutions (laws?) portrayed by many people who rely on the multiverse as part of the strong anthropic principle.

Moreover, research professor Ricard V. Solé, in his book *Phase Transitions,* asks a key question governing fractal chaotic systems (such as a universe): "What makes us label a collection of objects a 'system'? The most obvious answer is that there is some kind of link between different objects such that it makes sense to consider them as part of a larger structure".[656] Clearly, island universes are part of the larger system which has spawned multiple galaxies such as ours, all of which are governed by the laws of our system – our universe. Thus, having seen that Linde's chaotic inflationary universe implies 'island universes' with the same universal laws (but distorted by cosmic development), does M-Theory with its potential 10^{500} different solutions/universes imply more *physical* universes with different laws?

First, unlike Linde's theory implying single law universes based on evidence consistent with data from WMAP and COBE, M-Theory has not evolved into an evidence-based theory and thus the possible existence of these 'unilaw' or 'multilaw' universes remains a mathematical possibility unrepresentative of existing reality. (Recall 'The map is not the territory'[657] needs to be borne in mind.) Secondly, does a 'solution' have to mean a separate universe? Could it not mean a set of distorted states in one island universe corresponding to a set of solutions? Thirdly, Roger Penrose has observed that many of the 10^{500} possibilities (solutions) "turn out to be mathematically inconsistent. These constitute what has become known as the *swampland.*"[658] Fourthly, these potential 10^{500} different solutions involve a vast set of degrees of freedom (ie. the number of potential parameters governing each potential string system). Roger Penrose calculates these degrees of freedom to be the vast $\infty^{70\infty^9}$ for 'Ricci flatness'[659] in ten dimensions compared to the much smaller $\infty^{N\infty^3}$ in four space-time dimensions.[660] The problem is that these degrees of freedom do not seem to make an appearance in our reality, just like the six compact dimensions that are 'curled up' under string theory make no appearance (presumably because they are below the detectable limit in particle accelerators).

As strings are quantised, it is thought that there are constraints that will greatly reduce the degrees of freedom to a workable system like the four space-time dimensions. Things like the cosmological constant and the anthropic principle can also reduce these degrees of freedom as the M-theory mathematics must work with these constraints also. One of these constraints is identified by Andrew Chamblin of the University of Louisville as a 'no-go' theory which he describes as: "A no-go theorem, due to Gibbons, Maldecena and Nuflez, and also de Wit, Dass and Smit which basically asserts that if you compactify a string then you will never get de Sitter space… [ie. a de Sitter space is based on the cosmological constant in a model of our universe expanding exponentially using Einstein's relativity where $G=\Lambda$.[661]]… Since the universe is evidently past and future de Sitter (albeit with vastly different vacuum energies) this would seem to be a problem."[662] One of the possible solutions quoted by Chamblin include: "The internal space is described by certain scalar fields known as 'moduli'. These moduli describe the size, shape and other basic properties of the internal space. These moduli typically have exponential potentials, with the property that as you flow from the minimum of the potential the universe decompactifies… one can now imagine 'bouncing' the universe off of this potential."[663]

Professor Gubser describes the need to make M-theory consistent with the cosmological constant (but with less emphasis on the anthropic principle):

> *"In fact, observations over the last ten years seem to indicate that the expansion of the universe is accelerating in a way that's consistent (so far) with a very small cosmological constant. If we want to describe a world using string theory, then it seems we need to tie up the extra six dimensions so that they can't move at all, but leave the usual three with just the slightest tendency to expand and to accelerate their expansion. It's hard to figure out how to do that. But it does seem that the number of ways of tying up the extra dimensions is tremendously large… Our universe just happens to be one in which the extra dimensions are tied up in just the right way. If it weren't, the argument goes – intelligent life would probably be impossible; so we would not exist. Turning things around, our existence implies that the universe we inhabit has a small cosmological constant. Altogether, I find myself unconvinced that this line of argument is useful in string theory."*[664]

Another enigma from Andrew Chamblin of the University of Louisville is:

"How could the cosmological constant have been so large then [after the Big Bang], and so small now?"[665] Perhaps Professor Susskind's conclusion on the issue sums it up: [commenting on physicist Renata Kallosh's view, 'In science, the preference will always be given to non-anthropic explanations – unless there is nothing better,'] he said:

> ***"It is not enough to say, 'I hate the idea'. You have to say, 'Here's a better idea.' Every month or so somebody will come out with some screwball theory of why the cosmological constant is close to zero, but it won't last for more than a week. A legitimate controversy is when there are two or more or less equally good ideas which are in conflict with each other. The simple fact is there is no competition."***[666]

The anthropic principle was proven to work scientifically with Fred Hoyle's work with the origin of carbon-12, yet there seems to be the prejudice against it. If we defined this prejudice as one of Thomas Kuhn's paradigms[667] then one would probably define the paradigm as originating from the Enlightenment when God was banished from the laboratory. Despite Barrow and Tipler's multiverse 'gloss' to the principle (see Chapter 5), the reason is probably the link between the argument that God exists and the universe being fine-tuned for life, which has really been a major part of what the anthropic principle is all about. Hawking sees the anthropic principle as a 'counsel of despair', but it would seem the principle looms to solve the riddles of the universe after Susskind's comment. God figured strongly in the language of Einstein and Bohr as it did with Newton and Leibniz without hindrance to their spectacular scientific results. In fact, it seems it is hard to keep God outside scientific discussion! It would seem that Einstein was right as he saw God as the great riddle setter.

The cosmological constant (now thought to represent dark matter) has been an enigma that physicists cannot explain as to 'why it is so tiny' (it is of the order of 10^{-120})[668]. If the differing 'realisations' of the vacuum state in space vary in these island universes such that the constant in those universes also varies then it is thought this may explain why the cosmological constant is so tiny. Other constants may vary in those island universes also, perhaps according to deep underlying laws that we are yet to discover. In 1994, during a series of debates with Roger Penrose, Hawking said: "The only way to have a scientific theory is if the laws of physics hold everywhere, including at the beginning of the universe."[669] Presumably, this would also mean (in these island universes) – the megaverse. As appears below, it is only the purely mathematically based multiverses that permit variations of laws. Again the 'map is not the territory'.[670]

Having explained M-theory and its implied multiverse, it is perhaps important to take stock of how many parallel universe theories there are. Physicist Professor Max Tegmark in his book *Our Mathematical Universe*[671] refers to the various levels of multiverse. Level I – "distant regions of the universe which light..." (the 'wave of reality'[672]) "...hasn't had time to reach" yet; Level II Multiverse – the Linde megaverses described above with chaotic inflation giving a fractal universe; Level III multiverse, the classic 'many worlds' of Hugh Everett (see chapter 5) and the Level IV multiverse – a physical set of many parallel universes which Professor Tegmark postulates as equivalent to a mathematical universe – hence the title to his book which defines these multiverses.

The last multiverse mentioned is the most speculative and indeed subject to the same caution Professor Leonard Susskind expressed over the mathematical universe constructed by string theory – namely, no multiverse becomes accepted theory without experimental verification. Professor Tegmark does discuss what makes theories unscientific by referring to philosopher of science Karl Popper's definition as follows: "If it's not falsifiable, then it's not scientific... [Popper]" with Professor Tegmark commenting: "Physics is about testing mathematical theories against observation: if a theory can't be tested even in principle, then it's logically impossible to ever falsify it, which by Popper's definition, means it is unscientific." But he then states: "For a theory to be falsifiable, we need not be able to observe and test all its predictions, merely at least one of them."[673] This places many scientific theories into the 'unscientific category' of Popper. String theory is presently beyond the capacity of particle accelerators. Even the Big Bang theory seems to require a replication of the Big Bang, which would seem impossible. One of the rival theories to string theory is the simpler Loop Quantum Gravity hypothesis supported by physicist Carlo Rovelli. He describes it as involving dynamic quantum space or 'atoms of space' using the mathematics of Dirac's quantum mechanics and Einstein's relativity.[674] It awaits experimental verification from such sources as WMAP or COBE to detect cosmic gravitational background radiation.[675] Yet Popper's sweeping definition may be too harsh because Hawking's theory of black holes has been found to be difficult to verify experimentally, yet no one would state it is unscientific. It seems that Popper's principle should be applied philosophically on a case-by-case basis.

Professor Tegmark makes an interesting point about Linde's chaotic inflation theory, namely, if you accept it then you have to accept these island universes (otherwise, you are accepting a theory that could be falsified if you could disprove island universes).[676] Furthermore, he comments on the vital point of the laws and constants in Linde's megaverse/multiverse (Tegmark's Level II Multiverse) as follows: "[the Level II multiverse] covers regions that are forever unreachable because of the cosmic inflation of intervening space… [And] Whereas all the parallel universes in Levels I, Level II and III obey the same fundamental equations (describing quantum mechanics, inflation etc)." (His Level IV multiverse being mathematical is a 'multilaw' universe and aligns with the potential mathematical 'multilaw' string theory which Susskind cautioned against accepting when he said: "Mathematical existence is not the same as physical existence."[677] It is also worth restating that mathematics is an abstraction of methods to describe the real world. It is one of Wittgenstein's nets (see Chapter 23) adopted to describe reality but telling us nothing of what reality is, just as Newton's mechanics does not tell us what gravity is. If there is one thing that Chapter 13 demonstrates, it is that a perfect world is not possible because the mathematics cannot provide solutions to most physical problems. Proponents of a purely mathematical universe might claim, 'Well, a singularity is an unsolvable maths problem with its infinite curvature of space and time but this is handled by the universe in merely becoming a sink such as a black hole' – indeed, complex analysis has a way around singularities.

However, the analysis involving irrationals, infinities and super tasks in Chapter 13 indicates a far wider insolubility in physical systems occurs, showing that the packing of particles leads to mathematical roadblocks to a 'perfect reality'. Again, chaos is based on deterministic equations even if we cannot recover all the information in a chaotic system. However, mathematics cannot provide an exact solution to the square root of a number and can only provide ways around this insolubility in selected cases. In the chaos of physical systems, a tiny insolubility such as a square root can be amplified in terms of Lorenz's Butterfly Effect enormously as that system transitions to chaos. Mathematics also had great difficulty with the three-body problem in the age of Newton. Thus, for mathematics to be a complete universal system, it would have to have solutions for every event and system in the universe. The mathematical universe might become feasible if it was fully quantised, ie. being all discrete (not just quantum mechanics) and not continuous as has been suggested by Stephen Wolfram and in a 1981 paper by Richard Feynman.[678]

Newton's Reality has changed after Hawking in a number of ways. After quantum mechanics, the classical system of physics that Newton helped create is no longer a complete 'clockwork system'. Newton saw that chaos may produce incalculability but perhaps not the uncertainty that is inherent in Heisenberg's Uncertainty principle. Yet the hidden variable principle championed by Einstein and Bohm (yet made problematic by Bell) remains as a hope for classical systems. The strangeness of the quantum world is exemplified by what Einstein called 'spooky action at a distance' after the work of Bell and the testing of the famous Einstein-Podolsky-Rosen paradox by Alain Aspect, Jean Dalibard and Gerard Roger in Paris in the 1980s (which showed that 'spooky action at a distance' is real).[679] To use Carlo Rovelli's metaphor, Hawking's 'Rosetta stone' between quantum, gravitational and thermodynamic systems revolutionised cosmology. His brilliant insight was to see the instant of creation of the universe as a Black Hole singularity in reverse to solidify the Big Bang theory. His colleague Roger Penrose has also written of an extraordinary paradox at the moment of the Big Bang, namely, having on the one hand tiny entropy to obey the second law of thermodynamics, yet on the other hand the remarkable cosmic background radiation evidence to tell us there was enormous entropy.[680] In fact, Roger Penrose believes "we shall need to return to an examination of the very foundations of quantum mechanics... [to solve it]... for it is my strong opinion that these issues are deeply connected."[681] Perhaps a new Hawking-Penrose Rosetta stone will emerge.

Big changes have taken place in science. Newton was initially criticised by Hooke for not having a full hypothesis for his theory of light yet celebrated later for being rigorous in only relying on the results of experiment to construct a theory. Newton would go no further than was demonstrated by experiment to construct the theory and equations to clothe that theory. Yet now it seems that Hooke reigns in his approach because 'hypothesis first' has returned and the positivist approach of viewing metaphysics as 'nonsense' has been buried. This is a massive shift in a systemic scientific paradigm. Metaphysics, in the sense of hypothesis before experimental results, has been demonstrated to work. Gell-Mann and Zweig's work on symmetry predicted quarks before experiment showed their existence. The Higgs boson was theoretical before it was discovered in tests at the Large Hadron Collider. Roger Penrose's Twistor theory provides hope to unify relativity and quantum mechanics pending confirmation. Anti-particles were indicated by Dirac's equations before indicated in experiment, and even philosophy of science in the form of the anthropic principle works, as shown by Hoyle's stunning predictions on the resonances that needed to exist for the life-giving Carbon-12. Thus, M-theory with its strings and branes now thrives in metaphysical form awaiting validation (or will loop quantum gravity receive the accolades!).

Both Newton and Hawking have risen on the shoulders of giants and in tribute to both, it is worth repeating the profound words spoken by them:

> ***"I don't know what I may seem to the world but as to myself, I seem to have been only like a boy playing on the seashore, and diverting myself in now and then finding a pebble smoother or prettier shell than ordinary, whilst the great ocean lay all undiscovered before me."***[682] **[Isaac Newton]**

On Stephen Hawking's gravestone in Westminster Abbey between the graves of Newton and Darwin, the words appear: 'Here lies what was mortal' and it was reported that Hawking's advice was beamed towards a black hole: "Be brave, be determined, overcome the odds – it can be done." Let us hope that he does now truly know the mind of God.

NOTES TO PART II – MASTERS OF THE COSMOS

1. J. Borzendowkski, Marie Curie – *Mother of Modern Physics,* Sterling Publishing Co. Inc, 2009,pp55–56
2. J. Borzendowkski, ibid, p53
3. J. Borzendowkski, ibid, p3
4. J. Borzendowkski, ibid, p4
5. J. Borzendowkski, ibid, p7
6. J. Borzendowkski, ibid, p9
7. J. Borzendowkski, ibid, pp6–7
8. N. Pasachoff, exhibited in American Institute of Physics paper (Editor: S.R. Weart) based on her book *Marie Curie and the Science of Radioactivity,* Oxford University Press, 1996, p4 of paper.
9. J. Borzendowkski, ibid, p11
10. J. Borzendowkski, ibid, p11
11. N. Pasachoff, ibid, p5
12. J. Borzendowkski, ibid, p12
13. J. Borzendowkski, ibid, pp14–15
14. *Oxford Dictionary of Philosophy,* Oxford University Press, 2016, p94
15. J. Marías, *History of Philosophy,* Dover Publications, 1967 (English Translation), p355

16. J. Borzendowkski, ibid, p16–17
17. J. Borzendowkski, ibid, p18
18. J. Borzendowkski, ibid, pp 18–19
19. J. Borzendowkski, ibid, p20
20. J. Borzendowkski, ibid, p21
21. J. Borzendowkski, ibid, p22
22. J. Borzendowkski, ibid, p22
23. J. Borzendowkski, ibid, pp22–23
24. N. Pasachoff, ibid, p5
25. N. Pasachoff, ibid, p5
26. N. Pasachoff, ibid, p15
27. N. Pasachoff, ibid, p15
28. N. Pasachoff, ibid, p15
29. J. Borzendowkski, ibid, p29
30. N. Pasachoff, ibid, p16
31. N. Pasachoff, ibid, p16
32. N. Pasachoff, ibid, p17
33. J. Borzendowkski, ibid, p29
34. J. Borzendowkski, ibid, p34
35. J. Borzendowkski, ibid, p34
36. J. Borzendowkski, ibid, p37
37. J. Borzendowkski, ibid, p31
38. J. Borzendowkski, ibid, p41
39. J. Borzendowkski, ibid, pp63–65
40. J. Borzendowkski, ibid, pp66–67
41. J. Borzendowkski, ibid, p71, & N. Pasachoff, ibid, p47
42. N. Pasachoff, ibid, p48
43. N. Pasachoff, ibid, p51
44. J. Borzendowkski, ibid, pp80–81
45. J. Borzendowkski, ibid, p80
46. J. Borzendowkski, ibid, pp85–86
47. J. Borzendowkski, ibid, p82
48. N. Pasachoff, ibid, p57
49. N. Pasachoff, ibid, p58
50. J. Borzendowkski, ibid, p92
51. Quoted from N. Pasachoff, ibid, p59
52. J. Borzendowkski, ibid, p93
53. J. Borzendowkski, ibid, p93
54. J. Borzendowkski, ibid, pp93–95
55. J. Borzendowkski, ibid, pp98–99
56. N. Pasachoff, ibid, p64
57. N. Pasachoff, ibid, p63
58. J. Borzendowkski, ibid, p101
59. J. Borzendowkski, ibid, p102

60. N. Pasachoff, ibid, p68
61. J. Borzendowkski, ibid, p112
62. E. Curie, *Madam Curie – A Biography by Eve Curie* (originally published in 1937, Doubleday Doran & Co), this quote from De Capo Press Edition, Perseus Books, 2001, pxv
63. B. Marshall (himself a Nobel Prize winner) with Lorna Henry, *How to win a Nobel Prize*, Piccolo Nero, Schwarz Publishing, 2018, p36
64. N. Pasachoff, ibid, p72
65. J. Borzendowkski, ibid, p117
66. J. Borzendowkski, ibid, p108
67. E. Curie, Madam Curie – *A Biography by Eve Curie*, ibid, pxvii
68. *Oxford Dictionary of World History,* Oxford University Press, 3rd Ed, 2015, p527; *Longman Illustrated Encyclopedia of World History*, Ivy Leaf, 1991, pp730–731
69. *Longman Illustrated Encyclopedia of World History*, ibid, p730
70. *Oxford Dictionary of World History,* ibid, p383
71. *Oxford Dictionary of World History,* ibid, p383, pp527–528
72. *Oxford Dictionary of World History,* ibid, p457
73. *Oxford Dictionary of World History,* ibid, pp527–528, *Longman Illustrated Encyclopaedia of World History,* ibid, pp731–732
74. *Oxford Dictionary of World History,*, ibid, p457
75. *Oxford Dictionary of World History,* ibid, p457, *Longman Illustrated Encyclopaedia of World History*, ibid, p338
76. E. Curie, *Madam Curie – A Biography by Eve Curie*, ibid, p137
77. E. Curie, *Madam Curie – A Biography by Eve Curie*, ibid, p51
78. E. Curie, *Madam Curie – A Biography by Eve Curie*, ibid, p52
79. E. Curie, *Madam Curie – A Biography by Eve Curie*, ibid, p55, Marya Rakovska was a friend of Bronya, p52
80. J.Hawking, *Travelling to Infinity: My Life with Stephen*, Alma Books Ltd, 2007, pp138–139
81. I. Berlin, *The Proper Study of Mankind,*Vintage, 2013, p46
82. R. Scruton, *A Short History of Modern Philosophy,* Routledge Classics, 2002, p108
83. P. Watson, *Ideas, A History from Fire to Freud,* Weidenfeld & Nicolson, 2005, p883
84. *Oxford Dictionary of World History,* ibid, p121 & p527
85. B.Russell, *History of Western Philosophy,* Routledge Classics, 2004, p625
86. B.Russell, *History of Western Philosophy,* ibid, pp625–626
87. B.Russell, *History of Western Philosophy,*, ibid, p626
88. Will and Ariel Durant, *The Age of Voltaire – The Story of Civilisation Part IX*, Simon & Schuster, 1965, p 721
89. Will and Ariel Durant, *The Age of Voltaire – The Story of Civilisation*, ibid. p722
90. Will and Ariel Durant, *The Age of Voltaire – The Story of civilisation*, ibid, p722
91. Voltaire later softened this line and added twenty-eight more lines: Will and

Ariel Durant, *The Age of Voltaire – The Story of Civilisation*, ibid, p722

92. B. Russell, *History of Western Philosophy,* ibid, p627
93. B.Russell, *History of Western Philosophy,*, ibid, p626
94. B.Russell, *History of Western Philosophy,* ibid, pp628–629
95. P.Watson, *Ideas, A History from Fire to Freud*, ibid,, p879
96. B.Russell, *History of Western Philosophy*, ibid, p632
97. B.Russell, *History of Western Philosophy*, ibid, p632
98. P.Watson, *Ideas, A History from Fire to Freud*, ibid, p881
99. P.Watson, *Ideas, A History from Fire to Freud*, ibid, p882
100. P.Watson, *Ideas, A History from Fire to Freud*, ibid, pp881–882
101. P.Watson, *Ideas, A History from Fire to Freud*, ibid, p883
102. P.Watson, *Ideas, A History from Fire to Freud*, ibid, p884
103. P.Watson, *Ideas, A History from Fire to Freud*, ibid, pp884–885
104. Fyodor Dostoevsky *The Brothers Karamazov,* Vintage Books, 2004 Edition, p171
105. J. Marías, *History of Philosophy*, Dover Publications, 1967 (English Translation) p355
106. J. Marías, *History of Philosophy,* ibid, p352
107. J. Marías, *History of Philosophy,* ibid, p401
108. According to R. Scruton, *A Short History of Modern Philosophy,* Routledge Classics, 2002, p281
109. Per R. Scruton, *A Short History of Modern Philosophy,* Routledge Classics, 2002, p293
110. *Oxford Dictionary of Philosophy*, Oxford University Press, 3rd Ed, 2016, p421
111. I. Berlin, *The Proper Study of Mankind*, ibid, p567
112. I. Berlin, *The Proper Study of Mankind*, ibid, p573
113. I. Berlin, *The Proper Study of Mankind*, ibid, p570
114. P.Watson, *Ideas, A History from Fire to Freud*, ibid, p827
115. P.Watson, *Ideas, A History from Fire to Freud*, ibid, pp824, 828, 831
116. S. Okasha, *Philosophy of Science, A Very Short Introduction*, Oxford University Press, 2002, p84
117. Quoted from Bergson, *Creative Evolution*, London, Macmillan, 1911, p110 in R. Sheldrake, *The Presence of the Past*, Park Street Press, 2012 Ed, p374
118. A. Koestler, *The Act of Creation*, Pan Books Picador Ed. 1981, p iii – the author of this overview is not named.
119. E. Curie, *Madam Curie – A Biography by Eve Curie*, ibid, p76
120. J. Marías, *History of Philosophy*, ibid, p401
121. E. Curie, *Madam Curie – A Biography by Eve Curie,* ibid, p54
122. I. Berlin, *The Proper Study of Mankind*, ibid, p71
123. Nobel Prize winning physicist, Steven Weinberg has written: "The more the universe seems comprehensible, the more it also seems pointless." Quoted from M. Ruse, *A Meaning to Life*, Oxford University Press, 2019, p54. Science's grand scheme is not matched by a comparable religious grand scheme that satisfies the rationale and analytical mindset of scientists. Romanticism was a backlash against this vacuum of mindset.

124. For example, Dalton did not realise many gases were diatomic and it was only after Amedeo Avogadro discerned atomic masses more accurately that such misconceptions wer corrected – J. Lees, *Physics in 50 Milestone Moments*, Quad Books, 2017, p81
125. *Oxford Dictionary of Chemistry*, Seventh Edition, 2015, Oxford University Press, 2016, p161
126. *Oxford Dictionary of Chemistry*, ibid, p161
127. P. Bizony, *Atom*, Icon Books Ltd, 2017, p2
128. P. Bizony, *Atom*, ,ibid, p3
129. *Oxford Dictionary of Chemistry*, ibid, p352
130. N. Pasachoff, ibid, p11
131. N. Pasachoff, ibid, pp11–12
132. N. Pasachoff, ibid, pp12, 36
133. N. Pasachoff, ibid, p12
134. N. Pasachoff, ibid, p33
135. J. Latson, *The Shy Scientist Who Could See Through Skin*, January 5, 2015, Time
136. J. Borzendowkski, ibid, p45
137. J. Borzendowkski, ibid, p45
138. N. Pasachoff, ibid, p33
139. N. Pasachoff, ibid, p34
140. N. Pasachoff, ibid, p34
141. N. Pasachoff, ibid, p34
142. N. Pasachoff, ibid, p35
143. N. Pasachoff, ibid, pp35–36
144. J. Borzendowkski, ibid, p50
145. J . Borzendowkski, ibid, p51
146. N. Pasachoff, ibid, p36
147. N. Pasachoff, ibid, p37
148. N. Pasachoff, ibid, p37
149. N. Pasachoff, ibid, p37
150. N. Pasachoff, ibid, p38
151. J. Borzendowkski, ibid, pp62–63
152. J.Farndon, *The Great Scientists*, Arcturus Publishing, 2006, pp120–121
153. Heilbron, *The Oxford Illustrated Companion to the History of Modern Science*, Oxford University Press, 2003, p268
154. Heilbron, *The Oxford Illustrated Companion to the History of Modern Science*, ibid, p 268
155. J. Borzendowkski, ibid, p64
156. J. Borzendowkski, ibid, p64, who writes that one of her biographers described the discovery of radium as the birth of atomic energy and the key to unlocking the mysteries of the composition of the universe.
157. P. Watson, *Ideas, A History from Fire to Freud*, ibid, p176
158. *Oxford Dictionary of Philosophy*, Oxford University Press, 2016, p36
159. *Oxford Dictionary of Philosophy*, Oxford University Press, 2016, p36

160. *Oxford Dictionary of Philosophy*, Oxford University Press, 2016, pp36,179
161. *Oxford Dictionary of Philosophy*, Oxford University Press, 2016, p36
162. R. Iliffe, *Priest of Nature – The Religious Worlds of Isaac Newton*, Oxford University Press, 2017, p95 including preceding quote of Newton
163. Iliffe, Newton, *A Very Short Introduction*, Oxford University Press, 2007, p121
164. Quoted from Iliffe, *Newton, A Very Short Introduction*, p123
165. Quoted from Iliffe, *Newton, A Very Short Introduction*, p122
166. R. Iliffe, *Priest of Nature – The Religious Worlds of Isaac Newton*, ibid, p93
167. Quote from R. Iliffe, *Priest of Nature – The Religious Worlds of Isaac Newton*, ibid, pp93–94
168. R. Iliffe, *Priest of Nature – The Religious Worlds of Isaac Newton*, ibid, p93
169. R. Iliffe, *Priest of Nature – The Religious Worlds of Isaac Newton*, ibid, p93
170. R. Iliffe, *Priest of Nature – The Religious Worlds of Isaac Newton*, ibid, p93, including quote from Newton's Philosophical Questions
171. P. Bizony, *Atom*, ibid, pxi of Foreword
172. Heilbron, *The Oxford Illustrated Companion to the History of Modern Science*, ibid, p31
173. J.Farndon, *The Great Scientists, op.cit.* 2006, p123
174. Gorodeisky, Keren, '19th Century Romantic Aesthetics', *The Stanford Encyclopaedia of Philosophy*, Fall 2016 Edition, Edward N. Zalta (ed.), <https://plato.stanford.edu/archives/fall2016/entries/aesthetics-19th-romantic/>. Under heading 4.2 Bildung
175. D.Bodanis, *Einstein's Greatest Mistake*, Abacus, 2017, p6
176. J.Farndon, *The Great Scientists*, ibid, 2006, p123
177. J.Farndon, *The Great Scientists*, ibid, 2006, p123
178. A handwritten letter of Albert Einstein to Mr Eric Gutland known as the 'The God Letter' was reported by the BBC on 5.12.2018 to be sold for US$2.4 million and contained the following quote regarding Judiaism: "…like all other religions… [Judiaism]… is an incarnation of primitive superstition."
179. D.Bodanis, *Einstein's Greatest Mistake*, ibid, p7
180. Quote from Einstein in D. Bodanis, *Einstein's Greatest Mistake*, ibid, p7
181. D.Bodanis, *Einstein's Greatest Mistake*, ibid, p7
182. D.Bodanis, *Einstein's Greatest Mistake*, ibid, p7
183. D.Bodanis, *Einstein's Greatest Mistake*, ibid, p8
184. Quote from Einstein in D. Bodanis, *Einstein's Greatest Mistake*, ibid , p8
185. Quote from Einstein in D. Bodanis, *Einstein's Greatest Mistake*, ibid, p14
186. J.Farndon, *The Great Scientists*, ibid, 2006, p123
187. Quote from Einstein in D. Bodanis, *Einstein's Greatest Mistake*, ibid, p9
188. D.Bodanis, *Einstein's Greatest Mistake*, ibid, p9
189. D.Bodanis, *Einstein's Greatest Mistake*, ibid, p9
190. D.Bodanis, *Einstein's Greatest Mistake*, ibid, p11
191. D.Bodanis, *Einstein's Greatest Mistake*, ibid, p11
192. D.Bodanis, *Einstein's Greatest Mistake*, ibid, p12
193. A. Einstein, L. Infeld, *The Evolution of Physics – from early concepts to relativity*

and quanta, Touchstone (Simon & Schuster) reprint of 1938 work in 2007, ppxvii to xviii
194. D.Bodanis, *Einstein's Greatest Mistake*, ibid, p17
195. D.Bodanis, *Einstein's Greatest Mistake*, ibid, p17
196. D.Bodanis, *Einstein's Greatest Mistake*, ibid, p18
197. D.Bodanis, *Einstein's Greatest Mistake*, ibid, p19
198. D.Bodanis, *Einstein's Greatest Mistake*, ibid, p19
199. D.Bodanis, *Einstein's Greatest Mistake*, ibid, p19
200. J.Farndon, *The Great Scientists*, ibid, 2006, p123
201. D.Bodanis, *Einstein's Greatest Mistake*, ibid, p21
202. D.Bodanis, *Einstein's Greatest Mistake*, ibid, p21
203. A. Einstein, L. Infeld, *The Evolution of Physics – from early concepts to relativity and quanta*, ibid, pp xxi to xxii
204. "...like all other religions... [Judiaism]... is an incarnation of primitive superstition". See note 178 above for reference.
205. Calaprice, Alice, 2010, *The Ultimate Quotable Einstein*, Princeton: Princeton University Press, p. 325
206. Jammer, Max, 2011, *Einstein and Religion: Physics and Theology*, Princeton NJ: Princeton University Press, p. 75; originally published in Albert Einstein (1929). *Gelegentliches*. ['A Miscellany'] Berlin: Soncino Gesellschaft, p9
207. Viereck, George Sylvester, 1930, *Glimpses of the Great*, New York: The Macaulay Company, pp372–373
208. A. Einstein, L. Infeld, *The Evolution of Physics*, ibid, pp xxi to xxii
209. A. Einstein, L. Infeld, *The Evolution of Physics*, ibid, p xxii
210. *Science and Religion*, Edited by G. Ferngren, John Hopkins University Press, 2017, p360
211. P. Watson, *Ideas, A History from Fire to Freud*, Weidenfeld & Nicolson, 2005, p1006
212. S. Hawking, *A Brief History of Time*, Bantam Press, 1989, p50
213. Dr Pim van Lommel, *Consciousness Beyond Life – the Science of the Near-Death Experience*, 2007, Harper One
214. Dr Pim van Lommel MD, R. van Wees, V. Meyers, and I. Elffrerich, 'Near-Death Experiences in The Netherlands', *The Lancet* 358, 2001: 2039–45
215. N.E. Abrams, *A God that could be real – Spirituality, Science and the Future of our Planet*, Beacon Press, 2105
216. Dr Karl Jansen MD PhD. 'The Ketamine Model of the Near-Death Experience: A central role for the N-methyl-D-aspartate receptor', *Journal of Near-Death Studies*, 16(1) p5, 1997, Human Sciences Press
217. Karl Jansen Ketamine, *Dreams and Realities – Multidisciplinary Association for Psychcedelic Studies*, 2000, p134
218. Kenneth Ring, Sharon Cooper, *Mindsight: Near-Death and Out-of-Body Experiences in the Blind*, Moment Point Press, 1998, 2000, 2006 editions
219. S. Hawking, *A Brief History of Time*, ibid, p10
220. *Oxford Dictionary of Physics*, Seventh Edition, 2015, Oxford University Press,

2016, p431, & *Encyclopedia Britannica* Online under entry 'Photoelectric Effect'

221. J.Farndon, *The Great Scientists*, ibid, 2006, pp123–126
222. *Oxford Dictionary of Physics,* Seventh Edition, 2015, Oxford University Press, 2016, p59
223. A. Einstein, L. Infeld, *The Evolution of Physics*, ibid, p92, *Encyclopaedia Britannica* Online under entry 'Speed of Light'
224. A. Einstein, L. Infeld, *The Evolution of Physics*, ibid, p149
225. P. Parsons, *3-minute Einstein,* Murdoch Books, 2012, p24
226. B. Oakley, *A Mind for Numbers,* TarlerParlgee div Penguin Random House, 2014, p206
227. J. Farndon, *The Great Scientists,* ibid, 2006, pp124–125
228. D. Bodanis, *Einstein's Greatest Mistake*, ibid, p22
229. Quoted in L. Susskind & A. Friedman, *Special Relativity and Classical Field Theory, The Theoretical Minimum*, Allen Lane – Div. of Penguin Random, 2017, p25
230. *Oxford Dictionary of Physics,* Seventh Edition, ibid, p356
231. The invariance of the speed of light is now a central tenet of physics but in an article entitled *'Spatially structured photons that travel in free space slower than the speed of light'* by Giovannini Et Al. *Science*, 20 Feb 2015, Vol. 347, Issue 6224, pp 857–860 DOI:10.1126/science.aaa3035 It is claimed that the invariance of light applies only to plane waves and that under certain conditions different photons can vary in speed.
232. S. Hawking, *A Brief History of Time*, ibid, p135
233. S. Hawking, *A Brief History of Time*, ibid, pp43–44
234. *Oxford Dictionary of Physics*, Seventh Edition, 2015, Oxford University Press, 2016, p503. The exact quote is "…the universe is updated through a wave of reality, which emanates at speed c from the region in which the change took place."
235. *Oxford Dictionary of Physics*, ibid, pp502–505
236. Quoted from article 'Nature Editor Sir John Maddox Cries Heretic' on www.sheldrake.org as at 12 February 2019
237. Wikipedia. Entry of Grigori Perelman as at 11 February 2019
238. D. Bodanis, *Einstein's Greatest Mistake,* ibid, pp34–35
239. D. Bodanis, *Einstein's Greatest Mistake,* ibid, pp37–38
240. D. Bodanis, *Einstein's Greatest Mistake*, ibid, pp37–38
241. D. Bodanis, *Einstein's Greatest Mistake,* ibid, p51
242. J. Farndon, *The Great Scientists*, ibid, 2006, p125
243. A. Einstein, L. Infeld, *The Evolution of Physics,* ibid, p214
244. In Chapter 20, we will see that he referred to a system under acceleration as under a type of gravitational field akin to the artificial gravity that might be created in a rotating space station using centrifugal force.
245. D.Bodanis, *Einstein's Greatest Mistake*, ibid, pp54–55
246. D.Bodanis, *Einstein's Greatest Mistake,* ibid, pp62–63

247. J.Farndon, *The Great Scientists*, ibid, 2006, p123
248. S. Hawking, *A Brief History ofTime*, ibid, p32 – He also said: "A geodesic is the shortest (or longest) path between two nearby points." (p32)
249. D. Bodanis, *Einstein's Greatest Mistake*, ibid, p70
250. D. Bodanis, *Einstein's Greatest Mistake*, ibid, p70
251. D. Bodanis, *Einstein's Greatest Mistake*, ibid, pp71–71
252. D. Bodanis, *Einstein's Greatest Mistake*, ibid, p78
253. A. Einstein, *Relativity*, first published in 1916, Routledge Classics, 2001, pp102,103
254. P. Parsons, *3-minute Einstein*, ibid, p88
255. D. Bodanis, *Einstein's Greatest Mistake*, ibid, pp90–91
256. D. Bodanis, *Einstein's Greatest Mistake*, ibid, pp95–95
257. P.G. Fereira, *The Perfect Theory*, Little Brown, 2014, p19
258. D. Bodanis, *Einstein's Greatest Mistake*, ibid , p96
259. D. Bodanis, *Einstein's Greatest Mistake*, ibid, pp98–103
260. D. Bodanis, *Einstein's Greatest Mistake*, ibid, pp103–104
261. *Encyclopedia Britannica* Online
262. D. Bodanis, *Einstein's Greatest Mistake*,, ibid, pp104–105
263. R.C.Allen, *Global Economic History, A Very Short Introduction*, Oxford University Press, 2011, p27–32
264. R.C. Allen, *Global Economic History, A Very Short Introduction*, ibid, pp29–30
265. *1001 Days that Shaped the World* (edited by P. Furtado), Quarto Publishing, 2018, p565
266. *1001 Days that Shaped the World*, ibid, p573
267. *Oxford Illustrated History of Science* (edited by I.R. Morus), Oxford University Press, 2017, p357
268. Gorodeisky, Keren, '19th Century Romantic Aesthetics', *The Stanford Encyclopaedia of Philosophy* (Fall 2016 Edition), Edward N. Zalta (ed.), <https://plato.stanford.edu/archives/fall2016/entries/aesthetics-19th-romantic/>. Under heading 4.2 Bildung.
269. Gorodeisky, Keren, *19th Century Romantic Aesthetics*, ibid.
270. Gorodeisky, Keren, *19th Century Romantic Aesthetics*, ibid.
271. Arp and over thirty contributors, *1001 Ideas that changed the way we think*, Pier 9, Murdoch Books, 2017, p443
272. P. Parsons, *3-minute Einstein*, ibid, p28
273. P. Parsons, *3-minute Einstein*, ibid, p28
274. *Oxford Dictionary of Philosophy*, Oxford University Press, 2016, p94
275. A. Einstein, L. Infeld, *The Evolution of Physics*, ibid, pp xxi to xxii
276. P. Parsons, *3-minute Einstein*, ibid, p32
277. A. Einstein, L. Infeld, *The Evolution of Physics – from early concepts to relativity and quanta*, ibid, pp xxi to xxii
278. *Oxford Illustrated History of Science*, ibid, p357
279. See end of last chapter
280. Heilbron, *The Oxford Illustrated Companion to the History of Modern Science*,

ibid, pp352–353

281. Heilbron, *The Oxford Illustrated Companion to the History of Modern Science*, ibid, p354
282. Heilbron, *The Oxford Illustrated Companion to the History of Modern Science*, ibid, pp352–353
283. S. Hawking, *A Brief History of Time*, ibid, p187
284. *Oxford Dictionary of World History*, Oxford University Press, 2015, p712
285. P. Parsons, *3-minute Einstein*, ibid, p38
286. P. Parsons, *3-minute Einstein*, ibid, p38 including quote
287. P. Parsons, *3-minute Einstein*, ibid, p40
288. P. Parsons, *3-minute Einstein*, ibid, p42
289. P. Parsons, *3-minute Einstein*, ibid, p42
290. *1001 Days that Shaped the World*, ibid, p696
291. *1001 Days that Shaped the World*, ibid, p681
292. *Oxford Dictionary of World History*, ibid, p262
293. *1001 Days that Shaped the World*, ibid, p713
294. P. Parsons, *3-minute Einstein*, ibid, p50
295. E. Regis, *Who Got Einstein's Office?* Penguin Books, 1987, p37
296. P. Parsons, *3-minute Einstein*, ibid, p50
297. E. Regis, *Who Got Einstein's Office?*, ibid, pp4–7
298. E. Regis, *Who Got Einstein's Office?*,ibid, p xiv
299. E. Regis, *Who Got Einstein's Office?*, ibid, p39
300. E. Regis, *Who Got Einstein's Office?*, ibid, p34
301. E. Regis, *Who Got Einstein's Office?*, ibid, pp38–39
302. *1001 Days that Shaped the World*, ibid, p715
303. P. Parsons, *3-minute Einstein*, ibid, p52
304. S. Hawking, *A Brief History of Time*, ibid, p188
305. P. Parsons, *3-minute Einstein*, ibid, p56
306. S. Hawking, *A Brief History of Time*, ibid, p188
307. P. Parsons, *3-minute Einstein*, ibid, p58
308. *Oxford Dictionary of Physics*, ibid, p219
309. *Oxford Dictionary of Physics*, ibid, p502
310. *Oxford Dictionary of Physics*, ibid, p502 – this thought experiment is fully set out in Einstein's book *Relativity* in Chapter 9 (A. Einstein, Relativity, ibid.)
311. L. Susskind & A. Friedman, *Special Relativity and Classical Field Theory, The Theoretical Minimum*, ibid, p27
312. L. Susskind & A. Friedman, *Special Relativity and Classical Field Theory*, ibid, p27
313. *Oxford Dictionary of Physics*, ibid, p325
314. *Oxford Dictionary of Physics*, ibid, p383
315. A. Einstein, L. Infeld, *The Evolution of Physics*, ibid, pp189–191
316. *Encyclopedia Brittanica* Online – entry for Hermann Minkowski – as at 13 February 2019 – it was also noted in that entry that Minkowski was a colleague of mathematician David Hilbert who collected Minkowski's work in his *Gesammelte Abhandlungen*, 2 Vol. (1911, Collected Papers).

317. A. Einstein, L. Infeld, *The Evolution of Physics*, ibid, p213
318. Quote from Ernest Mach (1938–1916), *The Science of Mechanics*, quoted from Note 10, p268, Baggott, *Mass*, op.cit. *Oxford Dictionary of Physics*, op.cit. p331
319. A. Einstein, L. Infeld, *The Evolution of Physics*, ibid, pp31–32
320. A. Einstein, L. Infeld, *The Evolution of Physics*, ibid, p4
321. A. Einstein, L. Infeld, *The Evolution of Physics*, ibid, p33
322. A. Einstein, *Relativity*, first published in 1916, Routledge Classics, 2001, p66
323. See Chapter 18 – ref. D. Bodanis, *Einstein's Greatest Mistake*, ibid, p51 – note: Einstein calls reference-body a 'spacious chest' in his 1916 book Relativity (at p68 (ibid.) but calls it an 'elevator' in his book *The Evolution of Physics* (at p214 (ibid.))
324. A. Einstein, *Relativity*, ibid, p68
325. A. Einstein, *Relativity*, ibid, pp68–69
326. A. Einstein, *Relativity*, ibid, p70
327. A. Einstein, *Relativity*, ibid, p67
328. A. Einstein, *Relativity*, ibid, p71
329. Newton's first law of motion states: "A body continues in a state of rest or uniform motion in a straight line unless it is acted upon by external forces" – *Oxford Dictionary of Physics,* Seventh Edition, ibid, p383
330. A. Einstein, *Relativity*, ibid, pp12–13
331. A. Einstein, *Relativity*, ibid, p78
332. A. Einstein, L. Infeld, *The Evolution of Physics*, ibid, pp220–221
333. R. Penrose, *The Road to Reality, A Complete Guide to the Law of the Universe,* Vintage 2005, p641
334. *Oxford Dictionary of Physics,* ibid, p331
335. P. Parsons, *3-minute Einstein*, ibid, p86
336. P. Parsons, *3-minute Einstein*, ibid, p82
337. *Oxford Dictionary of Physics*, Seventh Edition, ibid, pp3–4
338. P. Parsons, *3-minute Einstein*, ibid, p90
339. P. Parsons, *3-minute Einstein*, ibid, p90
340. A. Einstein, *Relativity*, ibid, p63
341. Euclidean system here refers to Cartesian Co-ordinate system represented on a perfectly flat plane with no curved surfaces
342. A non-Euclidean system here refers to a Gaussian Co-ordinate system represented on curved surface – in general terms, it is as if a flat co ordinate system was distorted and pasted onto a sphere
343. See Chapter 25, Gaussian Co-ordinates – A. Einstein, *Relativity*, ibid, p87
344. A. Einstein, *Relativity*, ibid, p155 – and as a result of it being a special case it needs to satisfy the Riemann condition – see p156
345. *Oxford Dictionary of Physics,* ibid, p512
346. I. Stewart, *The Great Mathematical Problems*, Profile Books, 2014, pp195–196
347. P. Parsons, *3-minute Einstein*, ibid, p82
348. To borrow a term from biology where independent evolutions can occur

in nature – for example, the eye is thought to have developed six times amongst unrelated different species in a puzzling example of evolution which relies on randomised blind chance

349. P. Parsons, *3-minute Einstein*, ibid, p84. Parsons also notes that Poincaré also independently developed equations for special relativity ten years earlier but failed to recognise the revolutionary implications of the equations – p82
350. I. Stewart, *Calculating the Cosmos*, Profile Books, 2017, pp24–25
351. Carlo Rovelli, *Reality is not what it seems*, Penguin Random House, 2016, pp64–65
352. A. Einstein, *Relativity*, ibid, p151
353. A. Einstein, *Relativity*, ibid, pp156–157
354. 'Space' refers to space governing objects in the Special Theory of Relativity obeying equation $ds^2=dx_1^2 + dx_2^2 + dx_3^2 - dx_4^2$ from the Lorentz transformation
355. A. Einstein, *Relativity*, ibid, pp156–157
356. A. Einstein, *Relativity*, ibid, note 1 on pp156
357. A saying of Alfred Korzybski – a Polish American scholar
358. Carlo Rovelli, *Seven Brief Lessons on Physics*, Penguin Random House, 2016, pp40–41
359. J.C. Polkinghorne, *The Quantum World*, Longman Group Ltd, 1984, p5
360. Kricker & Butler, *Advanced Physics*, Angus & Robertson Ltd, 1968 (reprint 1969), pp125–126. This and subsequent references and quotes from this book is by kind permission of HarperCollins Publishers Australia Pty Limited
361. Max Planck: the reluctant revolutionary" by Helge Kragh *Physics World* December 2000. © IOP Publishing. Reused with permission. All rights reserved, pp 31-33.
362. *Encyclopedia Britannica* Online under 'Planck's Radiation Law'
363. J.C. Polkinghorne, *The Quantum World*, ibid, p5
364. P. Parsons, *3-minute Einstein*, ibid, p66
365. P. Parsons, *3-minute Einstein*, ibid, p66
366. J.C. Polkinghorne, *The Quantum World*, ibid, p5
367. P. Parsons, *3-minute Einstein*,, ibid, p68
368. The Ultrafast Einstein-de Haas effect, *Nature*, 565, 209–212 (2019), & P. Parsons, *3-minute Einstein*,, ibid, p72
369. D. Bodanis, *Einstein's Greatest Mistake*, ibid, pp175–176
370. Kricker & Butler, *Advanced Physics*, ibid, pp127–128
371. Kricker & Butler, *Advanced Physics*, ibid, p125
372. J.C. Polkinghorne, *The Quantum World*, ibid, pp8–9
373. J.C. Polkinghorne, *The Quantum World*, ibid, p9
374. Kricker & Butler, *Advanced Physics*, ibid, p129
375. Kricker & Butler, *Advanced Physics*, ibid, p129
376. R. Penrose, *Fashion, Faith and Fantasy*, Princeton University Press, 2016, pp55–56
377. Other anomalies noted by Kricker & Butler disclosed a mixture of classical and

quantum mechanics with contradictions of classical theory – specifically "The electron moving in a circular orbit is assumed to obey classical mechanics, and yet the quantisation of the orbital angular momentum is included. The electron is assumed to obey one feature of classical electromagnetic theory (Coulomb's Law) and yet not to obey another feature (emission of radiation by an accelerating charged body)". Kricker & Butler, *Advanced Physics*, ibid, p129

378. Quote from Kricker & Butler, *Advanced Physics*, ibid, p129
379. P. Parsons, *3-minute Einstein*, ibid, p68
380. J.C. Polkinghorne, *The Quantum World*, ibid, p7
381. Kricker & Butler, *Advanced Physics*, ibid, p130
382. D. Bohm & F.D. Peat, *Science, Order & Creativity*, Routledge 2011 (1st ed. 1987), p27
383. D. Bohm & F.D. Peat, *Science, Order & Creativity*, ibid, p29
384. M. Raymer, *Quantum Physics*, Oxford University Press, 2017, pp18 & 21
385. J.C. Polkinghorne, *The Quantum World*, ibid, p30
386. J.C. Polkinghorne, *The Quantum World*, ibid, p14
387. *Oxford Dictionary of Mathematics*, ibid, p135
388. *Oxford Dictionary of Physics*, ibid, p139
389. $t= |\psi|2$
390. J.C. Polkinghorne, *The Quantum World*, ibid, p14
391. *Oxford Dictionary of Physics*, Seventh Edition, ibid, p615
392. *Oxford Dictionary of Physics*, Seventh Edition, ibid, p615
393. J.C. Polkinghorne, *The Quantum World*, ibid, p14
394. D. Bodanis, *Einstein's Greatest Mistake*, ibid, pp180–181
395. D. Bodanis, *Einstein's Greatest Mistake*, ibid, pp176–177
396. D. Bodanis, *Einstein's Greatest Mistake*, ibid, p184
397. D. Bodanis, *Einstein's Greatest Mistake*, ibid, p184
398. L. Susskind & A. Friedman, *Quantum Mechanics, The Theoretical Minimum*, Penguin Random House, 2014, p xv
399. S. Hawking, *A Brief History of Time*, ibid, p57
400. The field equation was changed from $G=8\pi\gamma T$ to $G=8\pi\gamma T + \Lambda g$ (Λ was later discarded but scientists have re inserted it – Roger Penrose sees it as representing dark matter – R. Penrose, *Fashion, Faith and Fantasy*, ibid, p5)
401. D. Bodanis, *Einstein's Greatest Mistake*, ibid, p175
402. D. Bodanis, *Einstein's Greatest Mistake*, ibid, p166
403. Quoted from D.Bodanis, *Einstein's Greatest Mistake*, ibid, p186
404. D. Bodanis, *Einstein's Greatest Mistake*, ibid, p191
405. D. Bodanis, *Einstein's Greatest Mistake*, ibid, p193
406. D. Bodanis, *Einstein's Greatest Mistake*, ibid, p196
407. D. Bodanis, *Einstein's Greatest Mistake*, ibid, p118
408. D. Bodanis, *Einstein's Greatest Mistake*, ibid, p144–146, 166
409. *1001 Days that Shaped the World*, ibid, p680
410. R. Penrose, *Fashion, Faith and Fantasy*, ibid, p5

411. D. Bodanis, *Einstein's Greatest Mistake*, ibid, p201
412. D. Bodanis, *Einstein's Greatest Mistake*, ibid, pp202–203
413. D. Bodanis, *Einstein's Greatest Mistake*, ibid, p204
414. J.C. Polkinghorne, *The Quantum World*, ibid, p33
415. *Oxford Dictionary of Physics*, ibid, pp461–462, also see M. Raymer, *Quantum Physics*, Oxford University Press, 2017, pp138–139
416. M. Raymer, *Quantum Physics*, Oxford University Press, 2017, p140
417. J. Al Kalili, *Quantum, A guide for the perplexed*, Weidenfeld & Nicholson – Orion-Hachette, 2012, pp91–93
418. There is a mathematical impossibility that causes a problem with hidden variable theories outlined by Abner Shimony in a paper – *Conceptual Foundations of Quantum Mechanics* included in *The New Physics*, Cambridge University Press, 1989–1992 Ed.p384 in which he highlights a theorem: If the dimension of a linear vector space v is greater than two, there is no function which assigns to every subspace ε of v a number zero or unity in such a way that $m(v) = 1$ and $m(\varepsilon) = m(\varepsilon_1) + m(\varepsilon_2)$ if ε_1 and ε_2 are orthogonal subspaces and ε is the smallest subspace containing both ε_1 and ε_2. This basically means that the formality of vector spaces relies on a Boolean '1' for true and '0' for false like spin up or spin down represented in the linear equations in matrices. The only way out seems to be to depart from the strictly Boolean mathematics inherent in quantum mechanics that works so well.
419. L. Susskind & A. Friedman, *Quantum Mechanics, The Theoretical Minimum*, ibid, p36
420. L. Smolin, *Einstein's Unfinished Revolution – The search for what lies beyond the quantum*, Penguin Press, 2019
421. L. Smolin, *Einstein's Unfinished Revolution*, ibid, p89
422. L. Smolin, *Einstein's Unfinished Revolution*, ibid, pp93–94
423. T. Kuhn, *The Structure of Scientific Revolutions*
424. L. Smolin, *Einstein's Unfinished Revolution*, ibid, p142
425. L. Smolin, *Einstein's Unfinished Revolution*, ibid, pp137141
426. L. Smolin, *Einstein's Unfinished Revolution*, ibid, p215
427. D. Bodanis, *Einstein's Greatest Mistake*, ibid, p165
428. D. Bodanis, *Einstein's Greatest Mistake*, ibid, p29
429. Quoted from C. Zimmer, *A Dog and the Mind of Newton, National Geographic*, 11.8.2005
430. Newton was born on 25 December 1642
431. J. Farndon, *The Great Scientists*, ibid, 2006, p155
432. J. Hawking, *Travelling to Infinity: My Life with Stephen*, Alma Books Ltd, 2007, p49
433. J. Farndon, *The Great Scientists*, ibid, 2006, p155
434. S. Hawking, *A Brief History of Time*, ibid, p55
435. L. Susskind, *The Black Hole War: My Battle with Stephen Hawking to make the World Safe for Quantum Mechanics*, Back Bay Books/Little Brown and Co. – Hachette, 2008, pp174–175

436. S. Hawking, *My Brief History*, ibid, p36
437. S. Hawking, *My Brief History* ibid, pp33–36
438. S. Hawking, *My Brief History*, ibid, pp36–37
439. S. Hawking, *My Brief History*, ibid, p37
440. J. Hawking, *Travelling to Infinity*, ibid, p37
441. J. Hawking, *Travelling to Infinity*, ibid, p24
442. J. Farndon, *The Great Scientists*, ibid, 2006, p156
443. J. Farndon, *The Great Scientists*, ibid, 2006, p156; J. Hawking, *Travelling to Infinity*, ibid, p65
444. J. Hawking, *Travelling to Infinity*, ibid, p94
445. J. Hawking, *Travelling to Infinity*, ibid, p94
446. S. Hawking, *My Brief History*, ibid, p45
447. Einstein had inserted the Cosmological Constant (Λ) into his famous field equation in his theory of General Relativity namely, when $G=8\pi\gamma T$ became $G=8\pi\gamma T + \Lambda g$, which he later removed once he'd accepted Hubble's finding of an expanding universe over a static universe – see Chapter 21, 'Einstein's Reality'.
448. J. Hawking, *Travelling to Infinity*, ibid, pp54–55
449. S. Hawking, *My Brief History*, ibid, p46
450. S. Hawking, *My Brief History*, ibid, p46
451. S. Hawking, *My Brief History*, ibid, p63
452. S. Hawking, *My Brief History*, ibid, pp64–65
453. S. Hawking, *My Brief History*, ibid, p65
454. L. Susskind, *The Black Hole War: My Battle with Stephen Hawking to make the World Safe for Quantum Mechanics*, ibid, pp164–165
455. S. Hawking, *A Brief History of Time*, ibid, pp53–54
456. The Big Bang theory was originally proposed by Jesuit scientist George Lemaître in the 1930s – see *Oxford Dictionary of Physics*, ibid, pp42–43
457. J. Farndon, *The Great Scientists*, ibid, 2006, p157
458. J. Hawking, *Travelling to Infinity*, ibid, p214
459. J. Hawking, *Travelling to Infinity*, ibid, p214
460. J. Hawking, *Travelling to Infinity*, ibid, pp215–216
461. J. Hawking, *Travelling to Infinity*, ibid, p214
462. J. Hawking, *Travelling to Infinity*, ibid, p234
463. J. Hawking, *Travelling to Infinity*, ibid, p259
464. S. Hawking, *My Brief History*, ibid, p92
465. S. Hawking, *My Brief History*, ibid, p93
466. S. Hawking, *My Brief History*, ibid, p93
467. J. Farndon, *The Great Scientists*, ibid, 2006, p158
468. J. Hawking, *Travelling to Infinity*, ibid, p363
469. J. Hawking, *Travelling to Infinity*, ibid, p363
470. J. Hawking, *Travelling to Infinity*, ibid, p365
471. J. Hawking, *Travelling to Infinity*, ibid, p366
472. J. Farndon, *The Great Scientists*, ibid, 2006, p158

473. J. Hawking, *Travelling to Infinity*, ibid, p372
474. D. Hawkins, 'What made Hawking's A Brief History of Time so immensely popular?' *The Washington Post*, 14 March 2018
475. J. Farndon, *The Great Scientists*, ibid, 2006, p158
476. J. Farndon, *The Great Scientists*, ibid, 2006, p158; J.Hawking, *Travelling to Infinity*, ibid, back cover
477. S. Hawking, *A Brief History of Time*, ibid, p185
478. J. Farndon, *The Great Scientists*, ibid, 2006, p155
479. S. Hawking, *My Brief History*, Penguin Random House, 2013, p6
480. S. Hawking, *My Brief History*, ibid, p9
481. *Oxford Dictionary of Physics*, Oxford University Press 3rd Ed. 2016, p281 under entry for 'Logical Positivism'
482. Ibid.
483. Arp and over thirty contributors, *1001 Ideas that changed the way we think*, Pier 9, Murdoch Books, 2017, p816
484. Arp and over thirty contributors, *1001 Ideas that changed the way we think*, ibid, 2017, p816
485. J. Marías, *History of Philosophy*, Dover Publications, 1967 (English Translation), p400
486. Papineau D, *Philosophy Theories and Great Thinkers*, Shelter Harbour Press, 2017, p92
487. Quoted from Papineau D, *Philosophy Theories and Great Thinkers*, p92 from Rudolf Carnap, *The Elimination of Metaphysics through Logical Analysis of Thought*
488. Papineau D., *Philosophy Theories and Great Thinkers*, ibid, p92
489. J. Hawking, *Travelling to Infinity*, ibid, pp138–139
490. S. Hawking & L. Mlodinow, *The Grand Design*, ibid, p151
491. J. Marías, *History of Philosophy*, ibid, p352
492. Quoted from S. Flew, *There is a God, How the world's most notorious atheist changed his mind*, Harper One, 2007, p91
493. M. Epperson, *Quantum Mechanics and the Philosophy of Alfred North Whitehead*, ibid.
494. S. French, *Philosophy of Science*, Bloomsbury, 2016
495. L.Wittgenstein, *Tractatus Logico-Philosophicus*, 1921, Routledge Classics, 2001, pp81–82
496. L.Wittgenstein, *Tractatus Logico-Philosophicus*, ibid, p 82
497. S. Hawking, *A Brief History of Time*, ibid, pp x to xi
498. S. Hawking, *The Grand Design*, ibid, p227
499. Quoted from S. Flew, *There is a God, How the world's most notorious atheist changed his mind*, ibid, pp142–143
500. 'Lambda leads the Way' Landscape Theory Part 4, *Stanford News*, 10 September 2018
501. 'Lambda leads the Way', ibid, summarised comment by per Professor L. Susskind

502. See Chapter 5 generally – also note that evolutionary biologist Professor Stuart Kauffman believes that evolution cannot account for the extraordinary complexity and processes discovered in modern biology – see S. Kauffman *The Origins of Order – Self Organisation and Selection in Evolution* – Oxford University Press, 1993, p xiii – For a similar insight from a different viewpoint, see also former Cambridge biologist D. Axe, *Undeniable*, Harper Collins, 2016 – Professor of Computer Science and Applied Mathematics Leslie Valiant's analysis using evolutionary algorithms concludes that many biological features cannot have arisen by conventional evolutionary processes – see L. Valiant, *Probably Approximately Correct*, ibid, pp106–107
503. S. Hawking, *A Brief History of Time*, ibid , p50
504. Roger Penrose refers to this paradox at the moment of the Big Bang, namely, on the one hand there is a tiny entropy to obey the second law of thermodynamics, yet on the other hand the remarkable cosmic background radiation evidence tells us there was enormous entropy at the Big Bang – see R. Penrose, *Fashion, Faith and Fantasy*, ibid, p5
505. T. Kuhn, *The Structure of Scientific Revolutions*
506. Centre for Theoretical Cosmology – see http://www.ctc.cam.ac.uk/news/120829_newsitem.php
507. S. Hawking, *My Brief History*, ibid, pp66–67
508. S. Hawking, *A Brief History of Time*, ibid, p96
509. *Encyclopedia Britannica* Online – see definition of 'Schwarzschild radius'
510. S. Hawking, *A Brief History of Time*, ibid, pp95–96
511. S. Hawking, *A Brief History of Time*, ibid, p93
512. S. Hawking, *A Brief History of Time*, ibid, p93
513. S. Hawking, *My Brief History*, ibid, pp66–67 ; S. Hawking, *A Brief History of Time*, ibid, p90
514. S. Hawking, *A Brief History of Time*, ibid, pp86–87
515. S. Hawking, *A Brief History of Time*, ibid, p90
516. S. Hawking, *A Brief History of Time*, ibid, p108
517. S. Hawking, *A Brief History of Time*, ibid, pp107–108
518. *Oxford Dictionary of Physics*, ibid, p589
519. S. Hawking, *A Brief History of Time*, ibid, p108
520. S. Hawking, *A Brief History of Time*, ibid, pp109–110
521. S. Hawking, *A Brief History of Time*, ibid, p110
522. S. Hawking, *A Brief History of Time*, ibid, p110
523. S. Hawking, *A Brief History of Time*, ibid, pp112–113
524. S. Hawking, *A Brief History of Time*, ibid, p114
525. J. Farndon, *The Great Scientists*, ibid, 2006, p157
526. S. Hawking, *A Brief History of Time*, ibid, p114
527. *Oxford Dictionary of Physics*, Seventh Edition, ibid, p244, under Hawking process p243–244
528. L. Susskind, *The Black Hole War: My Battle with Stephen Hawking to make the World Safe for Quantum Mechanics*, ibid, pp165

529. L. Susskind, *The Black Hole War: My Battle with Stephen Hawking to make the World Safe for Quantum Mechanics*, ibid, pp174
530. L. Susskind, *The Black Hole War: My Battle with Stephen Hawking to make the World Safe for Quantum Mechanics*, ibid, p19
531. R. Penrose, *Fashion, Faith and Fantasy*, ibid, p384
532. L. Susskind, *The Black Hole War: My Battle with Stephen Hawking to make the World Safe for Quantum Mechanics*, Back Bay Books/Little Brown and Co. – Hachette, 2008
533. L. Susskind, *The Black Hole War: My Battle with Stephen Hawking to make the World Safe for Quantum Mechanics*, ibid, pp227–228
534. L. Susskind, *The Black Hole War: My Battle with Stephen Hawking to make the World Safe for Quantum Mechanics*, ibid, p262
535. L. Susskind, *The Black Hole War: My Battle with Stephen Hawking to make the World Safe for Quantum Mechanics*, ibid, pp250–251
536. L. Susskind, *The Black Hole War: My Battle with Stephen Hawking to make the World Safe for Quantum Mechanics*, ibid, See Chapter 21 generally which contains the summary of his analysis that follows this reference in the text
537. S. Gubser, *The Little Book of String Theory*, Princeton University Press, 2010, pp73–74
538. L. Susskind, *The Black Hole War: My Battle with Stephen Hawking to make the World Safe for Quantum Mechanicss*, ibid, p382
539. L. Susskind, *The Black Hole War: My Battle with Stephen Hawking to make the World Safe for Quantum Mechanicss*, ibid, p384
540. The equations for entropy in a collapsing black hole derived by J. Bekenstein and refined by S. Hawking are found at p270, R. Penrose, *Fashion, Faith and Fantasy*, ibid.
541. L. Susskind, *The Black Hole War: My Battle with Stephen Hawking to make the World Safe for Quantum Mechanics*, ibid, pp390–394
542. L. Susskind, *The Black Hole War: My Battle with Stephen Hawking to make the World Safe for Quantum Mechanics*, ibid, p419
543. J. Farndon, *The Great Scientists*, ibid, 2006, p158
544. S. Hawking, *My Brief History*, ibid, pp73–75
545. S. Hawking, *A Brief History of Time*, ibid, p122
546. S. Hawking, *A Brief History of Time*, ibid, p50
547. S. Hawking, *A Brief History of Time*, ibid, p50 & *Oxford Dictionary of Physics*, ibid, pp42–43
548. S. Hawking, *A Brief History of Time*, ibid, p121
549. S. Hawking, *A Brief History of Time*, ibid, pp124–125
550. S. Hawking, *A Brief History of Time*, ibid, p125
551. S. Hawking, *A Brief History of Time*, ibid, p125
552. S. Hawking, *A Brief History of Time*, ibid, pp127–128
553. S. Hawking, *A Brief History of Time*, ibid, p53
554. This is an interesting comment by Hawking because he would have been aware that mathematics does deal with infinity in clever ways such as by

series, limits and in complex analysis a technique known as contour line integrals and residue theory applied to imaginary numbers

555. S. Hawking, *A Brief History of Time*, ibid, p129
556. S. Hawking, *A Brief History of Time*, ibid, p129
557. Marcus Chown, *Life's Subatomic Secret*, New Scientist, 22 October 2016, p34
558. Friederich, Simon, 'Fine-Tuning', *The Stanford Encyclopaedia of Philosophy*, Winter 2018 Edition, Edward N. Zalta (ed.), URL = <https://plato.stanford.edu/archives/win2018/entries/fine-tuning/>.
559. Friederich, Simon, 'Fine-Tuning', *The Stanford Encyclopaedia of Philosophy*, ibid.
560. *Oxford Dictionary of Philosophy*, Oxford University Press, 2016, 3rd Ed, p23
561. S. Hawking, *A Brief History of Time*, ibid, p131
562. S. Hawking, *A Brief History of Time*, ibid, p132
563. S. Hawking, *A Brief History of Time*, ibid, p132
564. S. Hawking, *A Brief History of Time*, ibid, pp130–131
565. S. Hawking, *A Brief History of Time*, ibid, p132
566. S. Hawking, *A Brief History of Time*, ibid, p129
567. S. Hawking & L. Mlodinow, *The Grand Design*, Bantam Press, 2011, p151
568. S. Hawking, *A Brief History of Time*, ibid, p132
569. S. Hawking, *A Brief History of Time*, ibid, p140, he refers to using it as 'counsel of despair' to throw away any chance of understanding of the universe
570. S. Hawking, *A Brief History of Time*, ibid, p153, 159 & 174
571. S. Hawking, *A Brief History of Time*, ibid, p133
572. *Oxford Dictionary of Physics*, ibid, p106
573. S. Hawking, *A Brief History of Time*, ibid, p135
574. S. Hawking, *A Brief History of Time*, ibid, pp139–140
575. S. Hawking, *A Brief History of Time*, ibid, p140
576. Huffingtonpost.com – 11/16/2013 – Denise Chow Live Science
577. S. Hawking, *A Brief History of Time*, ibid, p140
578. S. Hawking, *A Brief History of Time*, ibid, pp141–144
579. J. Gribbin, *Hawking throws Higgs into Black Holes*, New Scientist, 2 December 1995
580. An article entitled 'Higgs v Hawking: a battle of the heavyweights that has shaken the world of theoretical physics' records the following comment by Professor Higgs which caused controversy: "It is very difficult to engage him [Hawking] in discussion, and so he has got away with pronouncements in a way that other people would not… his celebrity status gives him instant credibility that others do not have." S. Connor, science editor, *The Independent* (UK) 3 September 2002
581. Penrose refers to time asymmetry in relation to gravity's distortions in the early universe and believes it can be linked to quantum mechanics – see p819, Penrose (see below) – and in the early universe due to the General Relativity there can be no special choice of time co ordinates – p700 – and also see discussion of time asymmetry and second law of thermodynamics – p697, R. Penrose, *The Road to Reality, A Complete Guide to the Law of the*

Universe, Vintage 2005, p778

582. R. Penrose, *The Road to Reality, A Complete Guide to the Law of the Universe*, Vintage 2005, p778
583. It could be ellipsoid or some theories cast the black hole as cylindrical – they all allow a circular cross section topologically for the traveller to move forward in one direction forever with no 'edges' (as any circle would allow)
584. R. Penrose, *The Road to Reality, A Complete Guide to the Law of the Universe*, ibid, p861
585. FLRW refers to Friedman, Lemaître, Robertson Walker models of expanding universes – see R. Penrose, *The Road to Reality, A Complete Guide to the Law of the Universe*, ibid, p717
586. R. Penrose, *The Road to Reality, A Complete Guide to the Law of the Universe*, ibid, pp864–865
587. R. Penrose, S. Hawking, *A Brief History of Time*, ibid, pp718–719
588. R. Penrose, S. Hawking, *A Brief History of Time*, ibid, p774
589. Misao Sasaki, Dong-han Yeom & Ying Li Zhang, *Hartle-Hawking no-boundary proposal in dRGT massive gravity: Making inflation exponentially more probable*, Classical and Quantum Gravity 30(23):2001-December 2013. DOI: 10.1088/0264-9381/30/23/232001
590. M. Epperson, *Quantum Mechanics and the Philosophy of Alfred North Whitehead*, Fordham University Press, 2014, pp6–7
591. Carlo Rovelli, *Seven Brief Lessons on Physics*, ibid, pp61–62
592. S. Hawking & L. Mlodinow, *The Grand Design*, ibid, pp67–68
593. S. Hawking, *Brief Answers to the Big Questions*, John Murray (Publishers) – Hachette, 2018, p39 et. seq.
594. S. Hawking, *Brief Answers to the Big Questions*, ibid, p54
595. S. Hawking, *Brief Answers to the Big Questions*, ibid, pp55–56
596. S. Hawking, *Brief Answers to the Big Questions*, ibid, pp60–61
597. S. Hawking, *Brief Answers to the Big Questions*, ibid, p63
598. S. Hawking & L. Mlodinow, *The Grand Design*, ibid, pp151–152
599. S. Hawking, *A Brief History of Time*, ibid, p132
600. R. Penrose, *Fashion, Faith and Fantasy*, ibid, p87
601. Whereas functions map a point, say, x to f(x), a 'sheaf' will map a point to vector space via a rule called a 'sheaf' denoted with such notations as F(U) or F(U) for a set in the manifold under review –see p 156–157, E. Frenkel, *Love and Mathematics*, ibid.
602. E. Frenkel, *Love and Mathematics*, ibid, 221
603. 'The String Theory Landscape' Landscape Theory Part 1 – *Stanford News*, 10 September 2018
604. The gamma function involves factorials [eg Γ(x) = (x-1)! or say if x = 6 this means Γ(x) =5x4x3x2x1] suggesting cascades of shrinking values like ripples on a pond
605. P. Davies & J. Brown, *Superstrings, A theory of everything?* Cambridge University Press, Canto Edition 1992, pp67–68

606. *Oxford Dictionary of Physics*, ibid, p343 under 'mesons'
607. S. Gubser, *The Little Book of String Theory*, ibid, p126, pp151–152
608. *Oxford Dictionary of Physics*, ibid, p175, under 'elementary particles'
609. *Oxford Dictionary of Physics*, ibid, p575
610. S. Gubser, *The Little Book of String Theory*, ibid, p126, pp132–139, *Oxford Dictionary of Physics*, ibid, pp 574–575
611. S. Hawking, *A Brief History of Time*, ibid, p168
612. P. Davies & J. Brown, *Superstrings, A theory of everything?* ibid, p68
613. E. Regis, *Who Got Einstein's Office?*, ibid, p267
614. S. Hawking, *A Brief History of Time*, ibid, p173
615. E. Regis, *Who Got Einstein's Office?*, ibid, p264
616. 'A cosmic symphony of vibrating strings' Landscape Theory, Part 2 – *Stanford News*, 10 September 2018
617. E. Regis, *Who Got Einstein's Office?* ibid, pp265–267
618. S. Hawking, *A Brief History of Time*, ibid, p166
619. E. Regis, *Who Got Einstein's Office?* ibid, p267
620. Joël Swerk had sadly died
621. S. Hawking, *A Brief History of Time*, ibid, p171
622. 'A cosmic symphony of vibrating strings', ibid.
623. E. Regis, *Who Got Einstein's Office?* ibid, p269
624. E. Regis, *Who Got Einstein's Office?* ibid, pp268–269
625. E. Regis, *Who Got Einstein's Office?* ibid, p268
626. E. Regis, *Who Got Einstein's Office?* ibid, p269
627. Howard Georgi (Harvard University) E. Regis, *Who Got Einstein's Office?*, ibid, p269
628. Quote from chapter 'Edward Witten', in P. Davies & J. Brown, *Superstrings, A theory of everything?* ibid, p102
629. E. Frenkel, *Love and Mathematics*, Basic Books, 2014, p186
630. E. Frenkel, *Love and Mathematics*, ibid, p186
631. E. Frenkel, *Love and Mathematics*, ibid.
632. E. Frenkel, *Love and Mathematics*, ibid, p222
633. Prof. Susskind has a diagram of a Calabi-Yau space on p344 of his book L. Susskind, *The Black Hole War: My Battle with Stephen Hawking to make the World Safe for Quantum Mechanics*, ibid, p419; Roger Penrose has some simpler Calabi-Yau diagrams on p910 of this book, *The Road to Reality, A Complete Guide to the Law of the Universe*, ibid.
634. E. Frenkel, *Love and Mathematics*, ibid, p210
635. R. Penrose, *The Road to Reality, A Complete Guide to the Law of the Universe*, ibid, p910
636. *Oxford Dictionary of Physics*, ibid, p557
637. R. Penrose, *The Road to Reality, A Complete Guide to the Law of the Universe*, ibid, pp911–912
638. E. Frenkel, *Love and Mathematics* ibid, p210
639. E. Regis, *Who Got Einstein's Office?* ibid, pp270–271

640. E. Regis, *Who Got Einstein's Office?* ibid, p273
641. *Oxford Dictionary of Physics*, ibid, p575
642. S. Gubser, *The Little Book of String Theory*, ibid, p108
643. M. Kaku, *M-Theory: The Mother of all SuperStrings*, www.mkaku.org
644. S. Gubser, *The Little Book of String Theory*, ibid, pp89–90
645. S. Gubser, *The Little Book of String Theory*, ibid, pp114–115
646. S. Gubser, *The Little Book of String Theory*, ibid, pp114–115
647. S. Gubser, *The Little Book of String Theory*, ibid, p104
648. M. Kaku, ibid.
649. 'The fractal universe', Landscape Theory, Part 3 - *Stanford News*, 10 September 2018
650. 'The fractal universe', ibid – also see definitions COBE and WMAP in *Oxford Dictionary of Physics*, ibid, pp87 & 636
651. *Oxford Dictionary of Physics*, ibid, p636
652. 'The fractal universe', ibid.
653. 'A cosmic symphony of vibrating strings', ibid.
654. 'A cosmic symphony of vibrating strings', ibid.
655. 'The fractal universe', ibid.
656. R.V. Solé, *Phase Transitions*, Princeton University Press, 2011, p53
657. A saying of Alfred Korzybski – a Polish American scholar
658. R. Penrose, *Fashion, Faith and Fantasy*, ibid, p119
659. This is a measure between a Euclidean manifold and non-Euclidean manifold like the Einstein manifold modelling curved space – see https://en.wikipedia.org/wiki/Einstein_manifold which defines it as "If M is the underlying n-dimensional manifold and g is its metric tensor the Einstein condition means that: Ric = kg for some constant k, where Ric denotes the Ricci tensor of g. Einstein manifolds with k = 0 are called Ricci-flat manifolds."
660. R. Penrose, *The Road to Reality, A Complete Guide to the Law of the Universe*, ibid, p897 (and p380 for the notation)
661. R. Penrose, *Fashion, Faith and Fantasy*, ibid, p221; *Oxford Dictionary of Physics*, ibid, p132
662. A. Chamberlin, University of Louisville, *The Quest for a Realistic Cosmology in the Landscape of String Theory*, www.phys.isu.edu 2004.03.12 Jorge Pullin – references to Gibbons in GIFT Seminar on Supersymmetry, supergravity …World Scientific, 1984, Maldecena Nuflez Int J,Mod. Phys A16 (2001) 822-855 hep-th/0007018; B. d Wit, DJ Smit Nd.& Hari Dass, Nucl. Phys. B 283,165, 1987
663. A. Chamberlin, University of Louisville, *The Quest for a Realistic Cosmology in the Landscape of String Theory*, ibid.
664. S. Gubser, *The Little Book of String Theory*, ibid, p130
665. A. Chamberlin, University of Louisville, *The Quest for a Realistic Cosmology in the Landscape of String Theory*, ibid.
666. 'A cosmic symphony of vibrating strings', Landscape Theory, Part 5, *Stanford News*, 10 September, 2018

667. T. Kuhn, *The Structure of Scientific Revolutions*
668. Enormously below the expected calculation
669. Quoted from Chapter 5, p76 – series of lectures forming part of the debate between Stephen Hawking and Roger Penrose contained in *The Nature of Space and Time*, Princeton University Press, 1996
670. A saying of Alfred Korzybski – a Polish American scholar
671. M. Tegmark, *Our Mathematical Universe, My quest for the ultimate nature of reality,* Penguin Books, 2014, p321
672. *Oxford Dictionary of Physics,* ibid, p503
673. M.Tegmark, *Our Mathematical Universe,* ibid, pp124–125
674. Carlo Rovelli, *Reality is not what it seems*, Penguin Random House, 2016, pp144–145
675. Carlo Rovelli, *Reality is not what it seems*, ibid, pp190–193
676. M.Tegmark, *Our Mathematical Universe*, ibid, p125
677. 'A cosmic symphony of vibrating strings', ibid.
678. S. Weinberg, 'Is the Universe a Computer? A Review of a New Kind of Science by Stephen Wolfram' – www.nybooks.com
679. *Oxford Dictionary of Physics,* ibid, p21 & p183 for the EPR experiment
680. R. Penrose, *Fashion, Faith and Fantasy,* ibid, p5
681. R. Penrose, *The Road to Reality, A Complete Guide to the Law of the Universe,* ibid, p732
682. Quoted from Edward Dolnick, *The Clockwork Universe*, Harper Perennial, 2011, p234

SCHEDULE

SYNCHRONISATION OF BIOLOGICAL CIRCUITS IS AFFECTED BY CHAOS, IRRATIONALS AND INFINITIES

Schedule to above subheading in Chapter 13 illustrating model of biological system, states, circuits, bio equations and interactions to synchronise an organism

Modelling of cellular processes with respect to cellular replenishment of cartilage has been carried out involving metabolic pathways by setting up a matrix of over fifty external and internal metabolites (substances used by the cell) with their accompanying reaction rates (called fluxes).

Metabolites include substances such as ATP, glucose, H+ and water that are part of the cell's reactions. Homeostasis of the cell would provide a range of flux values and concentrations of metabolites that would 'work' in the cell – just as homeostasis of our blood pressure provides a range of values of body temperature, heart rate, etc. to produce an optimum range of normal blood pressure. The authors of the biological model referred to commented about values in this fifty-line (or more) spreadsheet/ matrix: "A system with a unique solution will have a stoichiometric of full rank. Most biological models are rank-deficient with an *infinite* number of solutions if the problem is feasible. To find a unique solution representing *empirically* calculated fluxes for a given experiment, we measure consumption and accumulation rates during the experiment to account for some of the unknown fluxes. This can make the system determined and therefore uniquely solvable" [my italics].

To continue our thought experiment, we need to abstract God's logical creative planning process and by doing so we can identify certain logical obstacles:

1. Out of the potentially hundreds of states in the human body that need to operate, a certain number of critical states – perhaps one hundred states – need to operate in synchronisation to sustain life and thus at least this same number is required to operate in perfect synchronisation if the human is deemed 'perfect'. ('Perfect synchronisation means at least free of disease, defects, etc.)

2. The hundreds of contributing subsystem bio equations from the thirteen main systems supply results data to this vast synchronisation before the process converts such data from the various equations' differing units of measurement and then synchronises all such data using state equations synchronising with equation results data from the other ninety-nine critical states to produce compatible 'synchronisation codes'.

3. These synchronisation codes for those operating parameters would then be used to 'reverse engineer' and create the 'bio hardware' for the perfect body with synchronised 'perfect' performance.

4. When the contributing subsystems pass results data to the states before synchronisation, three problems arise:

    (i) Some results of equations will be infinite values if there are more than two variables in any given equation (commonplace in biological equations).
    (ii) Some equations will produce rational finite results and some will produce irrational or infinite results making equations prone to infinite decimals. The Hippasus Dichotomy will prevent irrational

or infinite results being fed into state equations for synchronisation unless there are alternative means such as calculus, etc. and thus an insolubility problem arises.

(iii) On the conversion of units of measurement, some conversions will produce rational finite results and some will produce irrational or infinite results with the same insolubility problem.

(iv) Bio equations can change during the life of an organism, creating more suites of irrationals or infinities.

5. Without the possibility of an analytic solution in the metaphysical realm for a 'perfect human', the next possible method is whether a numerical solution in a simulated universe is available, ie. by iterating every permutation and combination of potentially infinite design possibilities. (This is analogous to philosopher Max Black's infinite machines described at the end of Chapter 11.)

6. However, these possible infinite iterations during a creative process to arrive at a solution or solutions is a super task because there can be no final 'iteration' (the same point raised by Max Black). Thus, the perfect human being (defined as a disease-/defect-free human being) is a super object and is thus analytically impossible to make.

INDEX

B

C

F

G

N

O

P

Q

R

S

T

U

V

W

Y

Z

www.ingramcontent.com/pod-product-compliance
Lightning Source LLC
Chambersburg PA
CBHW072010230526
45468CB00021B/1178

9781805142546